GARDEN SHRUBS

and their Histories

Nerium oleander

(OLEANDER)   Ferdinand Bauer (1819)

GARDEN SHRUBS
and their Histories

ALICE M. COATS

With notes by Dr. JOHN L. CREECH

SIMON AND SCHUSTER

New York London Toronto Sydney Tokyo Singapore

SIMON AND SCHUSTER
Simon & Schuster Building
Rockefeller Center
1230 Avenue of the Americas
New York, New York 10020

SIMON AND SCHUSTER and colophon are registered trademarks
of Simon & Schuster Inc.

Edited by Charles A. de Kay
Designed by Michelle S. Wiener

Produced by Smallwood and Stewart, Inc.
New York City

Printed in Singapore

1 2 3 4 5 6 7 8 9 10
First U.S. Edition

Library of Congress Cataloging-in-Publication Data
Coats, Alice M.
Garden shrubs and their histories/Alice M. Coats
    p.   cm.
ISBN 0-671-74733-9
    1. Ornamental shrubs—History.   2. Ornamental shrubs—Folklore.
3. Ornamental shrubs—Pictorial works.   4. Botanical illustration.
5. Watercolor painting.   I. Title.
SB435.C587   1992
635.9′76—dc20   91-20940
                                                 CIP

Contents

Foreword

The publication of the annotated American edition of *Garden Shrubs and their Histories* markedly advances the opportunity for gardeners to better understand one of the most important groups of commonly grown plants—woody shrubs. This delightful book, in which garden historian Alice Coats traces the origins of the most interesting species of our garden shrubs in more than a hundred genera, has been out of print for far too many years.

In *Garden Shrubs and their Histories* Miss Coats details the events leading up to each shrub's introduction into Western culture, particularly English gardens. Miss Coats had a keen understanding of the mutually dependent relationship between the collector and the botanists, gardeners, nurserymen and amateur plantsmen who received, propagated and tested these unknown plants from distant lands.

In order to make Miss Coats's erudite history of these shrubs more relevant to American gardeners, I have added specific information (both historical and cultural) that I hope will fill in what may appear as gaps from an American standpoint. It is not possible to offer a complete American historical overview and planting advice within the confines of this book, but there are many other works that provide comprehensive cultural information.

For my part, it has been a particular delight to write the notes to update *Garden Shrubs and their Histories* and to make it more applicable to American gardeners, as it is essentially an English book. I have inserted my ideas about the shrubs at the end of each of her plant histories. I am glad that the publishers asked that I take this approach to update the book. Miss Coats has a wonderful style and her text, while unquestionably factual, is so bright with its anecdotal material that I was reluctant to tamper with it directly. Any other method of addressing American gardening concerns might interfere with the enjoyment of the charm of the original.

Unfortunately, American horticultural history is not blessed with the meticulous record-keeping of the British plantsmen, and it is often impossible to determine with any certainty the exact year that a specific plant was first cultivated here. But the ties between amateur and professional gardeners in America and England have always been strong (in fact, even during the Revolution, plants were traded across the Atlantic by enthusiastic gardeners who could not be bothered with politics), that the English history of these shrubs effectively becomes snyonymous with the American experience with them. A common practice among Americans interested in botanic history is to take the British introduction date as the definitive one—using it as the date of *introduction into Western horticulture*. As a rule of thumb, though, one can argue that most plants made their way to the New World in a comparatively short span of time—usually within a decade—after their approval and introduction in the Old.

So intertwined, however, is American garden history with that of our allies across the ocean that many gardeners here would follow the fashions of English gardens in a way similar to say, clothing designers would wait to see what was popular in Paris. In point of fact, until very recently, American gardeners would regularly defer to their English counterparts for final judgements on most horticultural matters. During the heyday of the plant hunters, even the decision of which plants were best suited to the garden was posed first to the English. Americans looked across the Atlantic for guidance as to how to regard the latest plant introductions. Even American natives plants were often sent to England to be scrutinized and evaluated before being adopted by our own nurseries. Very often the official stamp of approval for ornamentals took the form of an award at the Royal Horticultural Society plant shows. (Please see the end of this section for more information on the significance and history of the RHS awards.)

The English gardener is very fortunate to have an environment that fosters the cultivation of such a broad range of shrubs which can be grown as companion plants in a single garden. In the United States, where we have so many different climatic zones, some of the plants featured in the book can be expected to flourish in one

Hibiscus syriacus

(ALTHAEA)   S. Edwards (1789)

region and to wither in others. I have added observations from my experience growing many of these shrubs in hopes that it may encourage adventuresome gardeners to try some of the more unusual shrubs that will thrive in their area. On the other hand, there are also some shrubs, Chinese holly (*Ilex cornuta*), Summersweet (*Clethra alnifolia*), Japanese Pittosporum (*Pittosporum tobira*), Redtip (*Photinia glabra*) and Japanese Ternstroemia (*Ternstroemia japonica*), for example, that have a place in our gardens but are not of historic or garden interest in England and so are not included here. The inclusion of these goes beyond the scope of my notes, which are restricted to topics discussed in the original book.

In many respects, though, I have felt that my work became almost a conversation with Miss Coats. Were it possible, I would dearly have enjoyed asking why some genera were omitted—*Ilex*, for example—and why some tree genera, such as *Cornus*, were covered, while others of similar habits were not.

As a historian Alice Coats's methods were exemplary. She had an unbridled enthusiasm for consulting primary sources; when she was working on a new project, she could be found at the Birmingham Reference Library, or the vast botanical libraries at Kew, the British Museum, the Edinburgh Royal Botanic Garden or the Royal Horticultural Society. Although physically afflicted with arthritis for most of her adult life, she was known for her persistence and good cheer. Miss Coats published six books on the art and history of garden plants during her lifetime and she will always be remembered for enriching the literature of garden history. Her scholarship and tenacity were rewarded in 1976 when she was awarded the prized Veitch Memorial Medal by the Royal Horticultural Society, just two years before her death.

A BRIEF LOOK AT THE HISTORY OF GARDENING IN AMERICA

The beginnings of ornamental horticulture in America found the original colonists preoccupied with survival and the economic necessities of life. Few ornamentals were among the plants to arrive in the early days of settlement in the New World. But as a more comfortable society emerged, around the mid-eighteenth century, landowners in the more affluent areas in Pennsylvania and Virginia began to develop large gardens; they imported trees, shrubs and other plants from England and other European homelands. In the deep South, where a gentle climate allowed for a style of living greatly influenced by European fashion, broadleaved evergreens—often species native to the Orient that had found their way into Europe much earlier—were introduced. Azaleas and camellias were commonly scattered among the native evergreen magnolias and live oaks. These early landowners created the image of what I consider the truly American garden.

Immigrants trained as gardeners and nurserymen took avantage of increasing interest in plants to grace improved urban dwellings, and the nursery industry was born. In 1730, Robert Prince established one of the first nurseries in Flushing, Long Island. Originally intended to introduce fruit trees, the nursery was offering trees and shrubs popular in England before long. Several generations of Prince nurserymen followed, each contributing new plants for American gardens. Other nurseries began to appear in and around towns along the eastern seaboard. Bernard McMahon, who began a nursery in Philadelphia in 1804, was successful in first growing the Oregon Holly-grape, *Mahonia aquifolium*. (See under Mahonia.) Other old nurseries that succeeded well into the twentieth century include Ellwanger and Barry (founded 1840) of Rochester, New York, and Fruitland Nurseries (1857) in Augusta, Georgia. These and many other early nurserymen, who had learned the arts of propagation, pruning and planting in their native countries, applied their skills in America before the Civil War.

These early nurserymen frequently used their contacts "back home" to introduce the latest varieties. They made great contributions to the wealth of garden plants, fruit and ornamental trees and shrubs available in the 1800s. It is said that by the mid-1800s there were about 150 varieties of European apples in cultivation. Our gardening forebears relied on these nurserymen because at that time we had no arboretums or botanic gardens to perform this service. The first of our great gardens, The Missouri Botanical Garden, was not founded in St. Louis by Henry Shaw until 1859. Our first major arboretum, Arnold Arboretum, in Jamaica Plain, Massachusetts was founded in 1872.

THE INTRODUCTION OF JAPANESE PLANTS

The opening of Japan by Admiral Perry in 1852 caused a tremendous, if belated, interest in Japanese gardens and flora. Incorrectly, many early plantsmen had maintained that China was the "mother of gardens" and that most plants of any interest in Japan had already been collected. Because of limited information and a seemingly insurmountable langauge barrier, it wasn't until the twentieth century that the West became aware of just what a botanic treasure-house Japan actually is. Japan can boast of over 6,000 native trees, shrubs and other plants, and of a tradition—more than 1,000 years old—of emphasizing native plants. When the Japanese landscape was finally fully explored by westerners, it was the Americans who led the way, surpassing for the first time the traditionally English-dominated field of plant hunting. (For a listing of some of the more important American-sponsored expeditions, see "Recent Ornamental Plant Explorations," page 219.)

Today, in many areas of the United States, more trees and shrubs in our gardens are Japanese natives or plants improved by Japanese horticulturists (over a period of centuries) than any other. A good number of specialized plant societies in the United States, such as those dealing with bamboo, camellia, chrysanthemum, hosta, iris, holly and peony, owe their existence to the plants of Japan. In addition, Japanese horticultural techniques have become among the most popular in recent years. The art of bonsai, an integral part of some Japanese gardens, for instance, became a fixture in the United States after the Second World War. The same is true for the recently popular art of Japanese flower arranging.

SOME THOUGHTS ON THE FUTURE OF HORTICULTURE

There are many exciting horticultural opportunities that bode well for future generations of horticulturists interested in discovering or developing new species of plants. The history of garden shrubs and other ornamentals continues to be written. One such opportunity is a unique and promising relationship between a number of groups of flora native to the eastern United States and that of eastern Asia.

Since the Perry expedition, botanists have become aware of the similarity of many Japanese and Korean plants to those from the southern Appalachian Mountains. There are perhaps more than ninety genera of trees and shrubs, native to the U.S., that share species with plants in Japan. Our better understanding of these plants not only offers us a fighting chance against some of the diseases that plague these genera but the chance to find ways to save weaker strains from disappearing altogether as well. When our native chestnut tree was devastated by a blight, for instance, an American plant pathologist, Dr. R. Kent Beattie, was sent to Japan and China in 1927 to collect Oriental chestnut seeds and seedlings to combat the disease. Unfortunately this program was discontinued after the Second World War and most of Beattie's collections were abandoned. But Beattie collected a large number of Japanese azaleas on the side; eventually these azaleas figured greatly in the development of the famous Glenn Dale azalea hybrids. Thus, as is often the case with scientific exploration, one cannot always predict from where the "pay-offs" of plant collecting will come.

More recently, during a winter expedition to Yakushima Island, Japan, I included the American introduction of a rare species of crape myrtle, *Lagerstroestroemia fauriei*, as one of my goals. Young trees were subsequently planted in the National Arboretum, where the late Dr. Donald Egolf discovered that it has a natural resistance to powdery mildew, the scourage of our common crape myrtle. In the 1970's, he transferred the resistance to an improved race of hybrids with remarkable success. These new hybrids have begun to displace the older, mildew-susceptible varieties.

It is now feared that our beautiful native flowering dogwood, *Cornus florida*, which is threatened by a serious new disease called "lower branch decline," might suffer the same fate as the chestnut and American elm. Several studies that are attempting to understand—and, if possible, control—the disease are underway. At the same time, horticulturists are looking at the Japanese counterpart dogwood, *C. kousa*, because it is similar in appearance and seems to be immune. Fortunately, early plant collectors first introduced *C. kousa* into cultivation some 120 years ago, and it became established soon thereafter;

so we know where it can be grown successfully.

Natural disasters like these diseases set the stage for intensive plant collecting, then scientific evaluation of the introduced material, selection of the best forms and their eventual distribution by nurseries. This can be a long process.

Another related issue is that time and time again we find the Japanese species being cultivated in our gardens to the exclusion of their native counterparts. Most gardens feature the Japanese pachysandra, *Pachysandra terminalis*, as a ground cover, for instance, while our own *P. procumbens* is scarcely known. We admire many of our native trees and shrubs in their wilderness setting but are more likely to purchase the Japanese counterparts for our gardens. The time is right for botanists to learn more about these overlooked natives and, perhaps, develop hybrids suited for ornamental purposes.

Finally, there is a very important development in recent American plant history, still in the making somewhat, that needs to be addressed. In Europe, a concern is emerging over the discarding of older cultivars from botanic garden collections and from the nurseries. While the same pattern of disposal is happening in the United States, the concern has yet to materialize here. Funds to maintain large cultivar collections are dwindling (try to justify maintaining just the vast number of named azaleas or camellias in public gardens) except where private financing is provided. Moreover, the emphasis in the large public gardens has shifted from the conservation of tree and shrub species and cultivars to landscape gardens and design schemes of a temporary nature. We are at a crossroads in the development of improved ornamental trees and shrubs concurrently with a call for widespread planting of trees throughout the nation. The nursery industry had relied on plant breeders in our federal and state experiment stations for this kind of work for decades. But the current trend is away from conventional plant breeding and toward basic technology. Few scientists these days are willing to pursue the tedious and long-term work of breeding new varieties of trees and shrubs. The loss of ornamental plant breeders combined with the apathy towards maintaining large collections of species and cultivars, which are the genetic resources essential for the future, should be a concern to all gardeners.

PLANT HUNTERS OF THE MODERN WORLD

Sadly, the romance associated with the intrepid plant explorers who journeyed for years, surmounting tremendous odds to bring back living plant collections is over. Swift means of transportation and effective lines of communication, as well as markedly improved methods of shipping collections (all praise to plastic bags!), have taken some of the adventure out of plant research. For today's collector, the information recorded by earlier explorers as well as observations of plant behavior in cultivation allow more discriminating selection of collecting localities.

In general, for woody trees and shrubs for most of the United States, Japan, Korea, China and the Soviet Union remain the most important centers of diversity. Instead of the long, lone expeditions of the past, now teams of horticulturists and botanists, often from several institutions venture forth for shorter durations. And with the nationalistic fervor in most countries to protect their natural resources, greater approval and participation by scientists from host countries is required. The institutionalization of what Thomas Jefferson called "the greatest service"—the bringing of a new plant to one's country—has changed the scene. Teams of experts may have replaced the solitary figure trekking across unknown habitats with a band of local porters. But the essense of the drive to capture the beauty of a universal garden is still there and we continue to reap the benefits.

JOHN L. CREECH
HENDERSONVILLE, NORTH CAROLINA

Author's Preface

The writer of a book such as this is obliged to make a number of more or less arbitrary decisions in order to fix the boundaries of the subject. When, for example, does a plant cease to be a shrub and become a tree? Many importations that are of quite modest dimensions in our gardens [in England] grow to a great size in their native lands. How are we to define the limits of "hardy" and "not hardy"? There seems to be a natural law that the most beautiful species in any genus is always the tenderest. Are evergreens to be included, or only flowering shrubs? Which species are sufficiently well known to be of general interest, and what of those that may become important in the future or were cherished favorites in the past? And what, above all, of the thorny problem of the rose?

A shrub is defined as a woody plant with a number of growths from the base, as opposed to the tree's single stem; height is no criterion, for many tall shrubs, such as lilacs, may overtop small trees. For our purpose, any multiple-stemmed woody plant that begins to produce its flowers when they are still at or below eye level, may fairly be regarded as a shrub, even though it may eventually attain a great height. In other respects, knowing it would be impossible to please everybody, I have pleased myself in the boundaries I have set. Thus I have included broad-leaved evergreens, but not conifers; climbers, but not rock-garden shrublets of less than a foot in height; and have added a few tender subjects considered indispensable by our ancestors, though less widely grown today. The rose, whose vast and complex history is difficult to reduce to any reasonable length, seemed yet impossible to leave out. I hope rosarians will be indulgent towards the small cupful of information I have dipped from the vast reservoirs available on this most specialized subject.

Where a plant family is named after a particular person (there are many examples, from Buddleia to Wisteria), biographical information concerning that person is given under the heading of the plant in question; the lives of those responsible for the introduction of numbers of plants, but whose names are not connected with any genus mentioned in this book, will be found in the biographical section at the end. Where footnote references to quotations give the author's name only, the quotation comes from the book mentioned in the bibliography; and in cases where two or more works by the same author are listed, they are differentiated in the footnotes by their dates.

The author wishes to express her gratitude to the staffs of the Birmingham Library, the Birmingham Reference Library, and the Lindley Library of the Royal Horticultural Society; also to the contemporary authors whose books she has consulted, and to the correspondents who have answered enquiries about particular subjects. A special mention should be made of Mr. Miles Hadfield, whose immense erudition has been of the greatest assistance.

A Note about the Royal Horticultural Society Awards

Alice Coats makes mention throughout the book of RHS medals awarded to various shrubs under discussion. These awards were often given soon after the plant's introduction into British cultivation—effectively a seal of approval for the shrubs use in the garden. The Award of Garden Merit (see below) best fits this category and is the award to which Miss Coats most often refers. All awards made by the society quoted in the book include the year in which the award was given, e.g. "A.M. 1984". Readers in America who are curious to learn more may wish to refer to explanatory notes below.

AWARDS TO PLANTS AT SHOWS

Awards, such as the Award of Merit (AM) and the First Class Certificate (FCC), are awarded by the society to group exhibits (i.e., a display of orchids), not to individual plants.

All awards are made either by the Council of The Royal Horticultural Society, or jointly by the council and an Executive Committee of one or more specialist societies. Awards made by the Council are, in most cases, made on the recommendation of apointed committees. The awards made by the Council and Executive Committees are given on the recommendation of Joint Committees composed of RHS representatives and representatives of the specialist societies concerned.

The awards given to plants, flowers, fruits and vegetables are:

FIRST CLASS CERTIFICATE (FCC). Instituted 1859. Given on the recommendation of the Standing and Joint committees to plants, flowers, fruits and vegetables of outstanding excellence.

AWARD OF MERIT (AM). Instituted 1888. Given on the recommendation of the Standing and Joint Committees to plants, flowers, fruits and vegetables which are of great merit.

CERTIFICATE OF PRELIMINARY COMMENDATION (CPC). Instituted 1931. Given at shows on the recommendation of the Standing and Joint Committees to a new plant of promise, whether a new introduction from abroad or of garden origin.

BOTANICAL CERTIFICATE (BC). Instituted 1878. Given on the recommendation of the Scientific Committee of plants of exceptional botanical interest.

THE AWARD OF GARDEN MERIT (AGM). Instituted 1922. Previously awarded to plants that best proved themselves to be of outstanding excellence for garden use (other awards are made at shows from pot grown or cut material and the awards do not necessarily imply *garden* excellence). The AGM was awarded by a separate committee who considered candidates on the basis of experience with plants rather than exhibits put before them. While the award has not been given since the mid-1980s, there are plans to relaunch it in 1992.

MEDALS

The following medals are available for award at shows

The society's	Gold Medal
	Silver—Gilt Medal
	Silver Medal
	Bronze Medal

These medals are available in the following ranges according to the type of exhibit:

THE GOLD METAL. Instituted in 1898. Redesigned in 1929. Awarded for exhibits of special excellence.

THE FLORA MEDAL. Instituted in 1835 and first struck in 1836 as the society's Large Medal replacing the original of 1811, renamed Flora Medal in 1836. Awarded for exhibits of flowers and ornamental plants.

THE BANKSIAN MEDAL. Instituted in 1820 in commemoration of Sir Joseph Banks, PRS, one of the founders of the society.

THE HOGG MEDAL Instituted in 1898 in commemoration of Dr. Robert Hogg, the great pomologist, sometime secretary of the society. Awarded for exhibits of fruit.

THE KNIGHTIAN MEDAL. Instituted in 1836 in honor of Thomas Andrew Knight, PRS, president of the society from 1811 to 1838. Awarded for exhibits of vegetables.

THE LINDLEY MEDAL. Instituted in 1866 in commemoration of Dr. John Lindley, PRS, sometime secretary of the society. Awarded for an exhibit of special scientific or educational interest. Individual plants are not eligible for this award.

Miss Coats also mentions the RHS *Dictionary of Gardening*'s star of excellence, which was an independent assessment of value that indicated a plant "of special merit".

Abelia

In 1816, Mr. Clarke Abel (1780–1826) was appointed as surgeon[1] to Lord Amherst's Embassy to the Emperor of China at Pekin. He was keenly interested in natural history and was anxious to make some collections on the journey through what was, to Europeans, an almost unknown country. This project was encouraged by Sir Joseph Banks, who arranged for him to be supplied with "a plant-cabin for the preservation of living specimens,"[2] a botanic gardener from the Royal Botanic Gardens at Kew, London, to look after it and collect seeds, and an assistant, Mr. Poole, who was Abel's brother-in-law. The diplomatic mission of the Embassy was a failure; the journey back from Pekin to Canton, made almost entirely by river and canal, was strictly supervised by the Chinese and afforded few opportunities for botanizing, and for part of the time Abel was confined to his cabin through illness. Nevertheless, he managed to procure a considerable number of plant-specimens; Hooper, the gardener, amassed more than 300 packets of seeds, and other friends and shipmates contributed collections of zoophytes, madrepores and geological samples. All were lost when the *Alceste* was wrecked (February 18, 1817) on the homeward voyage, on an uncharted reef in the pirate-infested Gaspar Straits; and though no lives were lost, Abel "had the mortification of hearing that the cases containing these seeds had been brought up on deck and emptied of their contents by one of the seamen, to make room for some of the linen of one of the gentlemen of the Embassy."[2] After establishing a precarious footing on shore, a party returned next day to the wreck; Abel found one collection of plants, seeds and minerals "still in a great measure uninjured, but only mocked the vexation of the owner, who saw no chance of preserving it."[3] This case was eventually put on a raft—which was burnt with all its contents by the hovering Malay pirates. All that could be retrieved from the ill-fated expedition was a small collection of plant specimens that Abel had given to Sir George Staunton at Canton, which was generously returned to him; and it was one of these that Robert Brown, Sir Joseph Banks' librarian and taxonomist, named "with friendly partiality" *Abelia chinensis*.

This pretty shrub was found by Abel at Takoo-tang in Kiang-si, in the neighborhood of Lake Po-yang; but it was not until 1844 that living plants were sent to

[1] There was also a physician, Dr. James Lynn. In all, counting servants, band and guard, the Embassy numbered seventy-two persons
[2] Abel, *Narrative of a Journey in the Interior of China*, 1819
[3] Ib.

England by Robert Fortune, who also introduced another Chinese species, *A. uniflora*, the following year. In the meantime, however, an Abelia had been discovered in the Cordillera of Oaxaca in Mexico, and introduced in 1841 as *A. floribunda*. This is the most showy of the family—one author describes the color of its flowers as a "furious magenta"—and it is no exception to the usual rule that the most ornamental sorts are sure to be the most tender; except in Cornwall and Ireland it cannot be grown out-of-doors without protection. *A. chinensis* is less delicate, and *A. uniflora* (now rare) is almost completely hardy; and these two species have been crossed to produce the most vigorous and one of the most popular of all the Abelias, the pink and white *A. × grandiflora*. The origin of this variety does not seem to have been recorded. It is not mentioned in the 1900 edition of William Robinson's *English Flower Garden*, but was well known by 1914. A fairly recent addition to the family is *A. schumannii* (syn. *A. longituba*), a dwarf species with pinkish-lilac flowers introduced from China by E. H. Wilson in 1908. It is slightly tender, but usually springs again from the root if cut down by frost.

The Abelias are members of the honeysuckle order, and are related to the Weigelas. Some of the species are fragrant.

NOTES: In America, the cultivar 'Edward Goucher', a hybrid between *A. grandiflora* and *A. schumannii*, has become popular. This hybrid was developed at the U.S. Plant Introduction Station, Glenn Dale, Maryland, in the early 1920s by Edward Goucher, when he was superintendent. The selection was later named by the Department of Agriculture (U.S.D.A.) in his honor.

It is a dense shrub with semi-evergreen leaves and lavender flowers. Like its parents (*A. grandiflora* × *A. schumannii*), 'Edward Goucher' is easy to cultivate, preferring semi-shade or full sun, and is hardy into New England. If someone were to hybridize the hardy Asiatic species with the tender *A. floribunda*, which is native to Mexico, the results might be spectacular.

Actinidia

This is a family of some forty hardy climbing shrubs, natives of Asia. Several of them are, or have been, cultivated in Britain, but only two, *A. chinensis* and *A. kolomikta*, are even moderately well known. These two, however, are very remarkable plants, well deserving a more extended renown.

Specimens, but not seeds or plants, of *A. chinensis* were sent to the Horticultural Society by Robert Fortune, and

the newly-discovered genus was named by Planchon in the *London Journal of Botany* for 1847. The plant was seen by Maries in Japan, but it was not introduced until E. H. Wilson sent seeds from Hupeh to Messrs. Veitch of Exeter in 1900. It is a vigorous climber, with leaves as large as one's hand, and clusters of handsome white flowers which turn an uncommon buff or chamois color as they age. They are succeeded by edible fruits the size of a walnut, containing a "green, subacid, palatable pulp, in flavor resembling the gooseberry."[1] These fruits are grown commercially in New Zealand, and experiments in their production, both in the open and under glass, have recently (1964) been carried out in the Netherlands. Unfortunately for the amateur, the plants are normally diœcious, and both male and female forms are needed to ensure good results; and few can afford wall-space for more than one specimen of a climber that may extend 30 feet in all directions. Occasionally individual plants bear flowers of both sexes, and good crops of fruit have been obtained in favorable seasons in the south of England. In one garden, it is said, the berries were swept up and fed to the ducks—until it was discovered that they had a pleasing flavor and could be sold for 5*s* a pound. After that, the ducks had to go without.

A. kolomikta is a much less vigorous climber, rarely exceeding 10 or 12 feet in cultivation; it is a native of the forest regions of Amur in Eastern Siberia, extending southwards to Korea and cultivated in Japan. Charles Maries[2] sent it to England in 1878 from Sapporo, on the Japanese island of Hokkaido. The Japanese called it the Cat's Medicine, for it affects felines even more powerfully than catnip or valerian. It is said that if charcoal-makers burned faggots that had been tied with this creeper, cats would come long distances to congregate around the fire, rolling on the ground and behaving as if intoxicated. A story is told of a former propagator in Boston who had all the young pot plants of this newly introduced species bitten down to the ground by the greenhouse cat before the culprit was discovered. It is advisable to protect it in gardens with wire netting. It is grown neither for its white fragrant flowers (much smaller than those of *A. chinensis*) nor for its fruits, eaten in Russia as "Amur gooseberries," but for the coloring of its leaves, which in the cultivated form are strikingly variegated with pink and white. It has been suggested that this variegated form is a male plant, and that that is why the fruit is so seldom seen here.

The generic name is derived from *actin*, a ray, because of the radiating styles of the flower. *Kolomikta* is the Amur name of the species.

[1]Wilson's description   [2]Then collecting for Messrs. Veitch

NOTES: *Actinidia chinensis* is the now-popular fruit found in most markets in the United States under the name Kiwi, reflecting its early cultivation and export from New Zealand. Originally introduced from China, it has been cultivated in California for years and called Chinese Gooseberry. It was ignored by the nursery trade and the general public until it acquired the simplified name Kiwi. Because of its recent popularity and the fact it is fairly hardy, the Kiwi is being tried experimentally in other parts of the country.

Another species, *A. arguta*, native to Japan and Korea, is occasionally seen in the trade because it is considerably hardier and has similar but much smaller fruits of no commercial value. *A. arguta* is sometimes used as an ornamental vine because it is vigorous and hardy in most regions of the country. In hopes of achieving a hardier "Kiwi" with larger fruits than *A. arguta*, a hybrid was created in 1923 and named *A. fairchildii* in honor of the "father" of federal plant introduction, David Fairchild. However, it never became popular. Another native of northern Japan is *A. polygama*, an ornamental vine with silver-white markings on the leaves of the male plant. Like *A. kolomikta*, it is also attractive to cats.

Aegle, see under *Poncirus*

Aesculus

Few would expect in a book on shrubs to encounter a horse-chestnut. We are apt to forget that this family of tall trees also contains a number of low-growing subjects, including one of the handsomest of late-flowering bushes, *A. parviflora*, unflatteringly called the Bottlebrush Buckeye. This plant was first discovered by William Bartram during his travels in Carolina, Georgia and Florida in 1773–78. He called it "a very curious nondescript shrub, which I observed growing in the shady forests, beneath the ascents, next bordering on the rich low lands of the river." He thought it a species of Aesculus or Pavia, "but as I could find none of the fruit, and but few flowers, quite out of season and imperfect, I am not certain." (A very old specimen, possibly planted by William, still grew in the Bartram garden in Philadelphia in 1930). The shrub was introduced to Britain by John Fraser (1750–1811) when he made his first trip to the American South in 1785. Fraser was a Scot who came to London about 1770, and began life as a linen-draper and hosier in Chelsea; but, stimulated perhaps by the proximity of the Physic Garden, he became interested in

Abelia floribunda
(MEXICAN EVERGREEN)   *Flore des Serres* (1846)

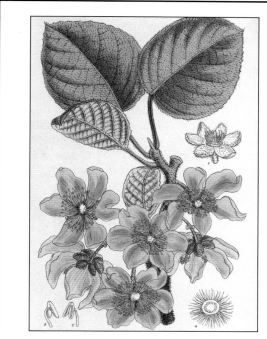

Actinidia chinensis
(CHINESE GOOSEBERRY)   Matilda Smith (1915)

Aesculus pavia
(CAROLINA BUCKEYE)   M. Hart (1826)

Aloysia triphylla   (now *Verbena triphylla*)
(LEMON VERBENA)   Sydenham Edwards (1797)

botany, and with the help of such eminent men as William Aiton and Sir J. E. Smith, embarked on a career as a plant-collector, starting with a journey to Newfoundland in 1780. His earlier finds were sent to the famous nursery firms of Lee and Kennedy, or Loddiges, or to private individuals; but later he started a rather unsuccessful nursery of his own in the neighborhood of Sloane Square. By 1820 the Bottlebrush Buckeye was "to be met with in most of our nurseries," but, in spite of its many garden virtues, which include good spring and autumn foliage color, as well as butterfly-haunted flowers in August, it is still unaccountably rare.

An earlier and perhaps more familiar introduction, *A. pavia*, the red or Carolina Buckeye, will grow into a small tree, but is often no more than shrub size and will flower when only 3 feet high. Seed was brought from Carolina in 1711, and by 1759 it was much cultivated in gardens. "There is a delicacy in this tree that makes it desirable," wrote William Marshall, who admired the effect of its red flowers at a season when so many of the plants in bloom were yellow. According to John Abercrombie, it merited a place in all ornamental plantations. "Some botanists have thought it proper to separate this from Aesculus on account of its smooth seed-husk and its four-, not five-petalled flowers; and it was grouped with a few similar species in a genus of its own, called Pavia after Peter Paaw (d. 1617), a Professor of Botany at Leyden; "but we confess," continues Loddiges' *Botanical Cabinet*, "we can discover no utility whatever in dividing a genus so very natural"—and modern botany apparently concurs.

The Buckeyes have a bad reputation among apiarists. The nectar secreted by the flowers is said to be harmful to bees, and "Buckeye poisoning," which affects the brood and young bees, is a well-known occurrence in the American South, where the shrubs grow wild. The flower of the ordinary horse-chestnut is quite harmless, and a valuable source of both nectar and pollen.

Aesculus is said to be derived from *esca*, nourishment, because a flour was made from the nuts of the horse-chestnut and of some other species.

NOTES: As is the case of many woody plants cultivated in the U.S.A., hybrids have become popular. In this instance, the hybrid *A. carnea* (*A. hippocastanum* × *A. pavia*), created in Europe over 100 years ago is the best for ornamental purposes and its cultivar 'Briottii' is an outstanding selection because it displays long pink-to-red flower clusters in May.

A. parviflora is frequently seen in gardens in the southeastern states in the vicinity of its natural habitat. Its large size (for a shrub) and suckering habit, which causes it to spread, do not make it a suitable plant for the small garden. One plant can easily encroach on the entire front lawn.

The Horse-chestnuts (*A. hippocastanum*) are generally hardy throughout the country. They were in fashion decades ago but because all the buckeyes are considered "dirty" trees—always dropping leaves, branchlets or husks—and because they are particularly susceptible to attack from various diseases and insects, Horse-chestnuts are no longer considered choice garden plants. In the garden, they are perhaps best placed in a mixed shrub border as an accent tree since they do attain tree-like proportions.

Aloysia

"This is among scented flowers . . . the masterpiece, and it should be grown in large quantities by the market-gardeners for London and all cities." Such was the opinion of a Victorian gardener on *A. triphylla* (syns. *A. citriodora, Lippia citriodora*), the Lemon Verbena. Perhaps it hardly qualifies for inclusion here, as it will thrive out-of-doors only in the most favored parts of England, though it luxuriates in Guernsey, and in Ireland has attained to a height of 25 feet. On the other hand, it strikes so easily from cuttings that it has often been used as a summer-bedding plant; and, even in districts where it is annually cut down by frost, with a little protection it will survive and spring again from the base.

The Lemon Verbena is a native of South America, and its early history was consequently much associated with Spain—though the botanists who discovered it were actually French. The first was Philibert de Commerson, who found it in Buenos Aires about 1768, when he was accompanying de Bougainville on his voyage around the world. From Buenos Aires the plant was introduced into Spanish gardens, and Professors Ortega and Palau of Madrid gave it the unpublished name of *Aloysia citrodora* in honor of Luis Antonio de Bourbon (1727–85), Prince of the Asturias and brother of Carlos III—*Aloys* and *Aloisio* being Provençal and Spanish forms of "Luis." (Prince Luis was created Archbishop of Toledo at the age of eight, but after the death of his father Philip V in 1746 he abandoned an ecclesiastical career in order to devote himself to the natural sciences and to music.) This respectable dedication was presently forgotten. The plant became popular all over Southern Spain as *Yerba Luisa*, and soon after it was confidently asserted that it was named after the Prince's niece by marriage, Queen Maria Luisa—made famous by Goya and notorious by

Godoy—who by all accounts was not a person with whom an innocent dimity flower with a clean, refreshing fragrance could suitably be associated.

Ortega sent seeds and specimens of the plant to L'Heritier in Paris, who described it in his *Stirpes Novae,* 1784, under the name of *Verbena triphylla;* and it was in Paris that John Sibthorp, Professor of Botany at Oxford and author of *Flora Graeca,* obtained the specimen which he introduced to England—according to the *Botanical Magazine*—on his return from Greece in 1784; but at this date he was only returning from some additional studies at the University of Montpellier and his first homecoming from Greece was in 1787, so the shrub may have come at either date. By 1797 the Lemon Verbena was already common in greenhouses in and around London, and it reached a high degree of popularity by the end of the next century, when market-gardeners were instructed how to manage their plants so as "to give a considerable proportion of stuff for bouquet-work."[1]

Meantime, in 1785, another consignment had arrived in Spain, this time brought from South America by another French botanist, Joseph Dombey (1742–96), one of the most devoted, painstaking and unfortunate of collectors. Arriving in Cadiz on his way back from Lima after an absence of eight years, his large collections were impounded by the Spanish Customs and left to rot in damp warehouses, while he suffered endless frustrations and delays, and could not even get permission for the seeds to be planted. The Lemon Verbena was one of the very few of his importations to survive, and Sir J. E. Smith, who wrote Dombey's story[2] in what was obviously a white heat of indignation at the treatment received by a fellow-botanist, thought it should become "a *monumentum aere perennis* with those who shall ever know his history."

At one time this shrub was allotted to the family of Verbenas called *Lippia,* in honor of Auguste Lippi, a French botanist of Italian descent who traveled in Egypt and died in Abyssinia in 1733; but from the assortment of available names *Aloysia triphylla* is favored at present. NOTES: The Lemon Verbena is not hardy and generally hard to find in the U.S.

Amelanchier

The Amelanchiers are very hardy trees and shrubs, members of the order of the Rosaceae, the natives mostly of North America and north Asia. There is

[1]Samuel Wood, *A Plain Guide to Good Gardening,* 1891 [2]In Rees' *Cyclopaedia,* 1819

one European species—the first to be recorded—*A. ovalis* (syn. *A. vulgaris*). This was herbalist John Gerard's (1545–1612) "dwarffe kind of Medlar growing naturally upon the Alpes," but it does not appear that he ever grew it himself; his rather scanty description, which omits to mention the flowers, was probably copied from the French botanist Matthias de Lobel (1538–1616), who called it *Chamaemespilum,* dwarf medlar. Thirty-four years later Gerard's editor, Johnson, added a picture and a description of the flowers, which "come forth in the Spring three or foure together, hollow and of a herby colour," but still made no mention of its being in cultivation, either here or elsewhere. It was certainly grown in England before 1730, but its actual introduction has not been recorded.

The best-known member of the genus, *A. confusa* (syn. *A. laevis* or *A. canadensis*), is a small tree of 20 to 30 feet, and was introduced by the "tree-monger" Archibald, Duke of Argyll, in 1746. In its North American home it rejoices in many names, such as Shad Bush, Serviceberry, Grape Pear or Bloody Choak-berry. *A. oblongifolia,* also from North America, was thought by some botanists to be only a variety of *confusa;* but it is definitely a shrub in habit, and spreads by means of suckers. Sir Joseph Hooker[1] (1817–1911) quotes Dr. Richardson as reporting that its wood (*Meesassquat-ahtic*) was prized by the Cree Indians for making arrows, and was thence termed by the Canadian voyagers *bois de flèche.* Its berries (*meesasscootoom-mena*) were used by the Indians either fresh or dried, and were considered the best fruit in the country. "They form a pleasant addition to *Pemmican,* and make excellent puddings, very little inferior to plum-puddings." Hence, probably, its other name of Swamp Sugar Pear. It was an early introduction to England, and was known to herbalist John Parkinson (1567–1650).

Until 1789, the Amelanchiers and the few known species of Crataegus, Pyracantha, Sorbus and Cotoneaster were all classed together in the family Mespilus; but, as more and more species were discovered, it became possible to divide this very mixed genus into more coherent groups, and it is now reduced to a single species—*Mespilus germanicus,* the Medlar. *A. ovalis* was named by Linnaeus *Mespilus amelanchier*—Clusius having written that the natives of Savoy called it *amélancier* because the berries tasted of honey—and this name was chosen for the new genus when it was created by Medicus in 1789. The name of Snowy Mespilus makes its first appearance, so far as I know, in 1778, when it was used by Abercrombie for *A. confusa;* since then it has been applied to several of the species, and the *Royal Horticultural Society* (R.H.S.)

[1]In *Flora Boreali Americana,* 1840

Dictionary of Gardening gives it exclusively to *A. ovalis*, which in the early books was only called the Alpine Mespilus or the Savoy or Shrubby Medlar.

NOTES: Because the several species native to North America overlap in their distribution, there is often confusion among them. However, *A. arborea* (syn. *A. laevis*) and *A. canadensis* are the most commonly cultivated, although they are prone to both insects and diseases. New cultivars are appearing and there is at least one cultivar with pinkish flowers. Eventually, the shadbush, shadblow or serviceberry, as it is commonly called, will

Amelanchier florida
(SHAD-BUSH)   M. Drake (1833)

gain popularity because of the improved forms. The "shad" common names are derived from the fact that the flowering time coincides with the running of the shad fish. The name serviceberry was coined by mountain people who used the flowers for weddings in early spring and for funerals when the service had to be delayed until spring.

In Japan, the counterpart species, *A. asiatica*, is often used as a bonsai plant. This is one example of the numerous vicariad, or counterpart species that exist between Asia and the Appalachian Mountains of eastern United States.

Today, there is new interest in the Amelanchier as garden shrubs. They expand the spring floral display because of their profuse sparkling white flowers and add color in the autumn when the leaves vary from yellow to red. In addition, they are excellent plants for attracting birds to the garden. Their cultural requirements are simple, and they are best located at the edge of a woods or in the background of the garden. The serviceberry has an enormous natural range and is hardy well into Canada.

Ampelopsis, see under Parthenocissus

Arbutus

A. unedo, the Common Arbutus or Strawberry Tree, can be either a shrub or a tree of considerable size, according to the environment in which it grows. Parkinson says it "groweth but low, or rather like a shrub tree then of any bignesse" but Gerard quotes a report that it grows to a great size in the valleys of Mount Athos, "where being in other places but little, they become great and huge trees." In England it usually takes the form of a large spreading bush, unless specially pruned to keep it to a single stem; but tall and handsome specimens have been recorded in Ireland.

Primarily a plant of the Mediterranean region and well known in classical times, the arbutus also occurs in the neighborhood of Killarney, where its presence is explained by the legend that it miraculously appeared (together with the London Pride and a few other members of the Mediterranean flora) to comfort the monk Bresal, homesick for the Spanish monastery in which he had spent many months when teaching the Irish style of choral music. The scientific explanation, that these plants spread by way of Brittany in the early post-glacial period, when the British climate was warmer than it is now, and survived in a few favorable places while becoming extinct in the intermediate stages of their range, appeals no less to the imagination. It is said that the existence of the arbutus in Killarney is now threatened by the rapid spread of *Rhododendron ponticum*, which smothers the young seedlings necessary for the continuation of the race.

Plants from the Mediterranean have been brought to England ever since Roman, or possibly Phoenician, times, and one would expect the conspicuously handsome arbutus to have been an early introduction; but although described by the Rev. William Turner (1508–

68) in 1548, there is no certain record of its cultivation in the British Isles before 1586, when an unknown correspondent in Ireland sent plants to Lord Leicester and Sir Francis Walsingham, who were both "very desirous to have some, as well for the fruit as for the rareness of the manner of bearing, which is after the kind of the orange, to have blossoms or fruit green or ripe all the year long."[1] An inventory taken in 1649 of the Manor of Wimbledon, "Late Parcell of the Possessions of Henrietta Maria, the Relict and late Queene of Charles Stuart late King of England," mentions "one very fayre tree, called the Irish arbutis, standing in the midle parte of the sayd kitchin garden, very lovely to look upon," and values it at 30s—a great sum for those days.

By the beginning of the eighteenth century the arbutus was sufficiently common for landscape gardener Batty Langley (1696–1751) to recommend its use for hedges. He says that it makes "the most agreeable hedge as can be desired" although it "will not admit of being clipped as other evergreens are." Philip Miller in 1759 described it as well known, "being at present in most of the *English* Gardens, and is one of the greatest ornaments to them in the months of *October, November*, and frequently great part of *December*." He listed double and red-flowered varieties, besides a number of kinds distinguished by variations in the leaf. Marshall, however, in 1785, regarded the red-flowered form as less valuable than the original white, because "the contrast is not so great between their fruit and them ... the colour approaching too near to a sameness"; and the doubleness of the other kind could only be perceived on a close examination, while its fruits were few, "so that a plant or two, to have it said that the collection is not without it, will be sufficient." The red-flowered kind, named var. *rubra* by Aiton in 1789, was found growing wild in Ireland in 1835, and received the R.H.S. (Royal Horticultural Society) Award of Merit in 1925.

The arbutus takes twelve months to ripen its fruit, and the large granular globes, either round or oval in form, being in color "like a pallide clarret wine"[2] and appearing in October or later with the new season's flowers, justify all the praises that generations of gardeners have so lavishly bestowed. The shrub is hardy in most parts of Britain, but it fruits best in mild districts, probably owing to the lack of insects in the colder parts during its unusual flowering season. Bees will visit the flowers when conditions permit. The honey which is obtained from them in some parts of the Mediterranean is said to be lemon-yellow and aromatic, but bitter in flavor,

[1]Letter quoted in *The Trees of Great Britain and Ireland*, Elwes and Henry [2]Parkinson, 1640

and Pliny says that the tree should not be planted where bees are kept. The fruit is edible, but at the best insipid; most authorities refer to it as unpalatable, though it is occasionally brought to market in southern European countries, and can be used to make wine,[1] spirits or jam. Parkinson described the taste as austere, and Gerard as "somewhat harsh, and in a manner without any relish," while William Turner summed it up as having no good properties "but that it delyteth some men for the diversyte, for it is evell for the stomache and ingendreth the head ake." Henry Phillips says that the fruit should

Arbutus unedo
(STRAWBERRY-TREE)   P. Miller (1755)

be left "as of old, for the bird-catchers, to entice their prey in the winter season;" blackbirds and thrushes are very fond of it, and the Corsican *pâté merle*, as Mr. Clarence Elliott points out, consists largely of arbutus-berries at second-hand.

It was from the color and size of the fruit, but not from its flavor or shape, that the arbutus got its name

[1]"In Tuscany, many years ago, a man gave out that he had discovered a mode of making wine from the Arbutus. His wine was very good; but upon his leaving the country, his wine-casks were found to contain a quantity of crushed grapes." *Flora Domestica*, 1823

of Strawberry-tree, which was suggested by Turner. The Latin *unedo*, one-I-eat, is usually taken to mean that no one tasting a berry would care for a second, though Turner interprets it rather differently: "It is a fruyt of small honor, and thereupon hath the name, that it bringeth forth but one alone by it selfe." Parkinson says that in Ireland it was called Cane-apple, "with as great judgement and reason as many other vulgar names are," but Caleb Threlkeld, who in 1727 published the first book on the flora of Ireland, points out that the Irish name was *Keeora caihne*. "His ignorance of the *Irish* language made him censure the Name, for Pliny calls the fruit *Pomum*, and the word *Cachne*[1] is *Irish*, so that Mr. Parkinson ought to have forborne his Fling upon the Word *Cane-apple*." Phillips spells it Cain-apple, and concludes "that this name was bestowed on it by superstition, whose terrible imagination alone was able to transform these beautiful berries into clots of Abel's blood." The Latin *Arbutus* is said to be derived from the Celtic *Arboise*, Austere Bush, and should correctly be pronounced with the accent on the first syllable. The English botanist John Martyn (1699–1768) gives a number of excerpts from Pliny, Virgil and Horace, and sternly concludes, "After all these classical quotations, I hope we shall no more have the classical ear wounded by pronouncing the last syllable but one of Arbutus long."

A few other members of the family are in cultivation, but *A. andrachne*, brought from the Levant before 1724, is tender, and rarely fruits in Britain; *A.* × *hybrida* (syn. *A. andrachnoides*), its hybrid with *A. unedo*, is somewhat hardier, but should be classed as a small tree rather than as a shrub; and the American Madrona, *A. menziesii*, is a tree of the largest size.

NOTES: In the U.S.A., the Strawberry Tree (*Arbutus unedo*) is limited to southern California and certain parts of warmer Washington where it forms a shrubby tree with several trunks. The bark on older trunks cracks open displaying the red inner bark. There are white flowers in early winter followed by brilliant red, strawberry-like, inedible fruit, hence the name "strawberry tree." The large leaves are evergreen and lustrous. Coming from Mediterranean Europe, it requires a dry climate and acid soil. Our own native, the pacific Madrone (*A. menziesii*) is a large tree and not considered here.

---

[1] Misprint for "Caihne?" It is now called *Caithne* in Kerry and *Cuince* in Clare

Aristolochia

A. *macrophylla* (syn. *A. sipho*), Dutchman's Pipe. This handsome, curious climbing plant was one of John Bartram's many introductions to Britain; he sent the seeds from the Ohio River to Peter Collinson in 1761, and Collinson gave them to the nurseryman James Gordon at Mile End to raise for him. (Aiton gave the date as 1763—probably the year in which it first flowered; W. J. Bean's, "1783," is probably a slip, unfortunately repeated in the R.H.S. *Dictionary*, Collinson and Bartram both being dead by then). Sims, writing in the *Botanical Magazine*, in 1801, considered the plant peculiarly adapted for arbors and trellis-work, where it would "form a canopy impenetrable to the rays of the sun, or moderate rain," and describes a particularly handsome specimen on the front of a greenhouse in the garden of Dr. Pitcairn, "to the top of which it ascends, and clothes the pier from bottom to top in a very beautiful manner." Dr. Pitcairn (1711–91), a physician renowned in his day, had a botanic garden of some five acres at his country retreat in Upper Street, Islington. His nephew, Robert, as a midshipman of nineteen, was the first to sight the land named after him Pitcairn Island, afterwards so famous as the refuge of the *Bounty* mutineers.

The Dutchman's Pipe is a plant of vigorous growth, and John Claudius Loudon (1783–1843) speaks of an energetic specimen in Cambridge Botanic Garden, which, after reaching the top of its wall, proceeded to climb a tree in the next garden, some 20 feet in all. It flowers freely in some localities, but seldom fruits in Britain, though it was recorded as doing so in the very wet summer of 1958; the curious, meerschaum-shaped flowers are beautiful chiefly to the eye of the botanist. One would like to think that the Dutchman of the name was the celebrated Rip Van Winkle, whose twenty-year sleep lasted approximately from 1765 to 1785; but the Catskill Mountains, in which the story is set, are a long way from the Ohio River, and in any case the name did not appear until a century or more after the plant was introduced. Collinson called it the Aromatic Vine—all parts of it are scented, the root smelling strongly of camphor—Loudon, the Pipe Vine, and the *Botanical Magazine*, Broad-leaved Birthwort. It is the best-known hardy garden representative of a very large and varied genus containing about 180 species, over fifty of which, mostly hothouse climbers, are described in the R.H.S. *Dictionary*. Their flowers are "of extraordinary forms and sometimes with a very disagreeable odour";[1] Loudon says that some tropical American species have grotesque and

---

[1] R.H.S. *Dictionary of Gardening*

gigantic blooms, big enough "to serve as bonnets for the Indian children." Three or four of the herbaceous kinds are natives of Europe, and have been valued for centuries as medicinal plants; the name used for the family by Dioscorides in the first century A.D.—from *Aristos*, best, and *locheia*, parturition—and the English name of Birthwort, indicate the principal use to which they were put. *A. serpentaria*, the Virginian Snakeroot, was reputed to cure the bite of the rattlesnake, as well as being "so offensive to these reptiles, that they not only avoid the places where it grows, but even flee from the traveler who carries a piece of the plant in his hand";[1] and Egyptian snake-jugglers are said to use a species of Aristolochia to stupefy their pets before they handle them.

NOTES: The Dutchman's Pipe is a major contribution of our native flora to European gardens. It is a choice wall plant with heart-shaped leaves sometimes a foot long. The vines will climb vigorously and cover walls densely, so it makes an excellent screen plant. Dutchman's Pipe requires common garden soil and sun or semi-shade. It is perfectly hardy well into northern states. It is seen occasionally on walls and patio screens. There have been no particular efforts to improve the wild type.

Arundinaria

"No man can live without a Bamboo tree in the immediate vicinity of his house, but he can live without meat."
Taoist proverb

From the very earliest times the Bamboo has been so indispensable in the economy of the Orient that the proverb quoted above probably holds some truth. In the case of the native of China and Japan, there is "not a necessity, a luxury, or a pleasure of his daily life to which it does not minister."[2] Besides serving for every conceivable use, it was early appreciated as a plant of ornament, and sixty-one varieties were recognized by Chinese gardeners before 1688, ranging from the Dragon Bamboo, whose stems were 7 feet in diameter and 20 feet between the knots, with leaves like those of the banana, to the Dragon's Moustache, no more than a foot high and cultivated in pots, with stems hardly thicker than strong pins. This last sounds like *Arundinaria disticha*, which Bean described as having zig-zag stems "about as thick as a lady's hatpin." The Chinese have six different names for the bamboo shoot or cane, each denoting a different stage of its growth.

There are many points of interest about these shrubby grasses besides their many uses and their long history.

[1]Freeman-Mitford   [2]Anne Pratt

Most of them flower very infrequently, and nearly all the species die after flowering, or survive only with difficulty, and have to be started again from seed. They are not the only plants with this habit, but few others carry it to such dramatic and spectacular extremes; for it is usual for all the bamboos of a given species to flower in the same year, all over the world, wherever they are growing, indoors or out, cultivated or wild, and often with such insistence that not only the mature canes but young shoots only a few inches high will also flower and die. Since many of the species cover large areas, a flowering-year can cause considerable economic loss, occasionally compensated for by the abundant seed, which can be eaten like rice, and has on more than one occasion averted an Indian famine.

Another spectacular habit of the bamboo is the extremely rapid rate of growth of some of the species; *Bambusa vulgaris* is said "to have grown 40 feet in 40 days at Chatsworth, the noble seat of the Duke of Devonshire,"[1] and 30 inches in twenty-four hours have been recorded of *Phyllostachys mitis* in California. The canes of some species contain so much silica that they emit sparks when struck by an axe, and can be used to make edged tools of great sharpness and durability—hence also their polished glassy texture. The underside of the leaf retains a silvery layer of air when immersed in water, and though left to soak for a week, will come out with the lower side still perfectly dry. The leaves of all the hardy species have a network of fine transverse veins connecting the main longitudinal ones, and any bamboo lacking this reticulation of the leaf is sure to be too tender for outdoor cultivation in Britain.

There are a great many different bamboos widely distributed in America, Africa and Asia, and botanists have found them difficult to classify, owing to the infrequency with which some of them flower—it is said in India that a man who has seen two flowerings of the bamboo must be about sixty years old. They have, however, been sorted out into a number of families, and the hardy species mostly belong to the groups of *Arundinaria, Phyllostachys* and *Sasa*. (The *Bambusae* or true bamboos are tropical in habitat and tree-like in size.) The canes of the Arundinarias are cylindrical, and have more numerous branchlets at the joints than Phyllostachys, whose stems are flattened on alternate sides; Sasa varieties are dwarf plants with large leaves. But all are unmistakably bamboos, and for convenience I will include them all here.

Travelers' tales of the bamboo were early brought to Europe—for example, by Marco Polo about 1298. Parkinson knew all about the "Huge great treelike Canes or

[1]Munro

Aristolochia hirta

(BIRTHWORT)   Ferdinand Bauer (1840)

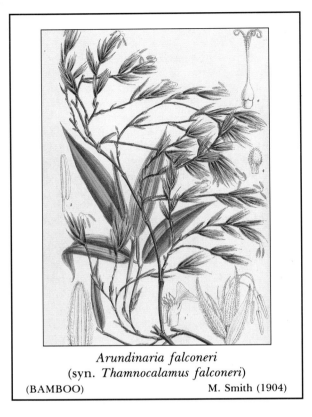

Arundinaria falconeri
(syn. *Thamnocalamus falconeri*)

(BAMBOO)   M. Smith (1904)

Reeds" which grew in Malabar, where they "bend them while they are fresh, that they may bee a little crooked and hollowish, to hange a Palankin." He saw, and carefully described, some broken pieces taken out of a Portuguese ship captured as a prize in 1593. But although known by reputation, it was a long time before any bamboo was cultivated in England. The first was *Bambusa vulgaris*, grown in a greenhouse by Philip Miller before 1730; but the hardy species were not introduced until the nineteenth century.

Arundinaria japonica, the Japanese Metaké or Female Bamboo (the "male" was *Dendrocalamus strictus*), was brought to Europe by the Bavarian naturalist and traveler Philip Franz von Siebold, and reached England in 1850. It is one of the hardiest and commonest garden species, and makes good screens and windbreaks, but is much exceeded in beauty by some of the later importations, for example, the Chinese *A. nitida*, which Reginald Farrer (1880–1920) called "the most lovely, delicate, vigorous and well-bred of all the hardy bamboos." Seed collected by Potanin in Northern Szechuan was sent to England by Dr. E. A. Regel of the St. Petersburg Botanic Gardens in 1889. It was raised in the Coombe Wood

nursery of Messrs. Veitch, and was at first thought to be *Bambusa khasiana*, the only kind then known to have black canes, but was later recognized as a new species and re-named accordingly. It is a graceful, 10-foot bamboo, less rigid in its growth than *A. fastuosa* which is one of the tallest kinds and has attained 25 feet in the southwest of England. This was imported from Japan by M. Latour-Marliac of Temple-sur-Lot about 1892, and came to England three years later. One of the best of the Arundinarias is *A. murielae*, similar to *nitida* but with green canes instead of black; it was found by E. H. Wilson in Western Hupeh in 1907, when collecting for the Arnold Arboretum, and was named by him after his daughter Muriel.

The first member of the Phyllostachys group to reach Britain was *P. nigra*, imported by the firm of Loddiges in 1823 or 1825, and listed in their catalogue of stoveplants as *Bambusa nigra* in 1826. It was soon found to be hardy, and a specimen in the garden of the Horticultural Society had reached a height of 7 feet by 1837; but it will grow much taller in favorable conditions, and Colonel Munro in 1866 spoke of "a very fine specimen in the Crystal Palace, 25 ft. high . . . but unfortunately it was one of the numerous splendid plants destroyed in the late disastrous fire." In Japan it has fifteen names and a great many uses, and its long creeping rootstock was the source of the flexible *wang-hai* or "whangee" canes which were an essential part of the equipment of the Edwardian "toff." The closely-related *P. henonis*, introduced from Japan in 1890, is first favorite with many bamboo-growers; graceful, luxuriant and hardy, its older stems bend almost to the ground with the weight of their foliage, while the young ones spring up, "arching and waving their feathery fronds, the delicate green leaves seeming to float in the air."[1]

P. mitis (syn. *edulis*) though very tall, and even in Britain a rapid grower, is not quite so reliably hardy as the foregoing species. This is the Moso, or Edible Bamboo, so much cultivated in the U.S. both in the east and in the southern states. The young shoots are cut when they are 8 or 10 inches high, peeled, and prepared in various ways for the table. The taste has been described as resembling that of sweet corn, combined with a flavor subtly oriental, but A. B. Freeman-Mitford (Lord Redesdale) says that though crisp and pleasant, like celery, "the flavour depends on the sauce." It was grown in Britain in 1890. Other species are also edible: Dr. David Fairchild tells a story about an elderly Frenchman who asked how he could get rid of a huge clump of the Japanese timber-bamboo, *P. bambusoides*, and who was much taken aback when Fairchild bawled down his ear-trumpet "Why don't you eat it?"

The largest leaves of all the hardy bamboos are borne by the dwarf-growing *Sasa tessilata*. Like those of the other large-leaved kinds, they were formerly used by the Chinese, sewn or skewered together, for packing tea; and a small curious glimpse of eighteenth-century London occurs in the *Universal Gardener and Botanist* (1778) where it is remarked that "we often see them in the streets, thrown out of the grocers' shops." This species has not been known to flower in Europe since its introduction in 1845. Unfortunately it is a terrible spreader and hardly to be trusted in even the wildest of wild gardens, if room is to be left for the owner. For it must never be forgotten that bamboos have the habit of couch-grass, and while some of the best only creep, there are many that run, and may be difficult to control. Some Japanese forests are rendered impenetrable by the density of their bamboo undergrowth, and high on the Andes another bamboo, *Chusquea aristata*, growing at an elevation equal to the height of Mont Blanc, makes thickets, called "Carizales," impassable even to animals. On swampy river-banks in the American South the North

American bamboo, *Arundinaria macrosperma*, grows in immense quantities and forms the "Cane-brakes" where runaway slaves used to hide. Gardeners must be cautious about encouraging certain kinds to make themselves at home.

In spite of this bad habit of spreading, the besetting sin of the family, the bamboo had, and still has, many admirers in England. In the 1890s it became for a time very popular—owing, perhaps, to the fashion for Japanese art set by Whistler, a decade or so before. The Bamboo-garden at Kew was planted in 1892, and in his monograph on the family published in 1896 Freeman-Mitford said there was at the time a "perfect craze" for hardy bamboos, and that "we need not despair of seeing in a few years miniature groves of bamboos, clothed in all their marvellous grace, and lacking no native beauty, save only at night the myriad darting lamps of the fire-flies, by whose light, as the pretty fable runs, Confucius and his disciples used to study." He did not, however, call his grove a Bamboozlem, the name given by Mrs. Mairi Sawyer to one particular section of her famous gardens at Inverewe in Ross-shire.[1] An appeal broadcast in 1959 by British zoo authorities for bamboo-shoots on which to feed the Giant Panda revealed that many species are still being grown in Britain. From the large variety of samples submitted the animal selected *Arundinariae anceps*, *japonica*, and *nitida*, and *Sasa tessilata*.

Apart from feeding pandas, British-grown bamboos are not particularly serviceable, except for furnishing garden-stakes—the canes do not ripen sufficiently well for other purposes, and are apt to split. Even the use for stakes is comparatively recent, for Colonel Munro, writing in 1866, mentions that "during the last summer" gardeners had been using, apparently for the first time, "a Bamboo which is sold abundantly in Covent Garden and elsewhere, for sticks for supporting plants, instead of the old-fashioned green ones." The bamboo furniture that some of us can remember was, of course, imported, but it is something of a surprise to learn that in the 1890s bamboo bicycle and tricycle frames were praised in the contemporary press for their strength, rigidity and comfort.

Arundinaria means reed-like, from *arundo*, a reed, Phyllostachys is derived from *phyllon*, a leaf, and *stachys*, spike, and Sasa is the Japanese name for a dwarf bamboo. In China, the thirteenth day of the fifth moon was the Day of the Drunkenness of the Bamboos, when a kind of clear eau-de-vie in which green bamboo-leaves had been steeped was drunk—but not by the bamboos.
NOTES: The name "bamboos" often strikes terror in the

[1]Robinson

[1]Now the property of the National Trust

hearts of the gardener because of the plants' insidious habit of appearing apart from their designated garden site. But the bamboos are regaining respect because of their versatility and charm. More frequently, they are being used in outdoor planter boxes, especially the ones with colorful culms (stems) such as the black bamboo, *Phyllostachys nigra,* or *P. viridis,* cv. 'Robert Young', which has yellow culms with green stripes. Young was the U.S.D.A.'s leading authority on the horticultural aspects of bamboo for several decades. *P. nigra* is a running bamboo attaining 25 feet in height with culms that turn black in the second year of growth. Running bamboos must be used with care. One way to use the black bamboo in a spectacular manner is in a container. The lower branches can be removed to display the shiny black culms and the plant kept in bounds by simply trimming the upper branches.

Sasa veitchii, although rampant, is a marvelous understory plant in light woods when the winter foliage displays pale bands at the leaf margins. The hardiness of the various bamboos is complex as they may be tropical, sub-tropical, or temperate in origin.

The American Bamboo Society in Springville, California, can provide information on culture and commercial sources of bamboo.

Aucuba

With characteristic perversity, we deny the name of laurel to the only member of that genus that we cultivate—*Laurus nobilis*—which we call the Bay, and bestow it on a number of totally unconnected shrubs such as the Common and Portugal Laurels (related to the Cherry and Plum), the Spurge Laúrel (a Daphne) the Laurustinus (a Viburnum) and, worst of all, on the Aucuba or Variegated Laurel, a member of the dogwood order, whose measled form is now so common that one hardly realizes that there is also an unspotted Aucuba, which can be quite a handsome bush.

A. japonica is diœcious, bearing its male and female flowers on separate plants. The first to be grown in Britain was a spotted female, introduced in 1783 by John Graeffer[1] from Japan, where it was cultivated for the sake of its handsome foliage—"for we do not know that it is applied to any use."[2] For some years it was grown

[1] A German and a pupil of Philip Miller, Graeffer was a partner with Thompson and Gordon in the Mile End nursery. After Gordon's death he went as gardener to the King of Naples at Caserta, and was murdered in 1816

[2] J. Sims, *The Botanical Magazine,* 1809

in the stove, and then in the cool greenhouse or conservatory. By 1809 it was being cautiously tried out-of-doors, though it was considered that the greenhouse specimens were more handsome. But lacking the essential cross-fertilization, the "Gold Plant," as it was called, produced no fruit, and one of the principal objects of Robert Fortune's visit to Japan in 1861 was to procure the male plant. As he said himself, "Let my readers picture to themselves all the Aucubas which decorate our windows and gardens covered . . . with a profusion of crimson berries. Such a result, and it is not an improbable one, would of itself be worth a journey all the way from England to Japan." At last he found an unvariegated male plant in the garden of Dr. Hall at Yokohama, and sent it to the nursery of Messrs. Standish and Noble at Bagshot, Surrey "where the mother plant was fertilized" and exhibited in fruit at Kensington in 1864. It created something of a sensation, *The Gardeners' Chronicle* calling it "this glorious shrub loaden with bunches of its large oblong berries of the brightest coral red." Two years later, while an unfruited plant could be bought for half a crown, a fruited specimen would cost from £15 to £20 and a male plant was priced at over a guinea a leaf, but could be hired for the purpose of cross-fertilization by enterprising nurserymen. (One almost expects to learn that a stud-book was kept.) Many varieties, of greater or lesser merit, resulted—twenty-one were given horticultural awards between 1861 and 1867—and the plant's popularity continued undiminished until the 1890s, when the Rev. W. Wilkes, then secretary of the R.H.S., wrote "You can hardly have too much of it. It is good in all stages, from the baby with only her six or eight mottled leaves, to the big spreading bush 4 ft. high by 5 or 6 ft. through, to fill a gap in your border." By 1914, plain and variegated plants of either sex were common, and Bean reports that "small plants in pots, with large crops of fruit, can be bought from costermongers in the streets of London." Artificial fertilization under glass is still practiced for the purpose of producing abundant berries, and if the plants should chance not to flower at the same time, pollen from the male can be kept for a week or two in a closed container until the female flowers are ready.

The name is a latinized form of the Japanese *Aokiba.* The Chinese too have a word for it, *Tao-yeh Shan-hu,* which means Peach-leaves Coral; but I suppose we shall go on calling it Variegated Laurel.

NOTES: This evergreen shrub became popular soon after its introduction in 1861 and remains so, especially the selections with golden markings on the leaves. The story of the introduction of the male type as described by Miss

Coats is an interesting one because Dr. Hall was so important for his early introduction of garden plants from Japan. One can visit Dr. Hall's estate in Barrington, Rhode Island, where some of his early plant introductions still exist. Although there are many forms of Aucuba with colorful leaves, none are more attractive than the wild type with plain green leaves when it bears large clusters of bright red fruits. For this to occur, a male plant must be present. There is a decumbent form, *A. japonica* var. *borealis* in northern Japan where it is able to survive by protection by the heavy snows. In cultivation, it forms a dense, low-growing plant suitable for planting on top of a low wall. I first introduced the *borealis* variety from Hokkaido, Japan, in 1955 and a demonstration planting is located in the Asian Valley at the National Arboretum.

The Aucubas are easy plants to grow, requiring semi-shade and a moist, acid soil with plenty of humus. They are perfectly hardy in the vicinity of Washington, D.C., and farther north in a protected locality.

Azalea, see under
Rhododendron

Benthamidia, see under
Cornus

Berberidopsis

The single species of which this family consists, the *B. corallina* or Coralberry, was introduced from Valdivia for Messrs. Veitch by their collector Richard Pierce[1] in 1862. It is a native of Chile, half shrub, half climber, and only half hardy, but worth attempting for the sake of its evergreen leaves and vivid red pendent flowers. It is also half-way between the Berberidaceae and the Lardizabalae and its discovery was very satisfactory to botanists, as it supplied a link between two natural orders which they had already surmised to be connected.

Berberidopsis means "berberis-like."

NOTES: Like most Chilean plants, the Coral Chilevine is hardy only in the most southern parts of California. It is not important in gardens elsewhere.

[1]Also spelt Pearce or Pearse

Berberis

. . . Conserves of barberries, quinces and such,
With sirops, that easeth the sickly so much . . .
Thomas Tusser, *Five Hundreth Points of Good Husbandrie,*
1573

The Berberis family is vast and complex. There are about 450 species in the genus, 170 of which are in cultivation, and they hybridize so freely that it is difficult to keep any of them true from seed. This is omitting the species with pinnate, not entire, leaves, which used to be included but are now separately classed as Mahonias; and although horticulturist James Shirley Hibberd (1825–90) thought the distinction served no useful purpose, at least it helps to relieve the congestion in a genus that has been much enlarged since his day. Most of them are natives of South America, Africa and Asia. There are a few European species, all but one of little garden value, and only two in North America, where British native *B. vulgaris*, introduced by early settlers, now almost outnumbers the indigenous *B. canadensis.*

B. vulgaris, the Common Barberry or Pipperidge Bush, may be either a British native or a very early introduction, for it was cultivated from time immemorial until the end of the last century as a fruit bush. The acid berries were pickled and used "to trimme or set out dishes of fish or flesh in broth . . . and many other wayes, as a Master Cooke can better tell than my selfe."[1] They were thought to be a good appetizer "for those that loath their meate,"[2] or as herbalist Nicholas Culpeper (1616–54) put it, "they get a man a good stomach to his victuals, by strengthening the attractive faculty which is under Mars." This "curious and very wholsome pickle"[3] was still a favorite garnish for dishes in the late eighteenth century, and Loudon, some years later, mentions the use of the fruit for jelly or "rob," and on the Continent as a substitute for lemons in punch, and in the famous *confiture d'épinette* of Rouen. For this a seedless variety was preferred, which was known to Gerard. Parkinson doubted its existence, but Loudon explained that it only became seedless when the plants were mature. The fruit was still in use in 1863—by which time five varieties were cultivated—but now an ominous comment appears, "tedious to gather"; a grave economic drawback, which the gooseberry has managed only with difficulty to survive, and which may account for the barberry's present disuse.

The leaves were also used for the table; they made a sour sauce similar to that of sorrel, which was good for

[1]Parkinson, 1629 [2]Parkinson, 1640 [3]Abercrombie

"a fainting hot stomacke and liver" and to repress "sour belchings of choller." Thus Parkinson, who also tells us Clusius' secret remedy of the inner yellow bark steeped in wine, which would "purge one very wonderfully." This bark was believed to be a remedy for jaundice, probably on account of its color; Kilvert records its use on a Herefordshire farm for the same purpose in 1878. It is very astringent, and was used in Poland to tan leather and to dye it a fine yellow at the same time. The root, or a lye made from the ashes of the whole plant, was used as a hair-dye, and various preparations from the juice of the fruit were valued for pestilential fevers from the time of the Ancient Egyptians till the mid-nineteenth century, when barberries were still used by herbalists, though discarded by the medical profession in favor of the currant. (No particular use seems to have been made of the spines.) The flowers have fascinated botanists on account of the irritability of their stamens, which will close in on the stigma if lightly touched at the base with a pin. Loudon described some rather unkind experiments which demonstrated that if the plant "is poisoned with any corrosive agent, such as arsenic or corrosive sublimate, the filaments become rigid and brittle, and lose their irritability; while on the contrary, if the poisoning be effected by any narcotic, such as prussic acid, opium or belladonna, the irritability is destroyed by the filaments becoming so relaxed and flaccid, that they can easily be bent in any direction." (Dr. Lindley thought that this indicated a rudimentary nervous system.) It was also discovered that the stamens would still contract after being cut from the flower, "just as the heart can be excited to action after removal from the body."

The common barberry was already in use for hedges in the early fifteenth century. Gerard says it was employed almost exclusively for the purpose in "a village called Iver, two miles from Colbrooke"; but it was early realized that it had some mysterious and baleful influence on corn, a belief which, Loudon said, "though totally unfounded, is of remote antiquity." Controversy raged on the subject during the eighteenth and nineteenth centuries, between practical men who demonstrated the fact by experiment, and scientists, many of whom denied it, quoting for example Saffron Walden, where the corn grew right up to the barberry hedges without ill effect. Sir Joseph Banks thought that an insect frequenting the bush generated a dust which blighted the growing corn; later scientists, who knew more, were actually further from the truth, for they examined the fungus on the shrub and the fungus on the wheat and flatly declared they were of different species. It was not then realized that the life-cycle of the fungus—*Puccinia graminea*, the Black Rust—went through two different stages, and that it was the intermediate or cluster-cup stage that was found on the shrub. Where climatic conditions are unsuitable for the growth of fungi, the berberis would be harmless. Unfortunately, in the panic engendered by this discovery, which was made between 1860 and 1865, the common barberry has been practically exterminated, and hardly dares show itself in a catalog,

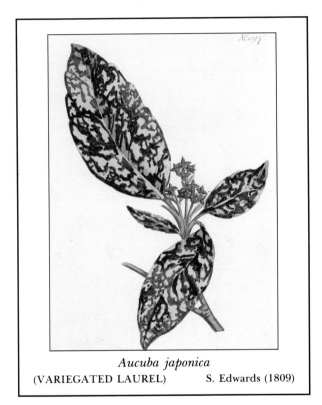

Aucuba japonica
(VARIEGATED LAUREL)   S. Edwards (1809)

though its purple-leaved variety is still occasionally to be found. This is a pity, for it can compete for beauty with any of the foreign species, and has been praised as an ornamental plant by discriminating gardeners from Batty Langley in 1728 to Mr. Haworth Booth and the late Captain Kingdon-Ward, and there are many gardens remote from cornfields where it might quite safely be grown. It is only *B. vulgaris* and its varieties that are subject to this blight—the other species are immune—but the whole race appears to be regarded as suspect in the U.S.A. and their importation forbidden.

The foreign species are excessively numerous; the R.H.S. *Dictionary* selects some eighty for full description, from *acuminata* to *zayulana*, almost every one with a thicket of synonyms, varieties, related species and hybrids springing up around it. Roughly speaking, the earlier ones came from the Himalayas and were followed first by a trickle and then by a spate from Upper Burma and western China, with sporadic additions from Chile from time to time. The majority of the Chinese species were introduced in the present century, Wilson, Forrest, Farrer, Kingdon-Ward and Ludlow and Sheriff's expeditions all making large contributions. (Farrer once

Berberidopsis corallina
(CORAL-BERRY)   *Flore des Serres* (1874)

in cultivation, which was found by Sparmann in Tierra del Fuego, where the natives used its wood for bows "on account of its great elasticity." But of far greater importance was another introduction from the same continent, *B. darwinii*. This was discovered by Charles Darwin himself, as a young man of twenty-six, in the course of the famous voyage of the *Beagle*, although it is not mentioned in his journal, which contains few references

Berberis cretica
(BARBERRY)   Ferdinand Bauer (1823)

described a mountain-slope in China as being covered with "Burberries"; having regard to the climate, he and his companions must often have wished that this was indeed the case!) Three useful and popular Chinese species, *B. candidula, gagnepainii* and *verruculosa*, were all introduced by E. H. Wilson for Messrs. Veitch, about 1904. The first had already been raised on the Continent by Messrs. Vilmorin from seed sent by the Jesuit missionary Paul Farges; the second, not showy but a good hedge-plant, was named after a French botanist, M. Gagnepain,[1] at one time in charge of the Herbarium at the Jardin des Plantes.

The earliest South American species to reach England was *B. ilicifolia* (1791), a handsome Berberis, now rare

[1] b. 1866

to plants.[1] It was introduced from the island of Chiloe about fourteen years later (in 1849) by William Lobb, a collector sent out by the firm of Veitch, and it is the parent of two of our most useful garden hybrids, *B. × lologensis* and *B. × stenophylla*. The former, a natural hybrid between *B. darwinii* and *B. linearifolia*, and almost worth growing for the luscious name alone, was found near Lake Lolog by H. C. Comber, who collected in the Andes on the Chile–Argentine border in 1925–27, and to whom we owe the introduction of about eight berberis species. *B. linearifolia* was among them, and plants raised from seed received in 1927 were awarded the R.H.S. First Class Certificate when exhibited in 1931. This spe-

[1] He presented all his specimens to Hooker, who named this shrub in his *Icones Plantarum*, 1844

cies has been called "a very Cleopatra" in its attraction for the pollen of other varieties; of seed collected from it in the wild only 50 percent came true, uncontaminated by the more abundant *B. darwinii*. The progeny is variable, and care must be taken to obtain a good form. The other important Darwin hybrid, *B.* × *stenophylla*, appeared as a self-sown seedling in the nursery of Fisher and Holmes at Handsworth, near Sheffield, about 1860, and is now perhaps the most popular of all the family, with a large progeny of varieties. Its other parent, *B. empetrifolia*, came from Chile in 1827, but is now little grown.

All these, and many others, are evergreen. The deciduous species are also very numerous, and have an even larger progeny of varieties and hybrids. The most popular of the early introductions was the Japanese *B. thunbergii*, first described by the Swedish botanist C. P. Thunberg in 1784, but not introduced here until 1883; it has several varieties, of which *atropurpurea* is probably the best known, and is one of the finest of all purple-leaved shrubs. It was first recorded in the French nursery of M. Renault during the First World War. Among the most prolific and popular of the deciduous kinds is *B. wilsonae*, another of those introduced from China by E. H. Wilson in 1904, and named in honor of his wife (*ae* is a feminine ending, and signifies that the plant in question is dedicated to a lady). Under this heading the R.H.S. *Dictionary* includes two other closely related species, and seven named garden hybrids, not including *B.* × *rubrostilla*, which has the honor of a separate entry, and has itself numerous descendants. It arose from one of two plants, supposed to be of *B. wilsonae*, received at the R.H.S. gardens at Wisley from Messrs. Veitch about 1909 or 1910, one of which showed signs of hybrid origin. Seeds raised from it gave varied results, and the best, selected and named by Mr. F. J. Chittenden, was awarded an F.C.C. in 1916; but the other parent of this chance cross is unknown. Wilson also introduced *B. aggregata* (1908)—another parent of many garden varieties, including the popular 'Barbarossa'.

There would, therefore, be no difficulty in finding enough to furnish a Berberis Garden, such as was enthusiastically advocated by Shirley Hibberd in 1898 when the genus was still something of a novelty. For my part I agree with Loudon, that their many suckers give them "a rough, inelegant appearance," and think their autumn color insufficient compensation for their lack of form and their tiresome prickles. (Only one nurseryman, in all the catalogs I have seen, has the courage and honesty to warn his clients that *B. wilsonae* has "frightful teeth.") Gertrude Jekyll has a curious passage in *Children and*

Gardens about her early hatred of the smell of barberries, and recalls a time "when to me it was so odious that it inspired me with a sort of fear; and when I forgot that the Barberries were near and walked into the smell without expecting it, I used to run away as hard as I could, in a kind of terror." Francis Bacon (1561–1626) seems to have disliked it too, and placed barberries in his heath garden "but here and there, because of the smell of their blossom." William Hazlitt's (1778–1830) recollections of the fruit, during his visit to America in 1783–87, were more favorable: "The taste of barberries, which have hung out in the snow during the severity of a North American winter, I have in my mouth still, after an interval of thirty years; for I met with no other taste in all that time at all like it. It remains by itself, almost like the impression of a sixth sense."

The Arabic name *berberys* or *barberis*, used for the fruit in the twelfth-century medical school of Salerno, was in turn derived from the medieval Latin; but the ancient Romans do not seem to have had a word for it, unless it was the "kind of thorny bush called Appendix" mentioned by Pliny, with "red berries hanging thereto which are likewise named Appendices' and which would stay the flux of the belly. The English name of Pipperidge-bush, which goes back at least to the sixteenth century (Turner in 1538 spells it "Pypryge"), may be derived from *pepin rouge*, which has a Norman ring about it, and hints that *B. vulgaris* may have an aristocratic descent from the Norman Conquest of England. Other dialect names include guild-tree, jaundice-berry, rilts and wood-sour. According to Sir John Mandeville, Our Lord was crowned with four successive garlands of thorns, one of which was made of "barbarynes"; the tradition is said to have arisen because the spines grow in threes, and therefore signify the Trinity.

NOTES: The taste for the many barberry species is not as developed in America as elsewhere, because of long-standing quarantine restrictions; many species have been found to act as alternate hosts for black stem rust of wheat. Of the few species commonly cultivated, the Japanese barberry (*B. thunbergii*) is one of the finest, especially the cultivar 'Crimson Pygmy', which is a rugged, low hedge plant with colored foliage. Miss Coats does not mention *Berberis julianae*, a hardy evergreen species with broad spined leaves and bright yellow flowers, making it the most popular of evergreen barberries in America. Use of *B.* × *mentorensis*, a hybrid of the two species mentioned above, as an upright hedge, is becoming popular throughout the states. In milder climates, it remains evergreen but is deciduous in the north. These are especially drought-tolerant shrubs of great garden merit. This hybrid was developed in 1924 by Elmer Schultz and Michael Horvath and patented in 1934 as a semi-evergreen, rust-resistant variety with extreme hardiness, withstanding low temperatures of −25 °F. without defoliation.

In many areas, *B. verruculosa*, the Warty Barberry, is also becoming popular as a dwarf evergreen shrub with a neat, compact habit. It is particularly useful in the rock garden and hardy throughout the eastern part of the country.

Bignonia, see under *Campsis*

Buddleia

The Buddleias are fairly widely distributed; species occur in Central and South America and in South Africa, but are most numerous in China. *B. globosa*, the Orange Ball tree, was the first to reach Britain, brought from Chile in 1774 through the agency of the famous firm of Lee and Kennedy, of The Vineyard, Hammersmith, London. Like most newly introduced exotics of the time it was first cultivated in the greenhouse, but it soon outgrew its quarters and was cautiously tried out-of-doors. It has since proved almost completely hardy, though it is at its best in warmer districts. Mr. Clarence Elliott quotes a well-intending nurseryman who tried to anglicize its name as "the Globose Buddlebush."

B. davidii, better known as *B. variabilis*, is the most familiar of the buddleias; in America it is called Butterfly Bush, and in its native China, Summer Lilac. It was one of the many plants found in the neighborhood of Ichang by Dr. Augustine Henry about 1887, and was named by him after the French Jesuit missionary also commemorated in Père David's Deer,[1] who had been the first to discover the shrub, some twenty years before. The first importation, via St. Petersburg, was of a poor form. A much better variety was raised by the French firm of Vilmorin from seed sent by another botanist-missionary, Père Jean André Soulié, in 1893, and reached Kew by way of the Jardin des Plantes in 1896. Two other handsome forms, vars. *magnifica* and *wilsonii*, were collected for Messrs. Veitch by E. H. Wilson in the first decade of the present century.

Burdened in late summer with Peacocks and Red Admirals, this incense-scented shrub is rendered almost

[1]David made several exploratory journeys in China between 1862 and 1867

more ornamental by the butterflies that it attracts than by its flowers; but in its native country it shelters a less innocent fauna, its thickets on the shingle-banks of the Satani River provided, according to Farrer, "famous harbourage for leopards." During the Second World War its windborne seeds found congenial conditions on the rubble of many of the London bombed-sites, where they grew into flourishing bushes.

At the start of the First World War, Mr. W. van der Weyer of Smedmore House, Corfe Castle, home on leave from France, found two belated heads of bloom on *B. globosa* and was able to cross them with *B. davidii* var. *magnifica* (not normally in flower at the same season) in the hope of procuring a plant that would have plumes of flowers similar to those of *davidii*, but yellow in color. The first results of the cross were disappointing, but the second generation of seedlings provided more variety, including a plant with upright panicles of balls of flowers described as "orange and yellow . . . shaded with pink and mauve," which was named *B. × weyeriana*. Unfortunately, the yellower forms of this cross proved infertile, and the experiment went no further. Since then the nurserymen seem to have concentrated only on raising handsomer and more colorful forms of *B. davidii*.

There are a number of other interesting species in the genus, of which *B. alternifolia* seems likely to become of increasing importance. It was discovered in north-west China as early as 1875 by a versatile Russian scientist called Pavel Jakovlevich Piasetski, but was introduced only in 1914 by Reginald Farrer, who described it as being "like a gracious small-leaved weeping-willow when it is not in flower, and a sheer waterfall of soft purple when it is." It is the most northerly of the species and is perfectly hardy, which unfortunately cannot be said of *B. colvilei*, whose large pink flowers are the showiest of the genus. This was found in 1849 by Sir Joseph Dalton Hooker, who called it "certainly the handsomest of all Himalayan shrubs"—high praise when it is remembered that some of our loveliest and most important rhododendrons were brought back from the same expedition. (The buddleia was not introduced until later, and first flowered in Ireland in 1892.) It is frequently misspelled with a double l. Hooker himself informs us that he named the plant after "my late friend, the Rt. Hon. Sir James Colvile, F.R.S." then Puisne Judge of the Supreme Court of Calcutta, who had presented him with a completely equipped *palkee* (a sort of palanquin) at the outset of his Indian travels. Other tender but less striking species are *B. salvifolia* (Miller, 1760) the South African Sage-wood, which, like *B. davidii* on the bombed-sites, is the first shrub to reappear in mountain forests after a

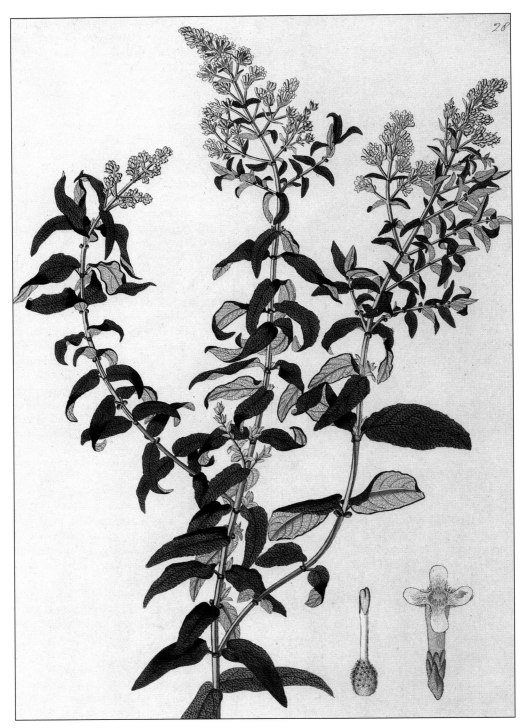

Buddleia salvifolia
(S. AFRICAN SAGE-WOOD) N. von Jacquin (1797)

fire, and whose wood was "esteemed by the Kaffirs for making assegai shafts,"[1] *B. lindleyana* (1844) a favorite garden plant in China, which Fortune admired growing wild in Chusan hedges, and whose crushed flowers thrown into water will stupefy fish, and *B. auriculata* (S. Africa, 1881), valuable in the gardens where it can be grown, for its winter-flowering and sweet scent.

The genus was named after the Rev. Adam Buddle (*c.* 1660–1715), one-time incumbent of North Fambridge in Essex. A keen amateur botanist and already an authority on mosses in 1687, he compiled before 1708 an exceptionally accurate and comprehensive herbarium of British plants, arranged according to a new system, of which he says, "I have jumbled Mr. Ray's and Mr. Tournefort's together (they are both dead). Some think I favor too much Mr. Tournefort, which is a reflection upon Mr. Ray which I am sure I do not design . . . but I find that he that would please everybody must never print." In fact his work was never published; after his death the herbarium and related manuscripts were left to Sir Hans Sloane, and through him passed eventually to the British Museum, where they were much consulted by later botanists, often without acknowledgment. "Justice was not done him by those of his immediate successors who profited by his labours";[2] and although the balance was somewhat redressed by the dedication of a genus in his honor (by Dr. William Houstoun, when collecting in South America in 1730–33), the name was so mangled by later botanists as almost to defeat its object. Houstoun himself spelled it Budleia; Linnaeus, publishing the new genus in his *Genera Plantarum*, made it Buddleja in the text, and Budleja in the index; and in a posthumous edition of Houstoun's work published by Sir Joseph Banks in 1781, three different spellings occur on the same plate—Buddleja, Buddleia and Buddlea. Other versions, such as Budlaea, Budlea and Buddleya have also been used, so that the name, as well as the commonest species, seems to have earned the epithet of *variabilis*.

NOTES: The buddleias are particularly useful in the garden because they grow vigorously and have no particular cultural requirements. The most commonly grown is the Fountain Buddleia, *B. alternifolia,* a large shrub with graceful arching branches and purple flowers borne on last year's branches in late spring. It is also the hardiest buddleia we cultivate. *B. davidii* is an upright shrub with terminal spikes of white, pink or purple flowers, depending on the variety. In colder climates it dies to the ground in winter and produces its flowers on new spring shoots. Consequently, it is more appropriate in

the perennial border than placed among other shrubs.

It is interesting that the improved forms of Buddleia are mostly of European origin, involving only the above named species. The most recent to attract attention is *Buddleia* 'Lochinch', an upright bushy plant of gray leaves and lavender flowers with an orange eye. The flowers are produced on the branches produced in the spring. Some growers advocate removing the dead flowers to extend the season of bloom.

Buxus

There is some doubt whether the common Box, *B. sempervirens*, is truly a native of Britain. Its distribution is very local, but tentative identifications have been made of box pollen from Interglacial, and of boxwood charcoal from Neolithic periods, and there are several records of it from Roman sites, including three of burials in coffins lined with sprays of box. Dr. H. Godwin in his *History of the British Flora* (1956) says there are no records of this practice of burying with box in Italy or elsewhere in Europe, and he deduces that the box must have been indigenous, as the Romans would be unlikely to have introduced it specially for this unusual funerary purpose. But he has perhaps forgotten how very fond the Romans were of topiary-work—to such a degree that at one time the terms "gardener" and "topiarist" (*topiarius*) were interchangeable. It is almost certain that a Roman owning a villa with any pretensions to a garden, in this remote outpost of their Empire, would have some form of topiary in it, and very probably of box—whether indigenous or not. The custom of placing a basin of boxsprigs at the door, so that persons attending a funeral might each take a spray which was afterwards thrown into the grave, still lingered in the north of England in the nineteenth century, and is mentioned by Wordsworth.

The principal locality for the "wild" box in Britain is at Box Hill in Surrey, which was a noted pleasure resort in the seventeenth and eighteenth centuries. The writer John Evelyn (1620–1706) in 1662 said it was frequented by the "Ladies, Gentlemen and other water-drinkers from the neighboring Ebesham[1] Spaw"; and half a century later it was remarked that it was "very easy for Gentlemen and Ladies insensibly to lose their company in these pretty labyrinths of Box-wood, and divert themselves unperceived . . ."[2] By 1759 the "delicious groves" had already been much cut down, but enough remained for the timber still to be sold at considerable profit in

[1]Bean   [2]*Dictionary of National Biography*

[1]Epsom   [2]John Macky, *A Journey Through England,* etc., 1714

the following century; the value of the wood marketed in 1715 being £3,000, and in 1815, £10,000. The wood was sold not by volume but by weight—"which renders it still more valuable to the cultivator."

The trees on Box Hill are of course of the tallest growing variety, which will reach in time a height of 30 feet. Its natural form, so elegant and graceful that in Persian literature women are frequently compared to box-trees, is rarely seen in gardens, where the box is almost invariably used for what Loudon called "verdant architecture and statuary." L. Liger tells us that it was formerly in demand for palisades about parterres, "but in *France*, where the people are so impatient that as soon as a thing is projected, they would immediately see the End of it, this Fashion of planting Box is quite out, because they can't stay till 'tis grown up, tho' 'twould contribute very much to the Ornament of a Garden."

There are many varieties. The first reference to a variegated sort is to be found in Parkinson's *Paradisi*, where he describes a "gilded" box whose leaves have "a yellow lift or garde about the edge of them on the upper side . . . which maketh it seeme very beautifull," and which, he tells us later in this *Theatrum*, "hath not been mentioned by any Writer before me." (Previous to this, in 1608, Sir Hugh Plat had forestalled nature by giving a recipe for gilding growing branches of box with gold-leaf, which he says "will carry . . . a long time fair, notwithstanding the violence of rain.") Eight varieties, with silver or gold stripes or edges, broad, narrow or curled leaves, and different habits of growth, were known by the end of the eighteenth century, and fourteen are listed in the R.H.S. *Dictionary* today. By far the most important of these is the dwarf variety *suffruticosa*. This goes back at least to the time of Gerard, and was considered by Parkinson "a mervaillous fine ornament" for edges and knots. It was much used for the embroidered parterres so fashionable in the grander gardens of the late seventeenth century, a style which survived until the nineteenth century in such places as the Vatican gardens, where "the name of the Pope, the date of his election, etc., may be read from the windows of the palace in letters of box,"[1] and in the gardens of Holland House, where the motto and crest of the family—a fox—were carried out in the same accommodating medium. (A box fox sounds like a pleasing conceit.) Almost all the early gardeners, from Evelyn onwards, are loud in praise of "this valuable little shrub" as an edging for borders. William Cobbett (1762–1835) is particularly eloquent on the subject, and concludes, "If there be a more neat and beautiful thing in the world, all I can say is, that I never

saw that thing." Only Batty Langley, in the eighteenth century, raises a dissenting voice; he warns us that it grows "more in its Roots than its Top, and is an open Robber of every Plant or Shrub that grows near it." And William Robinson (1838–1935), at the dawn of the twentieth century, points out that besides sheltering slugs and weeds, it requires a great deal of labor to keep it in really good order, "and not every garden workman can clip Box well."

Both tall and dwarf sorts were much used in the Dutch style of gardening that came in with William and Mary. Landscape gardener Stephen Switzer (1682–1745) thought the garden they laid out at Hampton Court was "stuffed too thick with box," and Queen Anne, who, it is well known, disliked the scent, after her accession cleared it all away, both from Hampton Court gardens and those of Kensington Palace. Long before her day, Gerard had called the smell "evill and lothsome," and earlier still, Henry Lyte had laid it down that Box was not only harmful if taken internally, but was "very hurtfull for the brayne when it is but smelled to." Nevertheless, Evelyn pointed out that society diverted itself on Box Hill "without taking any such offence at the smell, which has of late banished it from our groves and gardens," and added that water cast on it immediately after clipping "hinders all those offensive emissions that some complain of." It is a question of personal taste; Oliver Wendell Holmes found in it the fragrance of eternity, "for this is one of the odours that carry us out of time into the abysses of the unbeginning past; if ever we lived on another ball of stone than this, it must be that there was box growing on it."[1]

Various medicaments, apparently more than usually nauseous, were extracted from box-leaves, but were employed rather by "foolish emperickes and women leaches"[2] than by reputable physicians. As late as the nineteenth century, however, a decoction of boxwood was still being recommended for leprosies and "foulnesses of the blood," and Phillips says that "the quacks of ignorant ages" extolled the virtues of box "for diseases that delicacy forbids us to mention." (The leaves in fact yield an extract called Buxine, which was still used as a narcotic, sedative and purgative during the Second World War.) Parkinson, moreover, informs us that "the leaves and the dust of the wood boyled in lye, will make haires of an Aborne[3] (or *Abraham*) colour," and Phillips tells of a young woman of Silesia, who, having lost all her hair owing to a "malignant dysentery," washed her head with a decoction of boxwood, which resulted in the production of a fine mane of chestnut locks, "but having

[1]Loudon

[1]*Elsie Venner*, 1861   [2]Gerard   [3]Auburn

used no precaution to secure her face and neck from the lotion, they became covered with red hair to such a degree, that she seemed but little different from an ape or a monkey." He does not give the source of this horror-comic, and I would not vouch for its veracity.

The wood of both stem and root, so dense and heavy that it sinks in water, has been highly valued from the time of the Greeks, who used it for combs, musical instruments and other purposes. "Turners and cutlers," says Gerard, speaking of the root, "do call this woode dudgeon, wherewith they make dudgeon-hafted daggers"—which leads us straight to Macbeth's description of his visionary weapon: "And on thy blade and dudgeon gouts of blood . . ." (The *Herball* was published in 1597: *Macbeth* was acted in 1610.) The root was also used for cabinet-making and inlaying, furnishing "pieces rarely undulated and full of variety."[1] Cross-cut sections of the stem were employed for wood-engraving, and when this was the principal medium for illustration, before the advent of photographic reproduction, nearly 600 tons of the wood were annually imported from Turkey. It is still in use for this and other purposes, such as the making of mathematical instruments; the demand always tending to exceed the supply.

The name is derived from the Greek *puknos*, meaning close or dense. Boxes of the wood were called *pyxides*, the origin of the Roman Catholic "pyx," the sacred chest containing the Host. Martyn says that he has never known the dwarf box to flower; but the taller sorts will do so if allowed to grow freely, and produce a fruit which Gerard described as "having three feet or legs like a brasse or boiling-pot." The leaves are curiously constructed in two layers. Cut an old one across, and press the edges inward, and it will gape apart. They are poisonous to camels; the silly creatures love them, and in parts of Persia where box is abundant the caravans have to be restricted to horses, mules and oxen. Or so it is said.

NOTES: To update the history of common box (*Buxus sempervirens*), it should be noted that Edgar Anderson, an American botanist, searched along the Vardar River in Roumania in 1935 for hardier types. He collected several which were sent to the Arnold Arboretum. One has been highly successful and introduced into the trade as 'Vardar Valley'. Its main attribute is that it is especially hardy, remaining green throughout the winter where other boxwoods tend to be injured. In 1971, I collected boxwoods along the Vardar Valley in Yugoslavia, where the plant survives on barren calcareous outcroppings either in copses or as solitary plants in breaks in the

[1] Evelyn

rocks. As Miss Coats pointed out, horses and goats which browse everything else, studiously avoid the boxwoods.

The oriental boxwood, *B. microphylla*, because of its hardiness and somewhat distinctive habit, has an important place in American gardening. Native to China, Korea and Japan, it is considered to be hardier than common box but does turn brownish at the tips under severe conditions. It is a shrubby plant and some of the dense forms such as 'Kingsville Dwarf' and similar selections made at the old Kingsville Nursery in Maryland, are interesting small-leaved compact shrubs. They are suitable as rock-garden plants or low hedges. In Japan, the bright yellow, lustrous wood is highly prized for combs, seals, abaci counters and rulers. Small plants are prized for bonsai. Finally, the oriental boxwood does not have the unpleasant "foxy" odor of common boxwood.

There are several fine boxwood collections throughout the eastern states but the one at Blandy Farm, Virginia, headquarters of the American Boxwood Society, is the most complete.

Calluna

The tiny heath-flowers now begin to blow;
The russet moor assumes a richer glow;
The powdery bells, that glance in purple bloom
Fling from their scented cups a sweet perfume,
While from their cells, still moist with morning dew
The wandering wild bee sips the honied glue . . .
John Leyden, 1775–1811

It is true enough that heather-honey has a somewhat glutinous, or rather, jelly-like quality; it never granulates like other honeys, but retains the tiny air bubbles it acquires during extraction. Most people today consider it the best of all honeys, but this was not always so. The very first recorded description of heather was that of Dioscorides, thus translated by Turner: "Irica, sayth Dioscorides, is a busshy tre lyke unto Tamariske, but much lesse, of whose floures bees make noughty honey." Parkinson says that although called *mel improbum*, "we have not found any ill quality therein in our Lande, onely it will be higher coloured then in those places where no heath groweth." But right up to the end of the eighteenth century it was considered of less value than other kinds. The bees themselves find it indigestible, and taking hives to the heather is a hazardous and uncertain affair. The amount of nectar secreted depends on the type of soil on which the plant grows, as well as on the vagaries of the weather. On parts of the Continent it

Buxus sempervirens

(BOXWOOD)   J. Sowerby (1804)

Calluna vulgaris

(LING)   *Botanical Magazine* (n.d.)

is believed that thunder at once puts a stop to the nectar-flow, which will not be resumed till fresh flowers open.

Few shrubs monopolize such large tracts of land as this, the true Scots heather or ling (*C. vulgaris*); and what other flowering plant is responsible for a geological formation?[1] A heath is simply a place where heath or heather grows, and Carolus Linnaeus (né Karl Linné, 1708–78) tells of a Lapp superstition that there are two plants that will finally overspread and destroy the whole earth—heath and tobacco. The Erica and Calluna families were originally limited to the Old World, where they have an unusual longitudinal distribution, ranging from Norway, with few interruptions, to the Cape of Good Hope. But homesick Scots have introduced the ling to America, and it is now naturalized in parts of the Northeast, so a beginning has been made—though I have not heard of tobacco being widely grown in Lapland.

The impoverished economy of the moorland regions led to the discovery of many uses for this shrub, the dominant species over so many barren miles. The Highlander devised a method of using it for building. The thatch on his cottage was bound with heather ropes, he

[1]Peat

lay on a heather bed, and his wife swept the floor with a heather besom:

> Buy, buy, brushes,
>   Buy them when they're new;
> Fine heather besoms,
>   Better never grew.

It was used to tan leather, and to dye wool yellow, as a fuel to heat the bread-oven, and with malt, to brew the heather-beer. This was not the same as the legendary heather-ale, whose secret was lost when a Pictish father arranged for the execution of his own sons, knowing they might betray it to their Irish conquerors, while he himself never would. But the most important use of the ling is as fodder for sheep (also for deer, grouse, and the caterpillars of the Great Egger moth), and it is so superior to other heaths in that respect that on Lammermuir the ungallant shepherds used to call it "he-heather" while the worthless cross-leaved heath (*Erica tetralix*) was known as "she-heather." In favorable conditions, old plants can grow to the height of a man, and have hidden many a fugitive. But such bushes are little use to sheep or bees, so portions of the moor are annually burned by local farmers to ensure a plentiful supply of young

Calycanthus florida
(CAROLINA ALLSPICE)   J. Sowerby (1797)

growth; edicts regulating such moor burnings have been in force since 1401.

In spite of its many ancient uses, the ling has not a long history as a garden plant, and many people, even today, still believe that it cannot be grown away from its native moors. Bacon did not include any kind of heather when planning his six acre "heath garden," and I have seen no reference to its cultivation before the 1820s, when interest in the family had been aroused by the introduction of large numbers of heaths from the Cape of Good Hope. It is as a plant of the formal garden that it makes its first appearance. Loudon says that in Edinburgh it was used as an edging instead of box, standing clipping well, and affording less harborage for slugs. By 1838 there were already fifteen varieties in cultivation; the white heather had been known in 1597, and William Curtis (1746–99) in 1785 describes a hoary-leaved kind. Almost the only doubles among the Ericas occur in this family, and a double red form was found in Cornwall before 1855, when Anne Pratt reported that the botanist George Luxford had in his herbarium a specimen with long sprays crowded with double flowers, "and sweeter resemblances with wreaths of roses cannot be conceived." Since then, the plant has made up for the

lateness of its start by the number of its varieties, some seventy of which—tall and short, early and late, single and double, with variously colored foliage and flowers—have been introduced to cultivation. Some of these varieties have been found wild on moorlands, others are of garden origin. Among the older forms still grown are *alportii, hammondii,* and *serlei,* all of which date from before the present century. The white form of the common ling has long been superseded by improved garden varieties, but the double white is fairly recent; it originated in Germany, and received an Award of Merit in 1938, while the popular and excellent double pink, 'H. E. Beale', orginally found growing wild in the New Forest, southern England achieved the same distinction in 1942.

The reputation of white heather as a luck-bringer is old in Scotland, but seems to have been unknown in the south until brought by Queen Victoria from Balmoral. It is unusual in that the luck is transferable, and not limited to the finder of the flower. "On great occasions the table of a Highland chieftain would be poor indeed without its sprig of white heather . . . and it is considered the height of hospitality to present it to the stranger guest."[1] Moreover, sheep and grouse were supposed to spare it when they met it. White heather formed part of the bridesmaids' headdresses at the wedding of Prince Leopold in 1882, and when Victoria's youngest daughter, Princess Beatrice, was married three years later, a special party of Highlanders travelled to Osborne from Balmoral with a bouquet including the flower. Its commercial exploitation in relation to weddings and other auspicious occasions has now become a large and specialized industry. Mr. Chapple, in his book *The Heather Garden* (1952), says that it is grown chiefly in Derbyshire, and is sold by the pound, the annual produce running into tons; but labor costs for cutting are high, and the times are gone when a nursery could earn £400 for this product in a few weeks. The faded flowers of the common moorland heather keep their color a long time, and are often treated almost as "everlastings" by ignorant Sassenachs, incapable of distinguishing between the open, starry flower, and the closed fist of the faded one. The seeds will remain in the seedpods for as long as twelve months, which Curtis regarded as a special provision of nature to provide food for the grouse all year round; the neat little leaves are arranged one above the other in four ranks, to minimize loss of moisture on poor and droughty soils.

These starry flowers, consisting of a colored calyx

[1]*St. James's Gazette,* July 22, 1885. Quoted by Lean, *Collectanea, II,* pt I

rather than a true corolla, distinguish it so clearly from the other Ericas with their bell-shaped blooms, with which it was originally grouped, that in 1881 Salisbury removed it to a monotypic genus of its own, and christened it Calluna from the Greek *kallunein*, to sweep, in reference to its use for besoms. The same word also means "to beautify," so is doubly appropriate to a plant whose flowers, Curtis says, impart animation to the dreary waste of Britain. Old local names are Heth, Hather and, in Shropshire, Grig. According to the *New English Dictionary*, "Heather" is a comparatively recent word dating only from the eighteenth century. In Scotland the plant was called hadder or hedder, in the North of England ling, and in the Midlands and South, heath. The compound of hedder and heath would seem therefore to be a result of the Union of the Kingdoms in 1707. It is not, as some would have it, derived from "heathen," its flowers being colored by the blood of the Picts; on the contrary, heathens took their name because, uncivilized and unconverted, they dwelt on heaths. In the language of flowers, the ling signifies Solitude—"but from its frequent use as beds in the Highlands," says Loudon, "its sweet and refreshing smell rather recalls ideas of social enjoyment, and wild, though hearty, hospitality."

NOTES: The heaths are acid-loving bog plants of Europe, used frequently as groundcovers where these soil and moisture requirements are met. They also need plenty of sunshine but a cool climate. As for the heathers, many gardeners regularly struggle to cultivate them in this country, only eventually to give up on them.

Calycanthus

There are only four members of this family, all natives of North America, and their nearest relatives are the three species of *Chimonanthus*, which all belong to China. In each case there is only one of them that is of much value to gardeners, and in each case it is the fragrance, rather than the beauty of the flower, that is the chief attraction.

C. floridus, the Carolina Allspice or Sweet Bush, is the oldest and the best of its genus. It was introduced by Mark Catesby in 1726, from the country "some hundred miles on the back of Charles-Town in Carolina."[1] He describes, but does not name it, in the first volume of his *Natural History of Carolina* (1732): "The flowers resemble, in form, those of the Star-Anemony, composed of many stiff copper-coloured petals, enclosing a

[1] *The Botanical Magazine*, 1801

tuft of short yellow stamina. . . . The bark is very aromatic, and as odoriferous as cinnamon. These trees grow in the remote and hilly parts of *Carolina*, but no where among the Inhabitants." It remained very rare in Britain until about 1757, when Peter Collinson made a fresh importation from the gardens near Charlestown, where by this time it was being cultivated. In a letter to Linnaeus dated 1756, Collinson wrote "it is a charming suffrutex,[1] and grows in my garden in the open air, bearing flowers abundantly every year."

It is a shrub on which 200 years of cultivation have had no apparent effect; there are no garden varieties, and the curious, maroon-colored flowers, which almost look as though they were made of leather, appear much the same today as they did in Catesby's original illustration. Though not particularly small, their color and placing make them rather inconspicuous; but the leaves, wood and bark are all spicily fragrant, especially when dried, and Bowles described the flowers as having a strong scent of pineapple when newly opened and of grapefruit later, with an alcoholic fragrance like that of cider as they fade. Cobbett called the Calycanthus, "A hardy and exceedingly odoriferous shrub of Carolina . . . and blows a ruddy brown flower from May to August." Loudon quotes de Candolle as saying that if the terminal leaf-bud is removed, two new flower-buds are produced, and a succession of flowers may thus be obtained the whole summer.

Another species, *C. occidentalis*, the Florida Allspice, was introduced by David Douglas in 1831. Twenty-three years later it was illustrated in the *Botanical Magazine* with the comment that "it has not yet had the justice done it in the gardens of this country that it deserves." Its drawback seems to be that it does not flower very freely unless grown against a wall.

The family received its name, which may be translated "calyx-flower," because the sepals and petals are indistinguishable—in fact the flower has been described as consisting of "calycine folioles" rather than of true petals. It luckily escaped the name of *Basteria* which Miller tried to give it in 1763, "in honour of his worthy friend Dr. Job Baster, F.R.S., of Zurick Zee in Holland," through having been already named by Linnaeus in 1759.

NOTES: The Sweet Shrub is an old favorite in gardens in its native South. The aromatic leaves and reddish brown flowers with strap-like petals in late spring are the main qualities. It is fairly common in the mountains, and in the moist woods near my home I can find it growing along with native hollies and mountain laurel. It is of minor garden value today.

[1] i.e., Shrub, woody plant

Camellia

About the year 516 A.D. an Indian prince and mystic named Dharma or Daruma visited China "to instruct the natives in the duties of religion."[1] On one occasion, worn out with prayer and fasting, he was overcome by sleep when he had intended to meditate, and in expiation of this weakness he cut off his bushy eyebrows and flung them to the ground. There they took root, sprouted, and grew into the first tea-plants, a decoction of whose leaves has the property of promoting wakefulness and stimulating the mind.

This, then, was the origin of the first camellia; for though the tea-shrub was at first separately classified as *Thea*, later botanists found such "frittering down of useful genera into useless . . . too inexpedient . . . to be continued,"[2] and included it among the camellias as *C. sinensis*. There are a number of early examples of camellias being imported as accidental or intentional substitutes for tea-plants. John Ellis attributes this to the craftiness of the Chinese, who "when they sold the plants . . . for true tea-plants, had pulled off the blossoms"—but in some cases at least it may have been due to misunderstandings caused by the language difficulty. The Chinese word for tea is *Ch'a* or *Tch'a*, and the various species of camellia are similarly named—*Ch'a Wha* or *Hoa*, Flower of Tea, for *C. japonica* and its varieties, *Nan Shan Ch'a*, Tea of the South Mountain, for *C. reticulata*, and *Ch'a Yeao*, Oil-bearing Tea, for *C. oleifera*. Many of the species share the same properties to a lesser degree and have been called "possible tea-plants that nobody wants."[3] The true tea plant was brought to England before 1739 by Captain Goff of the East India Company, but it perished through the ignorance of a gardener, and was not reintroduced till 1768. It is not quite hardy enough to grow out-of-doors, though a plant survived in the open for several years at Kew, well sheltered by rhododendrons, until it perished in the winter of 1838.

The reputation of *C. japonica* as a cultivated ornamental plant of China and Japan preceded its actual introduction by many years. A dried specimen of a single red variety was sent by James Cuninghame from Chusan about 1700, and was described by the botanist Petiver in the *Philosophical Transactions* for 1702–03. A fuller account, under its Japanese name of *Tsubakki*, was given by the German botanist and traveler Engelbert Kaempfer, in his *Amoenitates Exoticae*, 1712. He said that the "Japan Rose" as he called it, grew wild in every wood and hedge, but that there were many superior garden varieties, and in his *History of Japan* he added that it had 900 Japanese names, "if it be true what the natives report." One of the names was *Dsisij*, which signified "the little dog . . . that we call the little lion" (i.e., the Pekinese), "a name which it gets from its pleasing flowers."

The first living camellia flowers to be seen in England were grown in 1739 at Thorndon Hall, Essex, by Robert James, Lord Petre, a keen plantsman, whose death from smallpox in 1743, at the age of twenty-nine, was lamented by Collinson as the greatest loss that botany and gardening had ever felt in Britain. His two camellias, a single red and a single white, were much admired by contemporaries for "the elegant brightness" of their flowers, and in August 1740 one of them bore "a most delightful crimsonish double flower of a Ketmia figure"[1]—perhaps a budsport, since in this way many *japonica* varieties have originated. Nobody knows where or how Lord Petre obtained his plants, and they are supposed to have perished shortly afterwards through being kept in too warm an atmosphere. But probably they had already been propagated since it is said that James Gordon, Lord Petre's gardener, who established a nursery of his own after his master's death, was the first to introduce the camellia to commerce. A branch that figured in a print of a Chinese pheasant executed in 1745 was sketched from a plant raised by Lord Petre.

After this, little more is heard of the camellia till 1792, when two new varieties, a double white and a double striped, were imported from China by Mr. John Slater of the East India Company, who was responsible for the introduction of a number of Chinese plants, "he having procured a catalogue to be printed of all the described Chinese plants in the language, with the descriptions translated, and by various hands transmitted it to that country."[2] Other varieties soon followed, brought by the East India Company's ships; a double red for Sir Robert Preston in 1794, and in 1806 a pale flesh-pink "which must be admired by everyone for the imbricated symmetrical distribution of its petals,"[3] for Amelia, wife of Sir Abraham Hume of Wormleybury, Hertfordshire—a variety later known as *incarnata*, or Lady Hume's Blush. (This lady, a keen botanist and gardener, was a pupil of Sir J. E. Smith, who named the Humea or Incense Plant in her honor.) By 1815 twelve camellias were being grown, and by 1819 twenty-five had bloomed and four more were being watched for their first flowers. This was the date of the first book on the subject, *A Monograph on the Genus Camellia*, by Samuel Curtis, a thin folio which compensates for the paucity of its subject matter by the

[1]Hooker, *The Botanical Magazine*, 1832   [2]*The Botanic Register*, 1815
[3]Hibberd

[1]Collinson   [2]Bretschneider   [3]Samuel Curtis. He was the cousin of William Curtis of the *Botanical Magazine*

magnificence of its size and the splendor of the five color plates on which ten of the varieties are shown. In the same year Alfred Chandler, a nurseryman of Vauxhall, saved "about half a peck" of seed from the anemone-flowered or Waratah camellia (introduced 1806) which he had fertilized with the double striped and other sorts, sowed them as soon as ripe, and thus started the raising of new varieties from seed. (One of these seedlings, which flowered in 1830 and was named *chandleri elegans*, is still among the twenty best camellias available today.) By this time Lord Petre's original single red, which struck remarkably easily from cuttings, was being used only as a stock on which to graft the newer kinds.

Early accounts of the introduction of camellias always give the names of the captains of the East Indiamen in which they were brought. Rightly, because without their active cooperation no plant could have survived the three or four months' voyage (the famous tea race of 1866 took ninety-nine days) with its two equatorial crossings and its stormy passage around the Cape, especially before the invention of the Wardian Case, first tried out in 1833. The first camellias, in 1792, were transported by Captain Connor in the *Carnatic* but afterwards the two chiefly concerned with their introduction were Captain Welbank of the *Cuffnels*, most of whose plants were consigned to Charles Hampden Turner, Esq., of Rooksnest Park, near Godstone, Surrey, and Captain Richard Rawes of the *Warren Hastings*, who brought a dozen or more on successive voyages between 1816 and 1824, for his sister and brother-in-law, Mr. and Mrs. Thomas Carey Palmer of Bromley, Kent. On one occasion both captains brought plants of a new double white variety in the same year. The one brought by Welbank for Turner flowered first, so it was given the name of *C. welbankiana*. Rawes is commemorated in the beautiful variety of *C. reticulata* which he introduced in 1820, and which is still known as 'Captain Rawes'. Another of his importations, *C. hexangulare* (probably a *japonica* variety), produced its first flowers on board the ship, during the homeward voyage—a horticultural happy event which has something of the pathos of a baby born at sea.

By the 1840s the camellia was at the height of its popularity, and big nurseries such as that of Messrs. Loddiges at Hackney were among the sights of London at flowering time. In Loddiges' greenhouses "many of the plants were thirty feet high, in splendid health and laden with blossoms. It was a perfect forest of camellias, tenanted with blackbirds, thrushes and other birds, which built their nests in the trees, passing in and out at pleasure through the open doors and windows."[1] On the

Continent the flower was equally popular. Some of us have heard of Marie Duplessis (1824–52), *La Dame aux Camellias*, whose sad and reprehensible story inspired a novel,[1] a play and an opera;[2] but few remember that about the same time a certain M. Latour Mézeray was known as *L'homme aux Camellias*, for the same reason—because he was never seen without one. It was estimated that during the nineteen years he lived in Paris his buttonholes must have cost him 50,000 francs. The camellia, being the most expensive of flowers, was considered the symbol of elegance; it also signified "Beauty is your only attraction" or, in the white form, purity—because of its lack of scent.

One of the Continental varieties of this period, the white-spotted *donckelarii*, deserves particular mention on account of its romantic history and its later importance as a parent of the hybrid 'Donation'. It originated in Japan, and was among a number of plants imported by Dr. Siebold, which arrived at Antwerp during the revolution of 1830. The place where the cases were put after being disembarked was shortly afterwards occupied by cavalry horses, which caused such extensive damage that hardly a leaf escaped unharmed and much was totally destroyed. This camellia survived the mêlée, and was rescued and propagated by M. André Donckelaer, then head gardener of the Botanic Garden at Louvain, who himself told the story. The name has been very variously spelled, even appearing as Don Klari. As the first of the "picotees," this variety immediately became popular, in spite of the fact that its informal flower did not comply with the exigent requirements of the florists of the day. It is still considered one of the best, and received a belated Award of Merit in 1960.

Towards the end of the nineteenth century camellia growing suffered a decline, perhaps due to a reaction against the excessive formality of the fashionable varieties. The trend showed itself in America in the 1860s, on the Continent a little later, and in England about 1880; references to the flower almost disappear from gardening papers and nurserymen's catalogues after 1888. A revival of interest in the more informal types as hardy plants for growing out-of-doors began at the turn of the century, but was checked for a time by the advent of the First World War. A pioneer in this movement was Mr. G. F. Wilson, who planted some camellias in his woodland garden, with arrangements for winter protection which afterwards proved to be unnecessary. After his death in 1903, his garden at Wisley was purchased and presented to the Royal Horticultural Society, and

[1]William Paul, quoted by Urquhart

[1]By Alexandre Dumas, fils [2]*La Traviata*, Verdi

Camellia japonica

(JAPAN ROSE)   N. von Jacquin (1786)

two years later some of the camellias were moved to sites where they had no shelter but that of deciduous trees. And there some of them are still to be found, in the Wild Garden at Wisley, bushes 30 to 40 feet in height, and flowering freely every spring. It is amusing to note that ever since 1827, when a paper on the subject was submitted to the Horticultural Society by a gardener called Joseph Harrison of Wortley Hall, horticulturists have never tired of telling us that the camellia is as hardy as a laurel—hardier indeed, for Harrison says that it survived frosts that much injured the common laurel; but apparently we still refuse to believe it.

Of late years the revival of the camellia has continued with ever-increasing impetus both in Britain and in America, and camellia societies and camellia experts flourish; the number of varieties of this one species alone has been estimated at over 3,000. They do not, however, include either a blue or a pure yellow. The blue camellia is a chimera as elusive as the blue rose, and growers attempting to raise one have only been able to achieve a dirty purple. The yellow has been sought since the early days of camellia culture. Rumors of the existence of such a flower had reached England early in the nineteenth century, and when the collector Robert Fortune was sent to China by the Horticultural Society in 1843, he was expressly ordered to look for "camellias with yellow flowers, if such exist." His first inquiries procured the reply, "No, my never have seen he, my thinkie no have got," but on his second journey in 1848 he found a plant of the coveted variety in a nursery at Shanghai, which he sent to Standish and Noble of Bagshot. It proved to be not a true yellow, but a creamy white flower of the "anemone" type, with a center of yellow petaloid stamens. Tender, difficult to propagate and "curious rather than handsome," it soon disappeared from cultivation, and was lost for the best part of a century until rediscovered in 1952, after a long hunt, by an American specialist, the late Mr. Ralph Peers, in a nursery in Portugal. Plants introduced to Kew and to America subsequently died, but more were procured, and it is now precariously re-established. It is thought to be a variety of *C. hiemalis*.

Although *C. japonica* has occupied so much of the limelight, there are other beautiful species in the camellia family. *C. sasanqua* had a neck-and-neck start with *japonica*, though it afterwards fell much behind in the race. (Both are natives of Japan, though cultivated also in China.) It too was sent home as a specimen by Cuninghame, though it was not recognized as a different species until later; and Kaempfer devoted almost as much space to the 'Sazankwa' as he did to the 'Tsubakki'. Several camellias thought to be of this kind arrived in England

in the East India Company's ships; but 'Lady Banks' Camellia' brought by Welbank in 1811, and Captain Drummond's double white in 1823, are now believed to have been forms of *C. oleifera*, while Captain Rawes' double pink (1818) was subsequently found to be a completely different species and was re-named *C. maliflora*. The true *sasanqua*, in double white and single rose varieties, was introduced by Charles Maries for Messrs. Veitch about 1879. It is hardy, but blooms from October to April, so its flowers are often damaged by frost unless given wall-protection; in countries where the climate suits it better, such as Japan and parts of America, it is a very popular plant, though its varieties do not rise above a paltry two hundred or so.

C. sasanqua and *C. oleifera* have often been confused. They are closely allied, and some botanists think *C. oleifera* merely the Chinese form of the Japanese *sasanqua*. Both have been long cultivated, and there are many intermediate varieties that seem to link the two species. In 1793 Sir George Staunton, accompanying Lord Macartney's embassy to Pekin, described "*C. sasanqua*"—as he then called it, knowing no better—both cultivated and wild; but Dr. Clarke Abel, with Lord Amherst's expedition twenty-two years later, considered it a different species, and chose its present name because it was cultivated on a large scale by the Chinese for the sake of its oil-bearing seeds. The typical form was introduced in 1820 by Captain Nesbitt of the *Essex*. These two, *oleifera* and *sasanqua*, unlike most camellias, have fragrant flowers, especially in certain varieties. Some kinds are dried by the Chinese, and used, like jasmine flowers, to perfume tea. Siebold said that in Japan windbreaks and hedges of *C. sasanqua* were planted in the tea fields, because it was believed that the fragrance of their flowers would be imparted to the tea.

It has already been mentioned that among the camellias introduced in 1820 by Captain Rawes was a semi-double variety, the magnificent *C. reticulata*. It first bloomed in 1826 and was described and named by Lindley the following year. This superb species had been cultivated in Yunnan since about 900 A.D., and by the eleventh century there were already seventy-two Chinese varieties. In England it proved slow to establish itself, and has always been scarce and expensive, but two specimens of 'Captain Rawes' planted in the conservatory at Chatsworth about 1840 are still in perfect health, cover a wall 24 feet high and bear thousands of blossoms every spring. The flowers often measure 6 inches, and occasionally 9 inches across, and have been compared to those of the tree-peony. Glenny despised them for their lack of formality and said that "no plant was ever more overrated," but all other authorities agree that this is the finest species of the genus. Unfortunately it is tender, and although it has been grown out-of-doors as far north as Yorkshire, it was only on a wall in exceptionally favorable conditions. In 1932, however, a plant of the original wild single ancestor of *C. reticulata*, raised from seed sent home from Yunnan by George Forrest in 1924, bloomed in the garden of Mr. J. C. Williams at Caerhays Castle in Cornwall, and it is hoped that seedlings from this wild type, some of which have already received horticultural awards, may prove hardier than the older garden sorts. Meantime, in 1949 Mr. Ralph Peers, the American specialist already mentioned, imported nineteen *reticulata* varieties from Kunming to Los Angeles, some of which have also been sent to Wisley and may presently become generally available.

This king of the genus has hitherto held itself somewhat aloof, being tender, very difficult to propagate, and reluctant to cross with other species. Although varieties were so numerous, hybrid camellias of any kind were virtually unknown until the present century, when the arrival of some new species from the wild caused a revolution in camellia-growing and gave it a fresh stimulus and a new direction. The first was *C. cuspidata*, as modest and hardy as *reticulata* is showy and tender. It was among the plants introduced by E. H. Wilson in 1900 for the firm of Veitch of Exeter, where it produced the first of its abundant, small white flowers in 1912. More important, though less hardy, was *C. saluensis*, which was found by George Forrest growing on the Salween-Shweli divide in western Yunnan, its wild-rose flowers varying in color from white to pink and crimson. Seed was sent home in 1924 and raised in several gardens, including Caerhays; and when the plants flowered, about 1930, Mr. Williams crossed the best of them with some *japonica* varieties, and produced the famous race of hybrids now called *C. × williamsii*. The one named after him, ' J. C. Williams' (F.C.C. 1942), has been called the best hybrid shrub ever introduced into our gardens. It has every garden virtue, including that of "burying its dead"—that is to say, it drops its faded flowers instead of retaining them, brown and unsightly, on the bush. (The latter habit, which we think a fault in camellias, may be considered a virtue in Japan, whose people, it is said, have inherited from feudal days a dislike for flowers that fall intact, like a head decapitated by a sword.) An even better variety was raised about 1941 from a similar cross (*C. saluensis* × *C. japonica* var. *donckelarii*) by the late Colonel Stephenson Clarke of Borde Hill, Sussex, and called 'Donation' (F.C.C. 1952)—a plant so striking in its beauty that a distinguished visitor to a Cornish garden in which it was growing was seen repeatedly taking off his hat to it. Hybrids from *C. cuspidata* are not yet numerous, but Mr. Williams' 'Cornish Snow' (*C. cuspidata* × *C. saluensis*) is valuable for the freedom with which it produces its small white flowers and for its great hardiness. There is also an almond-pink variety called 'Winton'.

The family is named after George Joseph Kamel (1661–1706)—latinized as Camellus—who had nothing actually to do with either the discovery or the introduction of the camellia, but deserves commemoration nevertheless for his services to botany. The botanist and Reverend John Ray (1627–1705) spoke of him as "being made, as I may say, for the advancing of natural knowledge."[1] Born at Brunn in Moravia, he became a Jesuit in 1682, and was sent as a missionary to the Philippines in 1688. In Manila he studied natural history, especially botany, and opened a pharmacy for the free treatment of the poor. He collected plants chiefly on the island of Luzon, including many cultivated by Chinese residents in their gardens, and sent specimens, drawings and copious notes to the British botanists Ray and Petiver. (His specimens included St. Ignatius' Bean, *Strychnos ignatii*, from the seed-coat of which strychnine is extracted— a useful product when legitimately employed.) Kamel's first consignment of drawings was intercepted by pirates; but his *Syllabus Stirpium in Insula Luzone*, in two parts, nevertheless occupied a ninety-six page appendix to the third volume of Ray's *Historia Plantarum* (1703), and a third section dealing with over two hundred climbers was published later by Petiver. Linnaeus gave the name of *Thea sinensis* to the tea-plant in the first volume of *Species Plantarum*, published in May 1753, and *Camellia japonica* to Kaempfer's *Tsubakki* in the second volume, which came out in August of the same year; so if it were not for the convention that counts the two volumes as simultaneous, the older name of *Thea* would have priority as the name of the whole genus, and the estimable Kamel would be forgotten.

NOTES: Much has been written about the history of the camellia. Important as an economic and ornamental shrub, camellias have made their way around the globe wherever they can be cultivated. But perhaps there has been more advancement over the past fifty years in the ornamental aspects of the camellia than during the several centuries prior. The application of advanced scientific techniques, particularly in hybridization and in bringing somewhat obscure species into the mainstream, has brought about marvelous new creations. Much of the credit for this accrues to American breeders. Among the many exciting advances is the discovery of a valid yellow-

[1]Quoted by Britten, *The Sloane Herbarium*

flowered species in China, *Camellia chrysantha*, now being studied in the U.S. and elsewhere. Nevertheless, the long-cultivated *C. japonica*, native to Japan, remains the most popular garden plant, followed by *C. sasanqua*. These two form the basis of the charm and character of the southern garden.

The cultural requirements for camellias become obvious when visiting a southern camellia garden. A mild, moist climate, acid soil with plenty of humus and shaded conditions are essential to their good growth. *C. sasanqua* is strictly a fall-flowering plant, while *C. japonica* will be found in bloom from late October to May with the peak of flowering in late winter.

We are still searching for the hardy camellia hoping to extend the cultural range farther north. In this respect, I have collected *C. japonica* across its entire range in Japan, particularly the rugged northern Pacific coast of Honshu, where it flourishes along the sea cliffs. There is also a desire to develop fragrance in camellias. One minor species with that potential is *C. lutchuensis* from Okinawa. I vividly recall first seeing the plants across a waist-deep stream, which I waded through in great excitement to collect seeds of this rare camellia. When I got across, my astonished local guide informed me to be careful of the poisonous snakes that infested the water. But with no harm done, I successfully established *C. lutchuensis* in cultivation in 1955, and it has been used by breeders in their improvement programs.

The oil from the crushed seeds of *Camellia japonica* is highly valued in Japan as a hair dressing and is also used for cooking.

Campsis

In the year 1700, the great French botanist Tournefort named a family consisting mostly of tropical climbing plants with showy flowers, *Bignonia*, "to express the extraordinary Esteem and Veneration he had for the illustrious *Abbé Bignon*," to whom he also bequeathed his collection of books on botany.[1] At that time it was already a large genus; Tournefort described fifteen species, and within the next hundred years a great many more were discovered. Botanists ultimately split the genus into no less than twenty-eight new ones, and the two hardy species eligible for inclusion here were allotted to the family called Campsis—from *kampe*, a bending—on account of their curved stamens.

Some difficulty in naming and classification was felt from the discovery of the very first species, *C. radicans* (syn. *Bignonia radicans*, Trumpet Creeper), which was

brought to Europe from the southeastern parts of America early in the seventeenth century. Parkinson says that the English settlers in Virginia called it at first a jasmine or a honeysuckle, and later a bell-flower, while he himself classed it as an Apocynum or Dog's Bane. In 1640 the plant, though introduced, had not yet bloomed in England, and he described its flowers at second-hand as "of a sad orange or yellowish red colour." He was very scornful of Cornutus, who was the first to describe and picture the plant,[1] for applying to it the name of *hederacea*, ivy-like; the French botanist was probably thinking of its self-clinging aerial rootlets, which Parkinson rather surprisingly omits to mention. Tournefort noted that the "Curls or Tufts, by which it lays hold of props, do not arise from the whole body of the plant, as in ivy, but only from the joints alone," and Abercrombie, that the plants will soon cover high walls, "they emitting roots at all the joints of the stalks, which strike into the joints of buildings and mount to the tops if ever so high, where they will endure many years, and flower annually." (Loudon, however, points out that old plants are "apt to get naked below.") In their native land, according to Mark Catesby (1682–1749), "the Humming Birds delight to feed on these flowers; and, by thrusting themselves too far into the flower, are sometimes caught." Cornutus compared the blossoms to the "little brass tubes with which women protect their fingers when sewing."

On the other side of the world, *C. chinensis* (syn. *Bignonia grandiflora*) was found by Kaempfer in Japan as early as 1691, but was not introduced to Britain until more than a hundred years later. In China, where it is called Sky-approaching Flower, it was cultivated as a medicinal plant as far back as records exist; but its pollen is thought to be harmful and it is not grown close to the house. Slighter and less hardy than *C. radicans*, it has no self-clinging roots, but larger and showier blossoms, "altogether" says Loudon, "a very splendid plant." The flowers of the type are tawny-orange outside and a "tolerably bright reddish-orange colour"[2] within. But improved varieties have been raised both of this and of the preceding species, and towards the end of the last century the brothers Tagliabue, nurserymen of Laniate near Milan, crossed the two kinds and produced a race of hybrids (*C. × tagliabuana*) in which the hardiness of *C. radicans* was combined with the large flowers of *C. chinensis*. The best of these hybrids was the red-flowered 'Mme. Galen', (about 1889)—a fine plant of which is to be seen at Wisley; this variety was given an Award of Merit in 1959.

[1]Bignon (1662–1743) was librarian to Louis XIV

[1]In *Historia Canadensis Plantarum*, 1635   [2]Loudon

Among the climbers formerly classified as Bignonias are the American *Doxantha capreolata* (syn. *Bignonia capreolata*, 1710), the Coral Vine or Cross-Vine, so called because its wood when cut transversely is marked with a cross, and the sweet-scented *Gelsemium sempervirens* (syn. *Bignonia sempervirens*, 1640), the Virginian Jasmine, whose high-clambering stalks, according to Marshall, "will form at a distance a grand figure from

improvement over either parent.

There are no particular cultural requirements, just a good garden soil and sunny position. As one nursery catalogue states, the Trumpet Creeper is a tenacious vine useful for shade on patios. Caution is advised, however, as it can destroy pergola supports. Also let it be noted that some consider the plant a nuisance because it drops flowers continually throughout the summer.

Campsis radicans
(TRUMPET CREEPER)   S. Edwards (1800)

Carpenteria californica
(TREE ANEMONE)   W. H. Fitch (1866)

the sway they bear." Almost the only plant that is left to the Abbé Bignon of his once large family is *Bignonia unguis-cati*, the Cat's Claw from South America, also mentioned by Marshall in 1785; today we consider it a greenhouse climber, and both the Doxantha and the Gelsemium are tender.

NOTES: Only the Trumpet Creeper (*Campsis radicans*), a native of the Southeast, is commonly grown here. Probably best seen rambling across fence rows on narrow winding roads of the South, the orange-red flowers in mid-summer are profuse and brilliant. There is also a yellow-flowered variety of the Trumpet Vine. The other native species, *C. grandiflora*, is less hardy and not often cultivated. The hybrid between these two species, 'Mme. Galen', has large scarlet flowers and is an

Carpenteria

C *californica*, the only species in this family, is like one of those husbands that are perfectly easy to manage, providing that they get their own way in everything. It is very handsome, and reasonably hardy if its requirements of shelter, plenty of sun and a clear atmosphere can be met. It was first discovered by Frémont during one of his expeditions to California, but it is uncertain whether it was that of 1843, 1845 or 1848. Colonel John Charles Frémont (1813–90) was an explorer who made a number of arduous and adventurous journeys in the service of the United States Topographical Corps, on all of which he collected largely, though he did not always succeed in getting his harvest safely home. In 1845 he set out for California, which then belonged to Mexico, with a "scientific survey" party of sixty men, which he secretly intended to transform into a military force if the expected war between Mexico and the U.S.A. should break out; and he played a prominent part in the subsequent conquest of California for the United States. According to Bean, it was on this expedition that the Carpenteria was found, though one would hardly think it a suitable occasion for picking flowers.

It is a rare shrub, even in the wild, and was slow to reach cultivation. The first British specimen was procured from Lemoine of Nancy in 1880, and grown in a garden in southern Ireland. In England, one bloomed in the garden of Miss Gertrude Jekyll near Godalming in 1885. The flowers are white and fragrant, but in shape and size resemble a single rose or a Cistus rather than the Philadelphus to which the Carpenteria is allied. It was named by the American botanist Torrey, in honor of Professor William M. Carpenter (1811–48) of Louisiana, "an assiduous local botanist." The Colonel is commemorated in the Fremontia, another Californian shrub—only semi-hardy in Britain. It was introduced by Messrs. Veitch in 1851, and has showy yellow flowers. NOTES: This shrub is native to California. Unfortunately, in America it is rarely cultivated outside its home state because it is difficult to transplant and limited in adaptation.

Caryopteris

A small family, with a relatively short history. The first of its kind was discovered by a German botanist, Dr. Alexander von Bunge, who accompanied an ecclesiastical mission to Pekin in 1830, and found this plant during an expedition to Mongolia in 1831. Soon afterwards, in 1833, the newly-named *Caryopteris mongholica* was introduced to France, but was lost a few years later and reintroduced from seed sent by Père David in 1866.

Meanwhile another species, *C. mastacanthus* (syn. *C. incana*, Blue Spirea) had been found by Robert Fortune growing wild near Canton, and introduced in 1844. Not being considered worthy of greenhouse space, and apparently (although reasonably hardy), not successfully cultivated out of doors, it too was allowed to die out, and was reimported by Charles Maries for Messrs. Veitch in 1880. It has been "in and out of cultivation and a kind of perpetual novelty"[1] ever since. *C. tangutica*, found in

western Kansu by Przewalski in 1880 and introduced by Reginald Farrer about 1915, is said to be the hardiest of the three, but is rarely to be found in nurserymen's catalogues, where the genus is usually represented only by the hybrid *C.* × *clandonensis*, which has superseded all the original species.

This was an accidental cross which appeared in the garden of Mr. A. Simmonds at Clandon, near Guildford. In 1930, wishing to propagate *C. mongholica*, Mr. Simmonds gathered seeds from a plant of it which was growing near one of *C. mastacanthus*. When the seedlings flowered two years later, about half of them proved to be hybrids, showing characteristics of both parents. Meanwhile a self-sown seedling had sprung up under the bush of *C. mongholica* (which it eventually smothered) and proved on flowering to be another hybrid, better in color than any of the seedlings that had been so carefully pricked out and potted up; so these were eventually discarded, and the chance-comer propagated. It inherited the deeper color of *mongholica* and the hardiness of *mastacanthus*, and proved more vigorous and floriferous than either; so it rapidly became very popular, winning all the medals of the R.H.S.—A.M. in 1933, F.C.C. in 1941 and A.G.M. in 1942. It is a valuable autumn shrub, for its blue flowers and greyish, silver-lined leaves make an admirable foil to the hot colors of late summer. But it seems to me to be insufficiently appreciated for another of its merits—the sweet aromatic fragrance with which it so charmingly rewards the gardener who weeds in its vicinity. The soft blue of the flowers does not tell at a distance, and it is essentially an intimate, or dooryard shrub. It sometimes dies in severe winters, especially in heavy soil, but is easy to propagate by cuttings or seed. The seedlings are variable, and improved forms have already been selected and named; it is said that those with darker colored flowers, resembling more strongly their *mongholica* parent, are less hardy than the paler varieties. It is curious that *C. mongholica*, the most northerly of the species, should be the most tender.

Caryopteris is from *karyon*, nut, and *pteron*, a wing, from its winged fruit.
NOTES: Although this is a somewhat useful plant for late summer bloom, it is not commonly offered in the nursery trade. The foliage is gray and the flowers are blue. Although a woody plant, Caryopteris is usually treated like a perennial and cut to the ground in spring, after which it reaches to about 30 inches by August when it flowers. It is a sun-loving plant and requires good, well-drained soil.

[1] Jason Hill, *The Contemplative Gardener*, 1940

Ceanothus

There are over fifty species of Ceanothus, evergreen and deciduous, all natives of North America, and ranging from small trees to the mat-like *C. prostratus* or Squaw's Carpet. They are most plentiful in California, forming a large part of the dense thorny brushwood known as the chaparral—an evocative word with pleasing overtones of cowboys and Indians; it is as a protection against the prickly ceanothus that the cowboy wears his heavy leather *chapareios*, or "chaps." Many of the species will thrive in hot and arid situations, and have proved useful for roadside planting and for checking soil erosion.

C. americanus, Red Twig, Red Root or New Jersey Tea, was the first species to be introduced in Britain. It is one of the few kinds native to the eastern states, and was grown by Bishop Compton in his famous garden at Fulham, London, before 1713. Afterwards it was lost for a time, and was reintroduced in 1751 by Peter Collinson. By the end of the century it was "pretty common in most of our curious gardens and nurseries,"[1] and was much admired. Abercrombie called it "a most elegant little flowering shrub . . . well calculated for the most conspicuous compartments of the shrubbery plantations," and Marshall praised it for the profusion of its white flowers, so covering the plant that "the leaves which appear here and there among them, serve as ornaments only, like myrtle in a distant nosegay." But we now have plenty of other handsome white-flowered shrubs, whereas blue-flowered shrubs are rare and highly-prized; and this ceanothus is valued today chiefly as a parent for hybrids. Collinson also noted that what he called "a very pretty tea" could be made from the dried leaves. This, known as "pong-pong tea," from the shrub's Indian name, was much used during the American Revolution. The twigs will dye wool "of a fine strong Nankin cinnamon colour."[2]

C. thyrsiflorus, the Californian Lilac or Blue Blossom, was the first of the Californian species to be recognized, and has proved one of the hardiest. It was discovered in 1815, probably near San Francisco, by J. F. Eschscholz, on the expedition during which he found the Californian Poppy "which perpetuates his unnatural name."[3] Seed was collected and sent to the Horticultural Society by R. B. Hinds, surgeon to H.M. survey ship, the *Sulphur*, in 1837. It is said that before the gold-rush days, this species formed thickets on the site of San Francisco, and that settlers clearing ground for houses often left specimens standing as "door-yard shrubs," some of which survived as late as the 1870s. It is tall; its bushes,

even in Britain, have reached a height of 30 feet.

Seeds of three or four more species were collected and sent to England by David Douglas in 1830, but it is uncertain whether any of them actually germinated. Theodore Hartweg sent a number of other species to the Horticultural Society in the years 1847 and 1848, but the most interesting introductions were those made by William Lobb, when collecting for the firm of Veitch of Exeter, between 1849 and 1863. He had a remarkable eye for a plant, for he found and sent home three new sorts, two of which, *C. dentatus* var. *floribundus* and the popular *C.* × *veitchianus*, have never since been found in the wild, and the third, *C.* × *lobbianus*, but seldom. The two last are thought to be natural hybrids.

All the blue-flowered evergreen ceanothuses (with the possible exception of *thyrsiflorus*) are tender, and, except in warm climates can only be grown against a wall; so horticulturists soon set to work to breed varieties that should combine the hardiness of the deciduous *C. americanus* and its cousin *C. ovatus* (which is "nowhere recommended as a garden ornamental of unusual merit"[1]) with the blue flowers of the more tender kinds. The first to be chosen as a parent was *C. azureus*, a Mexican species brought to France early in the nineteenth century by a Captain Baudin. Robert Brown, a colleague of Sir Joseph Banks, saw and admired it in the Empress Josephine's garden at Malmaison in 1816. Two years later it flowered in the nursery of Loddiges of Hackney, having been obtained by them "from our friend M. Parmentier of Enghein." It was on the Continent, however, that the first crosses were made between this species and *C. americanus*, resulting before 1830 in a group of hybrids (now called *C.* × *delilianus*) which included "Gloire de Versailles," still, probably, the most popular member of the family. Many other Ceanothus hybrids in various pink, white and blue shades, have since been raised from the same parentage, and one of them, 'Indigo' was crossed about 1930 with *C. dentatus* var. *floribundus* by Mr. A. Burkwood, of Burkwood and Skipwith, Kingston-on-Thames, to produce the hybrid "× *burkwoodii*." The same grower also raised 'Delight' from *C. papillosus* × *C. rigidus*.

Strangely enough, many of these hybrids, bred as they are for hardiness in a cool climate, are not a success in the native home of the race. The Botanic Garden of Santa Barbara in California possesses a large collection of ceanothus species, many of them unknown in Britain; but of some forty-five hybrids and varieties grown in Europe only a handful are cultivated in America, and those only in the eastern states, or further north in Brit-

[1]Miller-Martyn [2]Ib. [3]Van Rensselaer

[1]Van Rensselaer

ish Columbia, where *C.* × *veitchianus* was reintroduced as recently as 1938. Stranger still, several of the species that long since emigrated, have proved surprisingly reluctant to be repatriated, and in spite of several attempts at reintroduction, are rare or unknown to cultivation in their native land.

The genus was founded by Linnaeus in his *Species Plantarum*, 1753, and named after an unidentified prickly plant—certainly not this!—called *Keanothos* by Theophrastus. Robinson calls them "Mountain Sweets," perhaps an American name, which never seems to have become established in Britain. "As they are often natives of a charming climate" he further advises us, ". . . no one should attempt their culture except in a warm soil."

NOTES: Despite their lineage, the hybrids grow well only in California and similar Pacific coastal regions. One parent, *Ceanothus americanus* is widely distributed in eastern North America; nevertheless, it is rarely found in catalogues in the east because it is difficult to transplant.

Ceratostigma

This might be called a ladies' genus, since each of the two main species came to us by means of an eminent woman gardener, and was named after her, though one name was afterwards changed.

C. plumbaginoides is, frankly, no shrub at all, but a herbaceous plant; but with the authority of Bean's *Trees and Shrubs* for a precedent I will include it here, as it was the first species to be discovered and the one on which the genus was based. It was collected near Pekin by von Bunge (who also found the Caryopteris) about 1831, and it was from Pekin that Robert Fortune sent to England its seed thirteen years later; but whether because the seeds did not germinate or because the seedlings died, the plant failed to become established in cultivation. In 1846, a living specimen was sent from Shanghai to Sir George Larpent by a Mr. Smith, with the note, "Mr. Fortune tried to get a plant of it, but failed; yours is therefore the only one in England. It is very rare even in Shanghai, and I found it on the city wall. It is one of the most ornamental plants I have seen in China." This time Lady Larpent succeeded in growing it, and it flowered the following year (1847) in her garden at Roehampton, Hampshire. At that time it was believed to belong to the genus Plumbago, and it was therefore given the name of *Plumbago larpentae*, Lady Larpent's Plumbago. But very shortly afterwards Sir W. Jackson Hooker removed it from that family "because of some trifling peculiarities of structure," and reclassified

it as *Valoradia plumbaginoides*. He himself was inclined to apologize for the change. "We perform no enviable duty in restoring the original specific name (given to it as early as 1831)," he wrote in the *Botanical Magazine*, "for we know no lady who has deserved better of botany and horticulture than Lady Larpent. Her garden at Roehampton was long distinguished by high cultivation, and by the rarity and beauty of the plants." Later it was restored to Bunge's original name of Ceratostigma. Owing to the ease of its propagation it was "already common" in gardens by 1850, and there was much controversy about its horticultural merits.

C. willmottiana is a true shrub, and though it sometimes gets killed to the ground in cold winters, it will usually reappear, with buds as red as rubies, in late spring. It was discovered in 1908 by E. H. Wilson, in the semi-arid regions of the Min River valley in western Szechuan; and from the seed he sent to England, two seedlings were raised by Miss Ellen Willmott—all the rest failed to germinate. From these two plants all the stock of this shrub on the British Isles is believed to be derived. One of them was planted in Miss Willmott's famous garden at Warley in Essex, and the other in that of her brother-in-law at Spetchley in Worcestershire. By 1914 they had grown into bushes some 5 feet high.

Ceratostigma is a genus interesting to botanists on account of the peculiarity of its distribution; only about eight species are known, two from central Africa and the rest from Asia. It is a far cry from north China, where the first species was found, to Abyssinia, which produced the second. The name was taken from *Keras*, a horn, because the stigmas bear an excrescence supposed to resemble the antlers of a stag. This is not apparent in *C. plumbaginoides*, but an examination of herbarium specimens revealed that there were both long- and short-styled forms, and it has been surmised that all the plants in Britain have been derived from one short-styled clone—in fact, from Lady Larpent's original plant—which, in Britain at any rate, produces no seed. It therefore seems all the more unfair that Lady Larpent, through no fault of her own, should have lost her floral immortality, while Miss Willmott has retained hers. The unaccountable "Moustache Plant" seems to be the only name that the English genius for language has evolved.

NOTES: The Asian species, *C. plumbaginoides*, is a neat, reliable groundcover in the U.S. with a wide range of adaptation. It appears in many nursery catalogues that emphasize perennials and groundcovers. It is distinguished by its dark blue flowers in late summer and the maroon fall color of its foliage. No recent improvements have been made with this plant.

Caryopteris mastacanthus
(syn. *C. incana*)
(BLUE SPIRAEA)   Anne Barnard (1885)

Ceanothus azureus
(syn. *C. coeruleus*)
(RED ROOT)   S. Edwards (1818)

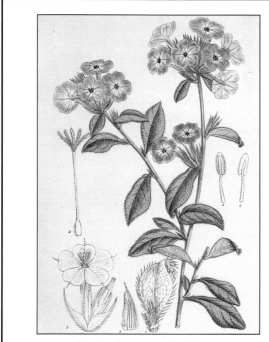

Ceratostigma willmottianum
(CHINESE PLUMBAGO)   M. Smith (1914)

Chaenomeles japonica
(FLOWERING QUINCE)   S. Edwards (1803)

Chaenomeles

This is the plant that the public still firmly calls "Japonica," as though no other flower ever came to us from Japan. No changes in nomenclature have been more justifiably resented by gardeners than those affecting this well-loved shrub; and yet the reasons for the changes are logical enough. The first species was found by Thunberg on the Hakone mountains in Japan, and was described by him in 1784 under the name of *Pyrus japonica*. In 1796 Sir Joseph Banks introduced a plant to Kew from China, which was thought (wrongly, as it afterwards proved) to be this *Pyrus japonica*, and it was illustrated under this name in the *Botanical Magazine* for 1803. In 1818 Robert Sweet noticed that it was not the same as Thunberg's plant, and renamed it *Pyrus speciosa*,[1] but by that time the name of "japonica" had become firmly established and nobody seems to have taken much notice of the correction. In 1869 the firm of W. Maule and Son of Bristol introduced Thunberg's original species from Japan; but as Sir Joseph Banks's Chinese plant was still usurping its name (in spite of Sweet's attempt at a rectification) another one had to be found, and it was christened *Pyrus maulei* by Dr. Masters in 1874. It has now been restored to its rights as the original "japonica" and Sweet's name of "speciosa" has been officially adopted for the Chinese plant. The other changes are explained by the fact that for a time botanists classed these shrubs as Quinces (Cydonia) rather than as Pears (Pyrus); then they were replaced among the pears, but that family, being inconveniently large, was split up into a number of more manageable sections, and Chaenomeles was chosen as the name for this particular group. So the old "Pyrus japonica" is now *Chaenomeles speciosa* and the old "P. maulei" is now *C. japonica*. How much simpler to keep to the Japanese name of *Boke*!

It does not seem to have been recorded how Banks procured the plant of *C. speciosa* that he introduced to Kew. The *Botanical Magazine* illustration shows a rosy-red, semi-double flower, the three blossoms and two buds of which "represented the whole specimen at the time it flowered." "This very rare plant" had been supplied, not from Kew, but from E. J. A. Woodford, Esq., "in whose collection at Vauxhall, we find a never-failing source of curious and rare articles." By 1838, Loudon could report that it was already common in gardens, grown usually as a bush or wall-plant, but occasionally trained as a standard, which had a "rich and striking

[1] A French botanist at about the same time gave it the name of *Cydonia lagenaria*, from its "lagenariaeform" or gourd-shaped fruit

effect" in spring. Unfortunately it would not graft easily on pear, hawthorn or quince, or it would have made, he suggests, "a most delightful little tree." He further quotes a Miss Twamley, authoress of *The Romance of Nature*—"a very elegant work on flowers"—who called the blossoms of this shrub fairy fires

> *That gleam and glow amid the wintry scene*
> *Lighting their ruddy beacons at the sun*
> *To melt away the snow . . .*

A number of varieties were soon in cultivation—a French firm listed over forty in 1869—but later there seems to have been a decline in interest; while Hibberd in 1898 mentions that several different kinds (whose origin he did not know) had been cultivated within the past few years, Robinson, very strangely, hardly mentions the plant at all, and only devotes a single sentence to it in the whole of *The English Flower Garden*. About a dozen varieties are available today.

It is said that the firm of Maule and Son came to ruin through speculating on a demand for their *C. japonica* (syn. *Pyrus maulei*) that never materialized. It may have arrived at an unfortunate moment, or it may simply have proved the less attractive shrub of the two, because its later blooms, though abundant, are less dramatic among the leaves than those of the early *speciosa* on its almost bare boughs. But it has an additional season of beauty in the autumn, when it frequently produces a crop of golden-yellow fruit, and occasionally a few belated flowers, just to show how effective the scarlet blossom can look in conjunction with the golden apple. This species too has its varieties, though they are less numerous than those of *speciosa*; and some hybrids have been raised between the two (known collectively as *C. × superba*), notably '*simonii*' and 'Knap Hill Scarlet'.

The fruit of *C. speciosa* is hard and green, but fragrant, and Loudon says it was sometimes used to put among clothes. The yellow fruit of *C. japonica* is still more strongly scented, and a few kept in a bowl will perfume a whole room. Both are hard and quite inedible when raw, but are said to make good jelly, and during the First World War, when every source of home-produced food was being investigated, it was decided to put it to the test. Fruits of *C. japonica* and of six different varieties of *C. speciosa* were supplied in 1917 by Mr. W. J. Bean of Kew, and made into jelly by the Rev. J. Jacob, of Whitewell Rectory, Whitchurch, Salop, using in each case exactly the same process, and a special tea party was arranged to sample the results. It was unanimously decided that the *japonica* jelly was much the best—as good, Mr. Jacob thought, as guava jelly from the West

Indies. The *speciosa* kinds varied in quality, the least successful being "hardly different from wild crabapple jelly." The late Mr. E. A. Bowles appreciated the flavor of japonica jelly as a condiment with meat, and transferred a number of plants to the kitchen garden for the purpose of fruit production.

Chaenomeles is from *chaino*, to gape, and *meles*, an apple. (I am not sure whether it is the fruit or the eater of it that is supposed to do the gaping.) The family includes a tree, *C. sinensis*, cultivated in China for its fruit, which is extensively candied; but it is not hardy here.

NOTES: As Miss Coats says, the names of the species are in constant confusion owing to the diverse pathways of introduction. *Chaenomeles japonica* is a Japanese plant and in the Kirishima mountains of Kyushu it can be seen in its true dwarf prostrate habit. *C. speciosa* (syn. *C. lagenaria*) is an upright plant, native to China. It is the hybrid between these two that is the basis of the many cultivars in the nursery industry. Nurseries rarely offer *C. sinensis*, but this tree has handsome flaked bark. A fine specimen can be seen in the Asian Valley at the U.S. National Arboretum.

Chamaepericlymenum, see under *Cornus*

Chimonanthus

A small family of Chinese plants, allied to the American Calycanthus, with which they were at first classified. The only one in cultivation in Britain is the Wintersweet, *C. praecox* (syn. *C. fragrans*). This has been grown for hundreds of years in the gardens of China and Japan for the sake of its winter flowering and the sweet scent of its waxy, yellowish, purple-concentrated blooms. It was first heard of in Europe through the letters of a Portuguese Jesuit missionary, Alvarez de Semedo, who mentioned it under its Chinese name of *La-Mei* in 1643; Kaempfer described it in 1712 under its Japanese name of *Obai*, but it was not introduced to England until 1766, when a specimen was sent to Lord Coventry[1] at Croome in Worcestershire. There it was planted in the conservatory, and by 1799 had grown into a shrub 16 feet high and 10 feet wide, whose fragrance when in bloom could be smelt 50 yards away from the

[1]His garden, laid out by Capability Brown, was much admired; a correspondent to the *Gentleman's Magazine* in 1792 went so far as to say "if there be any spot on the habitable globe to make a deathbed terrible, it is Lord Coventry's at Croome . . ."

building. It was illustrated in the *Botanical Magazine* (1800) from a spray supplied by Mr. Whitley, a nurseryman of Old Brompton, "His Lordship having presented most of the nurserymen about town with plants of it." Bushes propagated by layers or seed had already been tried out-of-doors with slight winter protection, and proved to be hardy. By 1838 it was "grown in most choice gardens for its flowers, a few of which are gathered daily, and placed in the drawing-room, or boudoir, in the same manner as violets." Loudon considered it "so very desirable that no garden whatever should be without it. . . . In all gardens north of London, it deserves a wall as much as any fruit-tree, at least judging from the measure of enjoyment it is calculated to afford." Gardeners with discriminating noses have echoed these praises ever since; E. A. Bowles went so far as to say that it should be grown "in any garden large enough to hold two plants"—the other being *Iris unguicularis*. "No nose that has ever smelt it in unsullied youth," says Mr. Haworth-Booth, "can ever forget it, even in extreme old age."

Some garden forms are in cultivation, with brighter and more substantial flowers than those of the type, which are rather subject to injury by frost. The first of these was the variety *grandiflorus*, which was grown by the Comtesse de Vandes at Bayswater, London, before 1819, and which has bright yellow flowers larger but less fragrant than those of the original. A later-blooming variety with flowers lemon-yellow throughout, lacking the purple inner petals of the previous sorts, was introduced sometime in the mid-nineteenth century, and was named var. *luteus*, but its exact date and origin are unknown, and it has always been rather rare. Another all-yellow variety with much larger flowers was raised by the firm of Vilmorin from seed collected in 1906 by R. P. Ancel in East China, and named by them, rather confusingly, var. *luteus grandiflorus*. It is probably this kind that appears in nurserymen's catalogues under the name of *luteus* today; it has a sweet scent, though less powerful than that of the original species.

In China there are at least five garden varieties, besides the type, which was domesticated during the Sung Dynasty (A.D. 960–1279) and had the first of many poems addressed to it in the eleventh century. In the temperate regions it blooms about the time of the Chinese New Year celebrations, and it has long been the custom for Chinese ladies to use the scented flowers to decorate their hair. The wood, too, is aromatic, and housewives both in China and Japan are said to tie the twigs and prunings into bundles to scent their linen cupboards and clothes closets.

The family is appropriately named from *cheimon*, winter, and *anthos*, a flower.

NOTES: The Winstersweet is a long-hoped-for, winter-flowering shrub in the United States. Unfortunately it defies vegetative propagation to a great extent. In Japan, Winterswet is grown by specialists for the delightful fragrance in November. The variety called 'Mangetsu' ("full moon") has large double flowers borne in profu-

have been unkindly described as smelling of furniture polish) and its white fragrant flowers sufficiently justify its name of Mexican Orange-Blossom. It does not seem to have been particularly valued in its native land, even by those very flower-conscious people, the Aztecs; the later Mexicans associated it with the clove, and called it *Hierbo* or *Flor del clavo* or *clavillo*. In 1825 a Spanish botanist suggested that it should be named *Juliana caryophyl-*

Chimonanthus fragrans
(syn. *C. praecox*)
(WINTERSWEET)   S. Edwards (1800)

Choisya ternata
(MEXICAN ORANGE)   A. Riscreux (1869)

sion. The National Arboretum brought plants from Japan in 1976 but it is still not available in the trade because it is so difficult to propagate by cuttings or grafting.

Choisya

C. *ternata* is the only representative in cultivation of a small family related to the Oranges, and is remarkably hardy considering that it is a native of Mexico—though in cold districts it is safest against a wall. It is not so close to the Citrus group as the Poncirus (q.v.) but its leathery, evergreen, aromatic leaves (which

lata, but it had already been christened, in 1821, by Humboldt and Bonpland, in honor of the Swiss cleric and botanist M. J. D. Choisy (1799–1859). Choisy was for twenty-eight blameless years Professor of Philosophy at the University of Geneva, and assisted the great botanist de Candolle in the preparation of his *Prodromus*.

The Mexican Orange was introduced to Britain in 1825, but was grown at first only as a greenhouse shrub, and was slow in becoming established. In 1880 it received a First Class Certificate, and the next year we hear of fine specimens grown as standards at a Fulham nursery, which had come satisfactorily through the previous severe winter; they had "mushroom-shaped tops" about 5 feet high, with scented snow-white flowers "scarcely inferior to those of the orange."

NOTES: The Mexican Orange is a delightful, aromatic shrub that, unfortunately, thrives only in the most southern regions.

Cistus

The members of this family have a reputation for "frailty in their conception of marital relations."[1] Nearly all of the numerous species are natives of the Mediterranean regions, where their generous flowers, widespread to the sun, offer abundant pollen to every passing insect; so it is not surprising that a great many hybrids occur, both in the wild and under cultivation. A specialist in the genus, M. Edouard Bernet of Antibes, who set himself the task of discovering how many different hybrids could be raised, achieved 234 between 1860 and 1875. This abundance of intermediate forms has led to great difficulties in naming and classification, many of the original "species" having since been found to be natural hybrids; few can be relied upon to breed true from seed, and even today, many of the forms offered by nurserymen are not purebred stock, but some closely associated mongrel. It was simple enough in Turner's day. He knew (in 1548) only three Cistuses; the red, which he called the Male, the white, or Female, and the Gum Cistus or Ledon (*C. ladaniferus*); and this comfortable division into Male, Female and Ledon persisted for the next hundred years—Parkinson still used it in 1640. Meanwhile the yellow-flowered kinds had come to light; Gerard cautiously does not commit himself as to their sex, but Parkinson puts them among the females. Helianthemums (not mentioned by Turner) were separately classed by both authors, and though united to the same genus by Linnaeus, the family was later split up into Helianthemum, with three-celled seed-vessels, Cistus, whose seed-vessels have five to ten cells, and whose flowers are white or pink, and Halimium, similar to Cistus, but with yellow flowers.

Another little failing of the race as a whole is that of being "very impatient of our cold climate."[2] One of the hardier kinds, says Parkinson, "will abide some yeares with us, if there be some care had to keepe it from the extreamity of our Winters frostes, which both this and many of the other sorts and kinds, will not abide, doe what we can." It is noticeable that the "female" of the species is tougher than the "male," the hardiest varieties being found among the earlier, white-flowered kinds. (It is only the white-flowered "female," too, that has sensi-

[1] Sir Oscar Warburg, in the R.H.S. *Journal*, Jan. 1930 [2] Gerard

tive stamens, which spread themselves out and down on the petals if lightly touched—the opposite movement to that observed in the berberis bloom.) In general they are short-lived plants, and most growers advise an insurance policy of keeping a reserve supply of cuttings in pots under glass; though "whether such labour would not be better bestowed on some family of shrubs quite hardy in our climate, may be worth considering save by those who

Cistus creticus
(syn. *C. incanus* subsp.)
(ROCK ROSE)   J. Hart (1825)

seek collections in the face of all difficulties."[1] Nevertheless, a well grown Cistus is worth a little trouble, and in the 1820s the family enjoyed a period of special popularity; Robert Sweet was able to illustrate 112 varieties in his monograph on the *Cistinae* in 1829 (including Halimiums and Helianthemums) twenty-nine of which came from one London nursery—that of Messrs. Colville of the King's Road, Chelsea, who supplied more specimens even than the Chelsea Physic Garden, with twenty-two.

C. salvifolius, the Female Cistus or Sage Rose, was the first species known to have been cultivated here; Turner in 1548 reported that it grew "in my lordes gardine in Syon"—Sion House, the home of the Duke of Somerset. It had perhaps been imported as a medicinal plant, for

[1] Robinson

its roots are still valued for wounds by the natives of Morocco. In Sweet's day many different cistuses were sold under this name, "scarcely any of the Nurserymen knowing the true plant" though it is very distinct. It is now rather rare in cultivation.

C. albidus would seem to be the "Male Cistus" grown by Gerard in 1597, though there is some uncertainty as to which of the pink-flowered species his figure and description represent. He says its petals were "somewhat wrinkled, like a cloth new dried before it be smoothed . . . the flowers for the most part do perish and fall away before noone, and never cease flowring, in such maner, from the moneth of Maie unto the beginning of September." The specific name has been a stumbling block to generations of cultivators—it refers not to the flower, but to the white downiness of the leaves.

C. ladaniferus, the Gum Cistus, appears on a garden list of plants grown in 1604 by Mr. William Coys of North Okington in Essex. The flowers, as Gerard says, "grow at the end of the branches like little Roses, consisting of five white leaves, every one decked or beautified toward the bottom with pretie dark purplish spots tending to blackness. . . ." It is the only spotted species, and any other so adorned is sure to be a hybrid of *ladaniferus* descent. The flowers are large and beautiful, with the almost inevitable corollary that it is a tender kind, and needs favorable conditions to succeed. "The beauty of this tree when in blow," says Marshall, "is often over, in very hot weather, by eleven in the morning; but that is renewed every day; and for about six weeks successively a morning's walk will be rendered delightful by the renewed bounties which they bestow."[1]

Parkinson described some twenty-seven Cistuses in his *Theatrum Botanicum* (1640), and is credited with the introduction of the variable, rose-colored *C. villosus*. Marshall, in 1785, knew of several forms of it, which had formerly been thought species, "but experience now teaches us better." One of the most distinct and important of them is *C. villosus* var. *creticus* (1731), a rather tender sort that needs protection near London, "which," Loudon says, "as it does not often receive, it is, in consequence, scarce." A number of Cistuses were grown by the younger John Tradescant in his garden at Lambeth; their names appear in the *Museum Tradescantianum*, 1656, and include, among others, *CC. monspeliensis, crispus* and *populifolius. C. monspeliensis*, with abundant but rather small white flowers, has the widest range of the genus, extending from the Canaries to Cyprus; it is not the

---

[1]All the cistuses, indeed, spread their flowers "like a laundress for a few hours to the sun, and then, careless of their delicacy, fling them to the winds." (Maund, *The Botanic Garden*)

handsomest nor the hardiest of the race, but is a prolific parent of hybrids. (Its pollen will fertilize most of the white and a few of the pink species, but the reverse crosses, using *monspeliensis* as a seed-parent, are ineffective.) *C. crispus*, with clustered small flowers of purplish-red, is tender and seldom true; the plant grown nowadays under this name is usually its hybrid *C. × pulverulentus. C. populifolius*, a bushy white-flowered plant with large poplar-shaped leaves, is much hardier and is indeed one of the more reliable kinds; in this case too the true plant is scarce. The hardiest of all the species, *C. laurifolius*, was a comparative latecomer; the earliest of several dates given for it is 1731. Loudon says that its profuse white flowers "with their bright red bracteas, are very ornamental before they expand, resembling at a distance the bursting buds of roses." It was one of the three sorts to survive the great frost that killed nearly all the cistuses at Kew in 1895.

The other survivors of the great frost were two of the numerous Cistus hybrids—both white flowered—*C. × corbariensis* and *C. × loretii*. The former, a natural hybrid, was grown by Tradescant before 1656; one of its parents is now known to be *C. populifolius*, whose rosy buds it has inherited, and plants of it, according to Sweet, "make very handsome snug bushes and have a lively appearance when in bloom." *C. × loretii* dates from the nineteenth century; two plants of slightly different parentage are grown under the name, the true form, a natural hybrid between *ladaniferus* and *monspeliensis*, and *loretii* "of gardens" (*C. ladaniferus × C. hirsutus*) which should properly be called *C. × lusitanicus* var. *decumbens*. Another fine hybrid of *ladaniferus* parentage is *C. × cyprius*, very like its famous papa, but with some of the hardiness of its other parent, *C. laurifolius*. It is supposed to have been introduced from Cyprus about 1800, and early in the nineteenth century often passed for *ladaniferus*, while the true plant of the name went as *C. salicifolius*. Bean considered it "the most beautiful of all the cistuses we can grow out of doors"; but it has a rival in 'Silver Pink', probably the best known of the family at the present time; a hybrid of unknown parentage (possibly *laurifolius × villosus*) which was raised in the nursery of Messrs. Hillier of Winchester, and gained an Award of Merit when exhibited at the R.H.S. in 1919. It seems reasonably reliable; but if anything emerges from a study of the literature of the genus Cistus, it is that no two authorities agree as to which species are genuinely hardy, and which are not.

Five of the species—*laurifolius, monspeliensis, salicifolius, ladaniferus* and *villosus* var. *creticus*, but especially the two last, secrete the sticky, aromatic gum or resin known as ladanum or labdanum. All five will perfume the air to a great distance in warm weather, even in England, with a warm spicy smell suggestive of pinewoods; in their native lands they can be smelt "the eighth Part of a *German Mile*."[1] Napoleon is reported to have said that he would know his native Corsica with his eyes shut, from the scent of these plants; and a Mr. Swinburn, visiting Sicily in the early nineteenth century, complained that they "exhaled so powerful an effluvium, when the sun had been risen for some time, that it quite overcame him."[2]

The gathering and export of the fragrant ladanum is a trade of very great antiquity. From the earliest times there seem to have been two methods of collection; either by brushing over the bushes with specially made instruments, or by combing or melting the gum from the hair of the goats that grazed on them. Both these methods were described by Herodotus, Pliny and Dioscorides—Turner deriving his account directly from the latter. "Whenas the gotes and gote buckes eat the leaves of cistus they gather manifestly the fatnes with their beardes, and carye away with their clammenes it that cleveth upon their heary and rough fete. The which the inhabiters of the countre combe of and streyne it, and make it in lumpes together and so laye it by; other pul ropes thorowe the bushes, and with them take of the clammines and make Ladanum of it." (Gerard disbelieved the goat story, calling it an old fable fostered by "lying munkes," but the process was observed by a Dutch traveler called van Bruyn as late as 1683, who said that the goats' beards were cut off once a year, and the gum extracted by heat.) The instrument usually described as being used was a tool like a rake, but with strips of untanned hide instead of teeth, which were drawn or trailed gently over the bushes till they became thickly charged with the resin, which was then scraped off with knives. An eminent Egyptologist, Mr. Percy Newberry, suggested that the "flail" or "scourge" carried by the god Osiris, and also by several other deities and by kings on certain occasions, was actually one of these "ladanisterions," and that the false beard worn by the god beneath the chin was one of the severed goats' beards—"in fact, the original goatee."[3] If this is the case, the traffic in ladanum must date back to the First Dynasty—from 3188 B.C. Two early Biblical references to myrrh, Genesis xxxvii, 25 and xliii, II, are actually mistranslations and refer to ladanum. The fragrance has been described as the nearest in the vegetable world to that of ambergris; it was used for incense, perfumes and cosmetics, and also for various medical purposes, both in Europe

[1]Tournefort [2]Phillips [3]Newberry, in *Journal of Egyptian Archaeology*, vol. XV (1929)

and the Orient. By the end of the nineteenth century the industry was mostly concentrated on the islands of Crete and Cyprus, which together exported nearly 10,000 pounds in the 1870s; Cyprus in particular being so famous for the product that a perfume in which it was largely used became known as "Chypre."

Cistus is derived from the Greek *kise*, a box, from the shape of the seed-vessels—the same root as our "chest" and the Scottish "kist." Turner in his *Names of Herbs* spells it "Cisthus," with a slightly Irish effect. In Gerard the flower appears as the Holly Rose, a name that aroused Parkinson's scorn—"but why *Gerard* should call it the Holly Rose I see no reason, having no resemblance unto Holly . . ." but Sir Thomas Hanmer spelled it Holy Rose, and it is possible that, like the Hollyhock, the plant had originally some association with Palestine. (The Rose of Sharon in the Song of Solomon was thought by Linnaeus to be a Cistus.) Later it shared the name of Rock-Rose with the allied Helianthemums (q.v.).

NOTES: Because of their Mediterranean origins, the Rock Roses are limited in their use to the Californian climate, and therefore seldom appear in general nursery catalogues. One can understand the English gardener's appreciation for the large, colorful flowers in spite of the cultural difficulties, such as their reluctance to take after transplanting and their tenderness to frost.

Clematis

This very large family flings its ragged tangles round the whole temperate world; there are more wild species of clematis even than of roses. Some are herbaceous plants, but the majority are woody climbers, most of them hardy and many of them ornamental. They include some of the most rapid growers in the vegetable kingdom, hauling themselves along hand over fist with the help of their leaf-stem hold-fasts; Traveller's Joy, native to England, is the nearest approach in the temperate zone to the lianas of the tropics. Some species are fragrant; many have ornamental seedheads (Farrer, indeed, refers to it as a "Struwwelpeter family") and the flowers of some have been compared to their not-very-distant cousins, the Anemones.

C. vitalba, the British species, is by no means to be despised. Gerard rightly praises "the beautie of the flowers, and the pleasant sent or savour of the same," and also the "goodly shewe" it makes in winter, "covering the hedges white all over with his fetherlike tops"; and it was he who named it Traveller's Joy. It is perhaps too rampant for the garden, though it may be "introduced into shrubbery and woodworks, here and there one, to run over the bushes."[1] It has its uses, too: the tough pliant stems make the best of withes for binding bundles of wood "and for this purpose it is always used in the woods where it can be got."[2] Short lengths of the woody stems were cut by old countrymen, lighted at one end and smoked—hence some of the local names, such as Smoking Cane, Gypsies' Bacca and Shepherd's Delight. The leaves are acrid and poisonous, and not even the early herbalists attempted to use the plant in internal medicine.

The earliest foreign species of clematis to be intro-

sad blewish purple colour," which produced no seed— "and therefore the tales of false deceitful gardiners and others" who offered seed of this sort "must not be credulously entertained." Marshall in 1785 speaks of four kinds, blue, red, purple and double purple, the latter, a particularly strong grower, being the best for arbors. Today we have six or eight named varieties, but the species is chiefly important as one of the parents of the large-flowered garden hybrids. It was also extensively used at one time as a stock on which to graft the choicer kinds.

Clematis viticella

(VIRGIN'S BOWER)   Ferdinand Bauer (1827)

duced to England were naturally the European ones; and the most important of these was also the first—*C. viticella*, which was grown before 1569 by Hugh Morgan, Apothecary to Queen Elizabeth the First. By 1597 two varieties, a blue and a red, were already being cultivated, and the name of Ladies' or Virgin's Bower had appeared. Parkinson in 1629 had a double, "of a dull or

[1]Abercrombie. Mrs. Loudon in 1846 said it was much planted in cemeteries   [2]Marshall

So far as is known, the next species, *C. flammula*, was first grown by Gerard, before 1597; he called it Biting Clematis or Purging Periwinkle (periwinkles at that time being included in the Clematis family). Miller says that a leaf of it bruised and smelt on a hot summer day, causes "a smell and pain like a flame." It is the most fragrant of the European kinds, the flowers being "of an exceeding sweete smell, much like the smell of Hawthorne flowers, but more pleasant, and lesse offensive to the head";[1] and Bean, who adds vanilla and almond to his analysis of the scent, calls it one of the greatest pleasures of the autumn garden. Some people find it overpowering, close at hand; no wonder, as the plant covers itself so profusely with its small white starry flowers, that we might be inhaling the perfume of the entire Milky Way. Phillips (in 1823) said that in the Royal Gardens of Paris this clematis was grown as a shrub in the parterres, tied to a stake and kept trimmed—"by which means it was very ornamental, being covered with white blossoms, and at the same time throwing the fragrance of May over the whole gardens." It was also popular as a pot-plant in the Paris markets. Two hybrids have been raised from it, *C.* × *rubromarginata* (*C. flammula* × *C. viticella*) and *C.* × *aromatica* (*C. flammula* × *C. integrifolia*)—a herbaceous species with blue flowers.

Gerard also grew the rather tender, winter-flowering *C. cirrhosa* and the herbaceous *C. integrifolia*; but one of the best of the European species had to wait until the eighteenth century for its introduction. This was *C. alpina*, whose air of fragile and delicate beauty belies its tough constitution; Robinson called it as "hardy as the Oak and tender in colour as the dove," and included it among the species sufficiently unobtrusive to be planted among azaleas and other choice shrubs, and allowed to ramble over them at will. It would seem to be the kind described by Miller in 1759 as an "Alpine Climber with a Crane's Bill leaf," except that he said the flowers were white (not, as usual, blue); he received it "from Mt. Baldus,[2] where it is in plenty." The *Botanical Magazine* dates its introduction with more certainty from 1784–85, when seed was obtained and raised by the firm of Loddiges.

Several Clematis species were introduced from America in the early part of the eighteenth century, but they were of little horticultural merit; it was the Asiatic ones that were really important, ushered in by the arrival of *C. florida* in 1776. This native of China had been cultivated in Japan since the seventeenth century, and several garden varieties were available by the time a plant reached Britain through the agency of Dr. John Fothergill (1712–80), a Quaker physician and keen plantsman, in whose garden at Upton, Essex, according to a contemporary, "the sphere seemed transposed, and the Arctic Circle joined with the Equator."[3] Fothergill's variety was probably a double form; the single white arrived at the beginning of the nineteenth century and the lovely bicolor known as var. *sieboldii* was imported by Siebold from Japan in 1837. All these were Japanese garden-forms; it was not until the 1880s that the original species

[1]Gerard [2]Monte Baldo, in Italy [3]Lettsom, *Memoirs of John Fothergill, M.D.,* 1786

was discovered by Augustine Henry, growing wild near Ichang in western China. It is rather a tender Clematis, but became of importance later as a parent for hybrids.

The vigorous *C. montana* (syn. *C. odorata*) or Great Indian Virgin's Bower, which will cover a wall 40 feet high, was the next in chronological order, but has always remained outside the mainstream of clematis development. It was brought by Lady Amherst, wife of the Governor-General of India, in 1831; the rosy form called var. *rubens* which replaces the white Himalayan type in western China was sent by Wilson to Messrs. Veitch in 1903, and many superior garden varieties have since been raised. Two important Asiatic species quickly followed: *C. patens*, originally a Chinese plant but introduced to Europe from Japan, together with some of its garden varieties, by Siebold in 1836, and *C. lanuginosa* (1850) one of Robert Fortune's introductions from China, which bears the biggest flowers—6 inches across—of any of the wild species.

The stage was now set for the production of the large-flowered garden hybrids. A pioneer had already begun—Mr. Henderson, of the Pineapple Nursery, St. John's Wood, who crossed *C. integrifolia* with *C. viticella* about 1835, and produced a plant that was named after him *C. hendersonii* (syn. *C.* × *eriostemon*) which, although a climber, inherited the herbaceous habit of its *integrifolia* parent, dying down to the ground every winter. In 1855, Mr. Isaac Anderson-Henry of Edinburgh raised a hybrid between *lanuginosa* and *patens*, patriotically named *C.* × *reginae* in honor of Queen Victoria, and several continental firms were also busy; but all were eclipsed when George Jackman and Son produced the first of their famous varieties in 1862, the result of crosses made in 1858 between *C. lanuginosa*, *C.* × *hendersonii*, and the red form of *C. viticella*. Many other kinds were subsequently raised, by the same firm and by numerous others, but the original *C.* × *jackmanii* has never been surpassed in its particular field, and is still one of the most popular of climbers.

Then began the golden age of the Clematis; more than 200 species and varieties were described by Jackman in his monograph on the subject, *The Clematis as a Garden Flower*, published in 1872 and dedicated to H.R.H. the Princess Mary, Duchess of Teck, "as treating of one of her favourite flowers." Gone were the days when the plant was relegated to arbors, woodworks or wildernesses. Jackman suggests the planting of a special clematis garden, which he calls a "Climbery" and which would "form a most striking episode in the pleasure-ground or flower-garden scenery." He thought the plant was also well adapted for growing on rockwork, or "amongst those grotesque arrangements of old tree-stumps to which the term Rootery is commonly applied," whose "picturesque irregularities," he continues, in prose as purple as his own flower, "serve as supporters of the gorgeous purple vestments of Queen *Clematis*, and become, as it were, the trainbearers who spread them out in all their rich exuberance and amplitude before the gaze of her admiring and astonished devotees." Clematis were grown in pots as exhibition or conservatory plants, and out-of-doors were trained on specially constructed pillars, pyramids and parasols of various materials; they were also used for bedding, carefully trained and pegged down over the surface of the soil, so that each plant covered the base of the next. Perhaps owing to the amount of work involved, there seems to have been a decline in the plant's popularity towards the end of the century; also, most of the varieties sent out by nurserymen at that time were grafted on stocks of *viticella*, which caused them, in Robinson's phrase, to "die off like flies," and gave them a bad reputation for delicacy. Few of the early hybrids survive today, although the firm of Jackman still flourishes, and still specializes in the flower.

Jackman and Clematis are not synonymous, however, nor are all the Virgin's Bowers blue, white or purple. The family contains species of many other forms and colors, that would hardly be recognized by those accustomed only to the garden hybrids. They include the "orange-peel" flowers of *C. orientalis*, found by Tournefort in the Levant and grown in James Sherard's Eltham garden in 1731; the scented cowslip-bells of *C. rehderiana*, sent to France from China by a Jesuit Missionary, Père Aubert, in 1898, and thence to Kew in 1904; the deep-golden Chinese lanterns of *C. tangutica*, sent to Kew via St. Petersburg in 1898, and its variety *obtusiuscula* (Wilson, 1908) admired by Farrer in Szechuan; and the scarlet pitchers of the American *C. texensis*, received at Kew in 1880 from Max Leichtlin of the Baden Botanic Garden, and now rare in cultivation, but the parent of some brilliant hybrids. Stranger still is the almost leafless, rush-stemmed clematis from New Zealand, *C. afoliata*, with its fragrant greenish flowers. Two valuable species from China are comparatively recent; the evergreen *C. armandii*, introduced by Wilson for Messrs. Veitch in 1900, whose bunches of white or rosy flowers are produced in April or earlier, on sheltered walls; and *C. macropetala*, seed of which was sent from Kansu to the same firm by William Purdom in 1910. This species had been known to science since 1742, but it seems rare even in the wild, and Farrer, who sent a second consignment of seed from the Da-Tung Alps in 1914, only found it once. He says it "rambles frailly through light bushes to a height of two or three feet, then cascades downwards in a fall of lovely great flowers of softest china-blue, so filled with petalled processes that they seem as double as any production of the garden." Botanically speaking, clematis flowers have no true petals; it is their colored sepals that form the inflorescence, but in certain species, such as *alpina, macropetala*, and some forms of *florida*, there is a ring of "petaloid staminodes" between the sepals and the true stamens, that give the effect of a double flower. These types are grouped in a subdivision of the genus called *Atragene*, and those with pitcher-shaped or closed flowers, under *Viorna*.

The name "Clematis" is derived from the Greek *klema*, meaning a twig or shoot of the vine. Both being climbers, the earliest botanists associated the clematis with the vine, and Turner's name for the Traveller's Joy was *Vitis sylvestris*—the wood or wild vine. Hence the specific names of *vitalba* (white vine) and *viticella* (vine-bower)—but not *viorna*: Gerard explains this, a name previously used by de Lobel, as derived from *vias ornans*, "of decking and adorning waies and hedges." (Later this name was inappropriately allotted to a most unornamental species, the American Leather-flower.) For English names Turner suggested "Heguine" (Hedge-vine) or "Douniuine" (downy-vine) but these seem never to have become established; Gerard's inspired "Traveller's Joy" for the wild plant and "Virgin's Bower" for the garden one (though he does not claim the authorship of the latter) were so obviously appropriate that they soon superseded all others. There is some difference of opinion about the Virgin or Lady of the bower—some giving the name a religious significance, and others holding that it referred to Queen Elizabeth I, "who, it is well known, liked to be called the Virgin Queen."[1] (So far as I am aware, the name does not appear in pre-Elizabethan times.) The Japanese name of Wire Lotus seems particularly appro-prite to some of the oriental kinds; in Europe wild clematis is said to have over two hundred names in various languages. The English ones include, besides those already mentioned, Old Man's Beard, Grandfather's Whiskers, Old Man's Woozard, Father Time, Hedge Feathers and Snow in Harvest, all on account of the downy seed; and less complimentary ones such as Hagrope, Bed- or Belly-wind, and Devil's Twister. The compilers of the Language of Flowers, too, are rather less than kind; they make the clematis stand for Artifice, because of the use formerly made by beggers of the acrid leaves of *C. vitalba*, to raise pity-compelling sores and ulcers. "Trust not the Clematis!" warns the French author A. Grandville, "which climbs slyly up the walls,

[1]Loudon

and shows her little head at the edge of the window, where young maidens go at evening to talk. The artful Clematis gets possession of their secrets . . ." though any other climber, one might think, could do as much.

NOTES: With some 200 species and cultivars to choose from, Clematis is one of the leading vine plants for flower beauty offered to the gardener. The large-flowered varieties exhibit a remarkable array of colors and are fairly easy to cultivate. Apart from these, *C. maximowicziana* (syn. *C. paniculata*) is more widely seen in the U.S. than any other species because it is the hardiest. This native of Japan, Korea and China produces a billowing mass of fragrant white flowers in the autumn, followed by flossy seed-heads. It has become widely scattered throughout the southeast and it is sometimes confused with the native Virgin's Bower (*C. virginiana*). But that species usually has leaflets in three's while the Japanese counterpart has leaflets in three's or five's. Our native species is not so spectacular and does not appear in general nursery catalogues.

Colutea

C. *arborescens*, the common Bladder Senna, was an early introduction from Europe, and was grown in England before 1568, probably as a medicinal plant. There seems to have been some confusion between it and the true Senna, and Gerard warns his readers that "they are deceived that thinke it to be Sena, or any kinde thereof: although we have followed others in giving it to name Bastard Sene, which name is very unproper to it." The leaves are purgative, but act "very churlishly, and with some trouble to the stomacke and bowells,"[1] but for such as can use them, they are said to be good against rheumatic pains.

"The Bladder Sennas cannot be called choice flower-shrubs, but they are very useful for poor hungry soils," says Robinson, and that seems to sum up the general attitude towards them; many of the gardening books and nurserymen's catalogues omit them altogether. From time to time they have had their admirers; Celia Fiennes observed in the gardens of Wolseley in Staffordshire (in 1698) "a fine sena tree that bears a great branch of yellow flowers"; Liger (1706) thought they were unfairly neglected as they had "so good an Effect" and were "very ornamental in the Borders of great Parterres"; and Abercrombie (1778) called them "elegant furniture for the shrubbery." In the eighteenth century two further species were introduced, both with coppery or red-

[1]Parkinson, 1640

brown flowers; *C. orientalis* (syn. *C. cruentus*) discovered by Tournefort in the Levant and grown in Britain in 1710; and the tender *C. istria*, grown by Miller before 1759 from seed collected by the Rev. Dr. Pococke[1] in Turkey. Neither is common in gardens today.

Miller recommends hanging lobster claws, or the bowls of clay tobacco pipes, on these shrubs, to trap the earwigs which eat into the seed-vessels and destroy the seeds; and a strange sight the bushes must have been, decked with these ornaments as well as their own pods, compared by the same author to "the inflated bladders of Fishes." It is for these seed-pods rather than for its flowers that the shrub is grown, and they have the additional advantage of bursting with a pleasing noise when pressed in the hand. "Children find amusement in dancing on, or pressing these little bladders, which make a considerable explosion as the air escapes," wrote Philips—which is perhaps why the same author made the plant stand for "Frivolous Amusement" in the Language of Flowers. (Goëthe, in the true spirit of scientific enquiry, solemnly ascertained that these "distended legumes" contained only "pure air.") A demand for these pods for winter flower-arrangements may revive interest in this still common, but disregarded, shrub.

Loudon says that the name is derived from *kolouo* to amputate, because the bush dies if lopped; the branches are brittle, and easily broken by high winds. Gerard says that pieces casually broken off "and as negligently prickt or stucke in the grounde, will take roote and prosper, at what time of the yeere soever it be done; but slipt or cut, or planted in any curious sort whatsoever, among an hundred one will scarcly grow . . ." However, it usually propagates itself readily enough from seed. John Ray observed it on Vesuvius, on the ascent to the crater, where hardly anything else could survive; so it is natural that it should have established itself on railway embankments in suburban districts of London. It will be interesting to see whether it disappears again, after having been deprived, by the advent of diesel and electric engines, of its favorite diet of cinders.

NOTES: In America, Colutea is not generally grown as an ornamental plant.

Cornus

The members of the Cornus or Dogwood family are remarkably diverse in appearance, and the half-dozen or so species most familiar to us are cultivated for quite different reasons. Besides the garden

[1]Richard Pococke, author of *Description of the East*, 1743–47

sorts we also have a native kind, *C. sanguinea*—Gerard's Female Cornel or Dog-berrie Tree—which has hardly attained the status of a cultivated plant, although it was mentioned as being "common in shrubberies" in the early nineteenth century, and a variegated form was occasionally grown. Its wood was formerly put to a number of minor uses, and its berries, "in taste unpleasant, and not cared for of the birds,"[1] were used, on the Continent at least, for the production of lamp oil. Some derive the name of dogwood from the fact that, as Parkinson says, this fruit "is not fit to be eaten, or to be given to a dogge."

The first foreign kind to be introduced, the European *C. mas* (Male Cornel or Cornelian Cherry) was, on the contrary, cultivated for its fruit, and in the early gardening books was always included in the section devoted to the orchard or fruit garden. Turner in 1548 had not seen it but in 1551 was able to report "I heare saye there is a Cornel tree at Hampton Courte here in Englande," and before very long we find Tusser advising the setting of "cornet plums" and barberries, as one of the tasks of January. Gerard says this tree was cultivated in the gardens "of such as love rare and dainty plants," including his own. The fruit was popular in the seventeenth century; Parkinson said that "by reason of the pleasantnesse in them when they are ripe, they are much desired," and that they were eaten "both for rarity and delight," and also for medical purposes, on account of their astringency. Evelyn claimed to have invented a method of pickling the fruit, and said that he had "frequently made them passe for olives of France"; but Turner, long before, had said they might be "kept in bryne, as Olives be." Charles Bryant, in 1783, said that the fruit was "seldom eaten fresh off the bushes," but was used to make "tarts and other devices," but from that time onwards we begin to find references to the tree having been "formerly" cultivated for its berries, and by the beginning of the nineteenth century they were so seldom seen "that many people do not know that this beautifully transparent fruit exists."[1]

It was not until the fruit went out of use that this dogwood began to be appreciated as an ornamental plant, and was transferred from the orchard to the shrubbery. The first real praise it receives, from Phillips in 1823 and from the *Botanical Magazine* in 1826, was for the decorative value of the fruit in August, which "has a very fine effect, as it hangs like cornelian drops from the branches."[2] Unfortunately it is sparsely produced, except on mature specimens and in favorable seasons; Loudon explains that *C. sanguinea* is called the Female Cornel because it fruits young, whereas *C. mas*, the Male Cornel, produces male flowers only, until it is fifteen or twenty years old. The flowers, in the early days, were little regarded; Gerard describing them as being born "in small bunches before any leaves do appeere, of colour yellow, of no great value (they are so small)." Philip Miller, 150 years later, accords them a grudging approval—"tho' there is no great Beauty in the Flowers, yet, as they are generally produced in Plenty, at a Season when few other Flowers appear, a few Plants of them may be admitted for Variety." It was only in the present century that these February balls of tiny acid-yellow flowers, which seem to have a special quality of gaiety about them, began to be fully appreciated; and the plant was given an Award of Garden Merit in 1924.

The Cornelian Cherry is a slow-growing, long-lived shrub or small tree, and has been long celebrated for its dense, iron-hard wood, which was used, it is said, to build the Trojan Horse. It also made good spearshafts. When Romulus was fixing the boundaries of the future city of Rome, he flung his spear toward the Palatine Hill; its cornel-wood shaft, stuck in the earth, produced leaves

Colutea arborescens
(BLADDER SENNA)   Ferdinand Bauer (1833)

[1] Gerard

[1] Phillips   [2] Ib.

and branches, "foreshadowing in its growth the spread and strength of the Roman state," [1] and became a shrub, afterwards regarded as sacred and solicitously watered in hot weather. Another story (related by Virgil) tells that when one Polydore was murdered, the myrtle and cornel-wood shafts of his attackers' lances sprang into trees and made a grove, and when Aeneas later tried to uproot one of them, the sapling dripped blood. This legend would be particularly appropriate to the species of Cornus that we cultivate for their red winter bark—and a very fine grove they would make, with the evergreen myrtles to back them. British wild dogwood, though it has been called *virga sanguinea* (which Gerard translates as Bloudy Rod) ever since classical times, is not the brightest kind; the best red-barked sort is confusingly named *C. alba*—the white cornel—from the color of the berries. Seed of this Siberian species was sent by Professor Ammann of St. Petersburg and cultivated by Philip Miller of the Chelsea Physic Garden in 1741. It is too rampant a spreader to be admitted into the garden, but it has produced two controllable varieties, the dwarf *sibirica*, origin unknown, first advertised in Loddiges' catalogue for 1836, and more recently, *C. alba* var. *atrosanguinea* (the Westonbirt Dogwood) with winter bark even more brilliantly scarlet than the type. Even these, however, are plants for the park rather than for the average-sized garden; but there are two beautiful variegated-leaved sorts which are less vigorous and would grace the smallest plot. These are *C. alba* var. *variegata* (sometimes called *elegantissima*, but not to be confused with another variegated form, *C. mas* var. *elegantissima*) with grey-green leaves broadly margined with white, and the gold-variegated *C. alba* var. *spaethii*, which originated in the nursery of Messrs. Spaeth near Berlin early in the present century. The plant described by Loudon as *C. alba* was actually the American *C. stolonifera*, the Red Osier Dogwood (called by the Indians *meenisan*, Red-stick Berry) which was grown by the younger Tradescant before 1656. It is even more intrusive and pushing than *C. alba* but for those who care to risk it there is a fine yellow-barked form (var. *flaviramea*) introduced by Spaeth in 1899.

These dogwoods depend for their beauty on their leaves and bark, their flowers and fruit being of little importance; but there are other species which are grown, not exactly for the flowers, but for the large and showy bracts which surround the clusters of small true flowers, and serve the same purpose of attracting the insects and us. The first, *C. florida*, was described and its introduction recorded by Mark Catesby in his *Natural History of*

[1]Skinner

Carolina, 1731. "In the beginning of March the blossoms break forth; and though perfectly formed and wide open, are not so wide as a sixpence; increasing gradually to the breadth of a man's hand . . . As the flowers are a great Ornament to the Woods in Summer, so are the berries in Winter: they remaining full on the trees usually till the approach of Spring; and being very bitter are little coveted by Birds, except in times of dearth. I have

Cornus mas
(CORNELIAN CHERRY)   John Curtis (1826)

observed Mockbirds and other kinds of Thrushes to feed on them. In *Virginia* I found one of these Dogwood Trees with flowers of a rose-colour, which was luckily blown down, and many of the Branches had taken Root, which I transplanted into a Garden. That with the white Flower Mr. Fairchild[1] has in his garden." (His illustration shows the pink form, and one of the mockingbirds, called by the Indians the Hundred-tongued Bird.) This fine shrub, which is usually grown as a tree in America, requires the extremes of summer heat and winter cold. It was not until thirty years after its introduction that it was first recorded as blooming in England, and the occasion was thought worth celebrating with a dinner party; on May 17, 1761, Peter Collinson was invited by a Mr.

[1]London nurseryman and author of *The City Gardiner*, 1722

Sharp, of South Lodge, Enfield Chase, to dine with him and see his *Cornus florida* in flower. Collinson found it just as described and figured by Catesby, and remarked that it was "the only tree that has these flowers amongst many hundreds that I have seen, and it began to bear them in 1759." In spite of disappointments, however, there were persistent attempts at cultivation in Britain thoughout the eighteenth century, perhaps stimulated by further travelers' tales of its beauty in America, such as those of Peter Kalm (1770) and William Bartram (1792). The pink form mentioned by Catesby (var. *rubra*) seems more floriferous with us than the type; it grows magnificently in some gardens in the southern counties, and received an A.G.M. in 1937.

C. nuttallii is a native of the western states of America; it was discovered by David Douglas in 1826, but was thought at first to be only a local form of *C. florida*, though its flat, six-bracted inflorescences are quite distinct from the four bracts of the eastern plant, each of which has a pinch or twist at the tip that gives something of the effect of a child's whirligig. Ten years later it was found again by Thomas Nuttall, a Yorkshireman who emigrated to America in 1808 and made several extensive trips of botanical exploration before publishing his *Genera of North American Plants* in 1818. He later became interested in ornithology, and produced a manual on American birds in 1832. He sent information about the tree, which he believed to be a distinct species, and about the pigeons that fed on its berries, to his friend J. J. Audubon, who published its description in the 1842 edition of his famous *Birds of America*, naming it after the finder. Audubon's plate shows two specimens of the Bandtailed Dove, "placed on the branch of a superb species of *dogwood* discovered by my learned friend THOMAS NUTTALL, ESQ., when on his march to the Pacific Ocean, and which I have graced with his name! ... Seeds of this new species of *Cornus* were sent by me to Lord Ravensworth, and have germinated, so that this beautiful production of the rich valley of the Columbia River may now be seen in the vicinity of London, and in the grounds of the nobleman just mentioned, near Newcastle-upon-Tyne." This handsome dogwood, called the noblest of the genus, in California attains a height of 100 feet and is so magnificent in autumn as to be "spared even by the settler;"[1] in England it is usually only a large shrub, but seems hardy and flowers freely when well-established, sometimes producing a second crop of smaller flowers late in the season, before putting on its autumn dress of silvery crimson.

C. kousa, which received its name from the Japanese

[1] *The Gardeners' Chronicle*, quoted in *The Botanical Magazine* (1910)

island of Kyushu where it was first found, is perhaps the best species for the average garden; its flowers, though individually less handsome than those of *C. nuttallii*, are produced in great profusion. It was introduced from Japan in 1875, and received a First Class Certificate in 1892. A superior form, var. *chinensis*, was collected by Wilson for the Arnold Arboretum in 1907, and was sent from there to Kew in 1910; it is said to equal even *C. florida* as it is in its homeland, and to begin flowering at an early age. It also fruits well, and was given an Award of Merit as a fruiting shrub in 1958. The fruit is said to be edible.

The family also includes a most beautiful evergreen tree, *C. capitata* (syn. *Benthamia fragifera*, Himalayas, 1825) which is unfortunately not hardy, and two small herbaceous species, the delight and despair of rock-gardeners, as they are equally hard to establish and to get rid of—*C. canadensis* and *C. suecica*. The latter grows wild in northern England and in Scotland; its Highland name is *Lus an Chraois*, Plant of Gluttony, because its berries are supposed to promote appetite. *C. canadensis* was introduced in 1758 by Collinson, who noted "grows all about Halifax and Newfoundland; called Baked Apples and Pears."

The name of Cornus, which dates back to Theophrastus, signifies "horn," and refers to the extreme hardness of the wood. Some authorities think that the translucent, tawny-red stone we call "cornelian" derived its name from its similarity to the fruit of *C. mas*.

The different groups in the family, long recognized as distinct by gardeners, have recently been officially separated by the botanists. *C. mas* keeps the original name; the running, red-stemmed species such as *alba* and *sanguinea* are now called Thelycrania; those with decorative bracts (*nuttallii*, *florida*, etc.) are Benthamidias, and the herbaceous ones return to their ancient name of Chamaepericlymenum.

One word of warning: if anyone handles cornel-wood (either from *C. mas* or *C. sanguinea*) until it grows warm, within twelve months of his recovery from the bite of a mad dog, he will have a return of his distemper. This was certified from personal experience by the Italian physician and herbalist Pierandrea Mattioli, who died of the plague in 1577.

NOTES: As Dr. Donald Wyman says, "there is a dogwood for every part of the country." They vary from multi-stemmed shrubs with bright red or yellow stems in winter to giants that reach 60 feet. First in line is our native East Coast dogwood, *Cornus florida,* so common and popular that it is the state flower (tree) in Missouri, North Carolina and Virginia. There are many cultivars, some with deep pink floral bracts, double flowers or a weeping habit. A serious disease (Anthracnose) now threatens this beautiful harbinger of spring; its Japanese counterpart, *C. kousa*, appears to be the dogwood of the future. Regardless, these two make a perfect combination for the garden because of their similar size and the fact that *C. kousa* flowers about three weeks later than *C. florida*. Although it has been around since before 1880, *C. controversa* is a newcomer to our gardens. The Giant Dogwood is a standard background tree in large feudal gardens of Japan, where it grows to 60 feet. Small flower clusters are borne across handsome layered branches. In winter the stems are a striking purple. It flowers as a young tree and roots readily from cuttings. There is a handsome form with variegated foliage that has been cultivated in Japan for centuries and is now available in the U.S. *C. nuttallii*, native to the Pacific coast, is a beautiful tree but unfortunately does not thrive in the eastern states.

Because of the current disease problems, considerable attention is being paid to our native Florida Dogwood. The future may rest on breeding programs utilizing the Asiatic species as well as selection of disease resistant strains of the Florida Dogwood itself.

Regarding *C. mas*, it is interesting to note that in the Soviet Union its fruits are regularly sold and prescribed in hospitals as a source of vitamin C.

Cotinus, see under *Rhus*

Cotoneaster

The Cotoneasters belong to the vast order of the Rosaceae, which contains so many useful or ornamental woody plants, and so few herbaceous ones. They are chiefly cultivated for the sake of their autumn display of berries, and differ from the Pyracanthas in their entire, not toothed, leaves and their lack of thorns. There are a few European kinds, but the great majority are natives of northern India and Tibet, extending eastwards through China, but not reaching Japan; and most are of comparatively recent introduction here. In the first quarter of the nineteenth century the number of known cotoneasters rose from four to twelve, thanks to those sent home by Dr. N. Wallich, Director from 1815 to 1846 of the Botanic Gardens at Calcutta; Asiatic introductions then steadily increased till the early years of the present century, when Chinese species arrived by the dozen—twenty at least before 1915, and many later.

There are now about sixty kinds known, more than half of which are in cultivation; some with black berries and some red, some evergreen and some deciduous, and ranging in size from prostrate shrublets to near trees— *C. frigidus*, if trained to a single stem, can reach a height of 30 or 40 feet and has been used as a street-tree. One cotoneaster has a precarious foothold as a native of Britain—*C. integerrimus*, which was cultivated by the younger Tradescant in 1656, but was not known as a wild plant until found on Great Orme's Head[1] near Llandudno by a Mr. J. W. Griffith in 1783.

The arrival of Wallich's Indian species in the 1820s gave quite a different aspect to a family which had previously been little regarded; some of them are still among the best in cultivation, including *C. frigidus*, *C. microphyllus* and *C. rotundifolius*. Seed of the first (the largest-growing member of the family) was brought by native collectors "from the mountains of that northern region of Nipal called Gossain Than," and was sent by Wallich to the Honorable Court of Directors of the East India Company, who in turn passed it on to the London Horticultural Society, in whose new Chiswick garden it was raised in 1824. This cotoneaster is said to have received its name of *frigidus* from the coldness of the locality in which it was found. Given room, it makes a superb large shrub, and is also of importance as a parent of hybrids; it received an A.G.M. in 1925, and has a yellow-fruited variety known as *fructu-luteus*. The evergreen *C. microphyllus* was sent in the same consignment; it is a prostrate species unless it meets with a wall, when it will walk up it in an obliging way that has earned it the name of Wall Cotoneaster or the Architect's Friend, as it will clothe unsightly masonry so densely that nothing can be seen. (Lindley thought that the snow-white flowers strewn over its deep glossy foliage had "so brilliant an appearance, that a poet would compare them to diamonds lying on a bed of emeralds.") The third species mentioned, *C. rotundifolius*, a medium-sized semievergreen shrub, also from the Gossain Than, arrived the following year (1825); it received an A.G.M. in 1927.

These early species came into gardens that were not, in a way, prepared to receive them. Naturalistic rock-gardens and informal woodland gardens were unknown, although there were both "rockeries" and "shrubberies"; and the uses proposed by Loudon for the cotoneasters in 1838 sound strange to us today. For three of the species—*integerrimus*, *microphyllus* and *rotundifolius*—he recommends grafting[2] on hawthorn as standards, to produce "singular" or "curious" drooping trees; the hardy *C.*

[1]Still its only British station [2]"Grafted stock should be avoided"— R.H.S. *Dictionary of Gardening*, 1956

rotundifolius, especially, "might be grafted standard-high in every hawthorn hedge in the north of Scotland." He further makes the suggestion that *C. microphyllus* might be used for topiary work. "To some, it may appear in bad taste to revive the idea of verdant sculptures; but such is the ardent desire of the human mind for novelty, that, we have no doubt, clipped trees and shrubs will, at no distant period, be occasionally reintroduced into

Cotoneaster frigidus

(ROCKSPRAY)   M. Hart (1829)

gardens. The contrast produced by beauties of this kind, in the midst of a profusion of natural and natural-like scenery, is delightful."

Later in the nineteenth century the Chinese species began to flow in—*horizontalis* in 1879, *franchetii* in 1895, *bullatus* in 1898. The first, the Fishbone or Herringbone Cotoneaster, is perhaps the most familiar member of the whole family, and the only one to have received a common name. Regular alternate branching on a level plane is characteristic of many of the cotoneasters, but is concentrated and exaggerated in this species into a striking, if rather rigid, pattern, which shows to great advantage when draped over a rock, or flattened, like a gigantic centipede, against a wall. Seed of this species was sent by Père Armand David to the Natural History Museum

of Paris, before 1874, and was distributed by M. Decaisne to various persons, so that its subsequent gradual introduction to horticulture (according to the *Revue Horticole*) was hardly noticed—certainly the means of its introduction to Britain does not seem to have been recorded. All the cotoneasters are good nectar-bearing plants, but *C. horizontalis* in particular is besieged in its season by bees and other insects, and is so alluring to the newly-emerged queen-wasps, that it has been suggested that it might be planted specially for the purpose of trapping them.

There are a number of other dwarf or prostrate species—beneath our notice except when they climb walls; so many, indeed, that Robinson suggested "Rock-spray" as the family name. A stimulus to the cultivation of the larger kinds was given in the present century by the arrival of many new species and varieties—*CC. divaricatus, henryanus, hebephyllus, hupehensis, lacteus, serotinus, wardii, salicifolius* and *salicifolius* var. *rugosus* were all introduced from China before 1914—all excellent shrubs from 6 to 12 feet in height, except the last two, which are rather taller. The valuable *C. salicifolius* was another of Père David's discoveries, but although introduced to France in 1869, it does not seem to have been grown in England until seed collected in the same locality by Wilson in 1908 was distributed by the Arnold Arboretum to various centers, including the Glasnevin Botanic Gardens at Dublin. *C. divaricatus* and *C. henryanus* were also introduced by Wilson, but *C. lacteus* and *C. serotinus* were grown from Forrest's seed; the latter is particularly valuable because its berries ripen late, and stay on the trees till spring. For it must be admitted that the cotoneaster does not produce its fruits primarily for our benefit, and the most glorious display of color in the garden may be stripped by peckish blackbirds almost overnight. The depredations of birds vary greatly with the season and the species—some kinds seem unpalatable and are left untouched except in emergencies. *C. serotinus* is outstanding in this respect, and so is one of the most recent and most successful introductions, *C. conspicuus*. This fine shrub was discovered and collected by the late Captain Kingdon-Ward at Gyala in southeast Tibet, in 1924; there is both an upright[1] and a prostrate form (var. *decorus*) the latter described by its finder as looking like "a bubbling red caldron of berries" when seen from the cliffs above. By Rima in the Lohit Valley, called by Kingdon-Ward "a rather unbuilt-up village," the berries remained on the bushes till they were taken by migratory birds in April; and at Nymans in Sussex, where Lieutenant-Colonel L. C. R. Messel was among the first to grow

[1]Now, I believe, classed as *C. microphyllus* var. *conspicuus*

it, the berries have been seen still looking bright and attractive among the azaleas the following June.

The tall *C. frigidus* is the parent of three or four handsome and vigorous hybrids, dating approximately from the 1920s—*C. × cornubia, C. × St. Monica* and *C. × watererii*. The first occurred as a chance cross between adjacent plants of *C. frigidus* var. *vicarii* and *C. glabratus*, in the garden of Mr. L. de Rothschild at Exbury in Hampshire; it is supposed to have the largest fruit of any cotoneaster. The second, a poor orphan of unknown parentage (probably *frigidus × salicifolius* var.) was "raised accidentally at St. Monica's Home, Bristol" by Dr. Bauer, and distributed by one of the trustees, Mr. Hiatt Baker. *C. × watererii* was brought up in comfort in a nursery—that of Messrs. John Waterer, Sons and Crisp, of Bagshot, Surrey, and is the result of an alliance between *C. frigidus* and *C. henryanus*. In each case one parent is an evergreen species, and the progeny may keep their leaves in the winter in mild seasons or situations and lose them in cold ones. Unfortunately it has recently been discovered that both *C. frigidus* and *C. salicifolius*, together with *C. simonsii, C. × cornubius* and possibly other species, are susceptible to the dreaded disease of Fire Blight. Any tree affected with this extremely infectious bacterial disease must immediately be grubbed up and burnt, so that there is a danger that many of the handsome cotoneasters of our gardens may vanish as quickly as they came.

The name, according to Loudon, is a barbarous word meaning quince-like, from the Greek *Kotoneon*, a quince, and the Latin *ad istar*, similarity or likeness; its invention was attributed to the Swiss botanist Konrad Gesner (1516–65). It has recently been decreed that this word should be regarded as masculine, not feminine, and the endings of all the specific names have been altered accordingly. If the names used above seem unfamiliar, they are nevertheless quite correct—at the moment.

NOTES: Peerless among shrubs for use on banks and as rock garden plants, there are about six species in common use despite the large number of both deciduous and evergreen species. One should look for the improved forms. One cotoneaster that has been neglected is the evergreen *Cotoneaster conspicua* 'Decora', often called the "necklace" cotoneaster because of the profusion of small pinkish flowers followed by strands of scarlet fruits. There are some fine plantings of this form in the National Arboretum.

The Cotoneasters are relatively easy to grow, requiring only good garden soil and a sunny position in the garden. Some of the more prostrate ones tend to sprawl and must be pruned regularly to keep them in bounds. They are often used in foundation plantings as well as for bank covers and in rock gardens.

Crinodendron, see under *Tricuspidaria*

Cydonia, see under *Chaenomeles*

Cytisus

. . . And pypës made of greenë corne
As han thise litel herdegromes
That kepen bestës in the bromes . . .
Geoffrey Chaucer, *The House of Fame, c.* 1384

Cytisus, Genista and Spartium have long been favorite puss-in-the-corner families among botanists, the members of which have frequently been transferred from one genus to another; the only sure way, we are told, of distinguishing a Cytisus from a Genista is by the presence in the former of a small protuberance called a strophiole on the seed—and even that is not an infallible guide. The Laburnum, once classified as a Cytisus, has long since been removed from the family, and our own common broom (*C. scoparius*) does not fit very comfortably into it; some botanists, after trying to include it first in Genista and then in Cytisus, have decided that this particular piece of the jigsaw does not belong to either puzzle, and have allotted to it a genus of its own, under the name of Sarothamnus.

C. scoparius (or *Sarothamnus scoparius*) has an "Atlantic" distribution, and will not grow in the interior of the Continent, where the winters are too cold for it; and what to us is a common weed is vainly coveted elsewhere. Gerard sent plants and seeds of both broom and furze "to Elbing, otherwise called Meluin[1] . . . where they are most curiously kept in their fairest gardens . . . being first desired by divers earnest letters;" and in the eighteenth century Collinson sent seeds to John Bartram in Pennsylvania, but there too the plant perished in the hard winters. It has a bad reputation for impoverishing the land on which it grows, but it used to afford valuable winter fodder for sheep,[2] and in Scotland whole fields of it were formerly sown for fuel. Probably they were rocky slopes where little else would grow; and the "broomy knowes," wild or cultivated, made a favorite trysting place for lovers, often mentioned in Scottish

[1]In Poland [2]It also afforded an ointment "to salve poore mennes shepe, that thynke teere [tar] to costely"

song. (Warwickshire too had its broom-groves "Whose shadow the dismissed bachelor loves, Being lass-lorn."[1]) There seems to have been a belief that the scent of the flowers was soporific; in the ballad of *The Broomfield Hill* a girl has made a tryst to meet her lover among the broom, and then regrets her rashness, until a witch-wife tells her how she may keep both her promise and her maidenhood. The hour was to be in the afternoon, and on arrival she would find the young man asleep:

> *Tak ye the bloom frae aff the broom,*
> *Strew't at his head and feet,*
> *And aye the thicker that ye do strew*
> *The sounder he will sleep . . .*

A faint, faint breath from a broom-grown river-bank still lingers in the name of Glasgow's "Broomielaw."

The plant has been put to a great variety of uses. The young flower buds, pickled in salt, were "used in Sallades, as Capers be, and eaten with no lesse delight:"[2] the seeds have been used as a substitute for coffee, and the wood, when it attained sufficient size, to supply the cabinetmaker with "most beautiful material for vaneering."[3] Broom has also been used, in Britain and elsewhere, for thatching, cloth- or paper-making, tanning, dyeing and in the manufacture of potash, and above all, for the making of brushes—so much so that it has given its name to the implement, which now has rarely anything to do with the Cytisus family; "in modern Gibbridge, a Beesom made of Birch is called a *Broom*."[4] "The greene new brome sweepth cleene" was already a proverb in 1562, and Culpeper said it was "generally used by all good housewives almost through this land to sweep their houses with." The itinerant broom-seller was once a familiar sight:

> *There was an old man and he lived in the wood,*
> *And his trade was a-cutting of broom, green broom . . .,*

and until quite recent times gypsies used to cut broom and make besoms, which were used for sweeping out brick bread-ovens because they did not readily take fire.

Broom was probably first cultivated, in districts where it was not sufficiently abundant wild, as a medicinal plant; a decocotion of the green twigs was valued for centuries as a diuretic and purgative, though Gerard warns us that "it doth by reason of his sharpe qualitie manie times hurt and fret the intrailes. . . . That woorthie Prince of famous memorie *Henrie* the eight King of England was woont to drinke the distilled water of

[1]*The Tempest*, Act IV, Sc. I   [2]Gerard   [3]Miller-Martyn   [4]Threlkeld, *Synopsis Stirpium Hibernicarum*, 1727

Broome flowers against surfets and diseases thereof arising." Turner speaks of broom twigs steeped in vinegar being used for the sciatica, but with a caution unusual for the period, remarks "I thinke it were better to mix it with oyle, and so to lay it upon the greved place, then to take in inwarde, except the pacient were very stronge." A drug extracted from the plant is still valued for kidney and liver complaints, and broom tops were collected for this purpose during the Second World War.

Writing in 1728, Batty Langley said that the cultivation of the common broom was "not yet considered" in the garden, as it was so plentiful wild. Miller considered it equal to any as a decorative shrub, and Loudon pointed out that it has the largest flowers of the genus, "and were it not so common, would doubtless be considered the most ornamental." Phillips (1823) makes the suggestion that it might "judiciously be placed at the foot of towering trees, where it will shine as gay in the gloom as a gypsy's fire in a forest." (But it does not thrive in the shade.) Its status as a garden plant was much improved by the discovery by M. Edouard André (1840–1911) a celebrated French landscape gardener and horticulturist, of a variety in which the wing petals were brownish-red. He found it about 1884, growing among wild broom in Normandy, and it was propagated and named after him *C. scoparius* var. *andreanus*. A near-white variety, and the sulphur-colored Moonlight Broom were also by that time in cultivation; but there were no broom hybrids until the 1890s, when Mr. George Nicholson, then Curator of Kew, persuaded the late Mr. W. Dallimore of the Kew Arboretum, to try to raise one by means of artificial fertilization. This was a task hitherto thought to be impossible, as broom pollen ripens while the flower is in bud, and it is an extremely delicate operation to cut away the stamens without injuring the pistil, when the flower is still so small. Mr. Dallimore isolated a plant of var. *andreanus* in a greenhouse, and after several failures managed to emasculate two flowers successfully, one of which was fertilized with pollen from *C. albus*. Four seeds resulted, all of which germinated; two of the seedlings grew well, flowered, were found to be of little value and were discarded. The remaining two grew very slowly; it was not until 1902 that one of them produced a promising-looking flower, and it was only with difficulty that a shoot was obtained large enough to graft on a stock of laburnum, after which it grew and flowered freely. This was the lovely pink-and-crimson *C.* × *dallimorei*, which has never proved a very robust plant, but whose fertile seeds have provided many beautiful varieties. (The other seedling eventually opened yellow flowers, but produced a grandchild with cream blooms touched with rose.) Catalogue descriptions of a dozen or so of these varieties read like a list of racing colors; but the bicolored brooms, though lovely close at hand, do not show up so well at a distance as the plain forms.

Several foreign species were introduced during the eighteenth century, (natives, with one exception, of southern Europe) including *C. monspessulensis* (1735), *C. hirsutus* (1739), *C. purgans* (1750), *C. austriacus* (1759) and the tender *C. canariensis*, the popular "Genista" of the florists, which was brought home by Francis Masson from the Canary Islands in 1777, and pot plants which are, or were, often sold on costers' barrows in the spring. But the most important introductions for general garden purposes were the White, the Black and the Purple brooms—*CC. albus, nigricans* and *purpureus*.

C. albus, the White Spanish or Portugal broom, was known to Gerard and later authors through the descriptions of Continental botanists, but does not seem to have been grown in Britain until about 1752, when it was cultivated by James Gordon, the celebrated nurseryman of Mile End. Loudon considered it a very ornamental lawn plant even when not in bloom, "by the varied disposition and tufting of its twiggy thread-like branches" and spectacular in flower. "Trained to a single stem, its effect is increased, and grafted on the laburnum, a common practice about Paris, it forms a very remarkable combination of beauty and singularity." Phillips' praises were even more fulsome. He described it as being "clad like a virgin bride in pearls" and "rather studded with flakes of snow than bedecked by Flora's hand ... while its graceful waving bend so well accords with the chastity of its colour." Even the R.H.S. *Dictionary*—not usually given to eulogy—calls it an "upright and lovely shrub"; and though it has stooped so far as to become the parent of two important hybrids, in the case of *C. × dallimorei* at least, it was through circumstances beyond its control. Bees approve of it as much as gardeners, finding its flowers more convenient in size and more welcome in their earliness than those of the common broom.

C. nigricans is not really black—most authorities say that the name was given because the flowers turn black on drying. It too was known to botanists long before its introduction. Miller said it was formerly cultivated in some of the curious gardens, but was lost until reintroduced by him in 1731; also that it formed "a low shrubby bush which is with difficulty raised to a stem" but which made nevertheless "a fine appearance." Needless to say, Loudon grafted it as a standard; Maund on the other hand said it was "well-suited for mingling with low American plants." It is a shrub of a very different pattern from the elegant *albus*, but its slender upright spires

of yellow flowers are bright in July and August, and deserve to be better known.

C. purpureus was described by Sims in the *Botanical Magazine* (1810) as "a humble shrub with weak stems" introduced about 1790, "we believe, by Mr. Loddiges"; it was found, says Miss Kent, by Pallas "in the Wolga desert." Loudon considered that this was the best broom of them all—grafted, of course, as a standard; and this practice seems to have persisted, for Robinson, sixty or seventy years later, protested that its place was among boulders in the rock garden, and not, as it was generally seen, "grafted mop-fashion on Laburnum stems." On one occasion this custom had an unexpected result. In 1825 a scion of *C. purpureus* grafted on a laburnum stock in the nursery of D. Adam at Vitry near Paris, was accidentally broken off, leaving a small piece of it behind. Both the stock and the remaining scrap of the scion subsequently grew, the latter inside the former, producing what is known as a "graft-chimaera," which the utmost skill would probably never have achieved on purpose. The result is a curious small tree, which produces for the main part purplish or coppery-pink laburnum-like flowers, but which occasionally sends out a branch from its outer tissues of pure unadulterated laburnum, or from its inner ones of pure purple cytisus. This pathetic mixture is still in commerce under the name of *Laburnocytisus × Adamii*. Left to itself to grow in its natural fashion, and not imprisoned like Ariel inside his cloven pine, *C. purpureus* will attain a height of about 2 feet; but there are some out-and-out prostrate brooms whose upstanding height is limited to a few inches.

With so much of the material already assembled in the eighteenth century, it is remarkable (and says much for the chastity of the family) that no broom hybrids appeared until late in the nineteenth—with the exception of *C. × versicolor* (*C. purpureus × C. hirsutus* or *ratisbonensis*) which dates from about 1850. There are five of these hybrids—all what we fondly call accidental, meaning not designedly made by man—besides the intentional *C. × dallimorei* already mentioned. Two of them, *C. × kewensis*, and *C. × beanii*, are rock garden shrubs, sharing as a common parent the 6-inch *C. ardoinii*; they appeared in the nursery at Kew about 1891. The tender Porlock Broom (*C. racemosus × C. monspessulensis*) was raised in the early 1920s by Mr. Norman G. Haddon in his garden at West Porlock in Somerset. The popular *C. × praecox* (*C. purgans × C. alba*) appeared as a seedling in a bed of *C. purgans* in the nursery of Messrs. Wheeler of Warminster, about 1867. Unfortunately its beauty, resembling a waterfall in sunlight, is marred by a disagreeable smell, which can only be imputed to original sin, as it

Cytisus hirsutus

(BROOM)   Ferdinand Bauer (1833)

does not seem to be inherited from either parent. Fortunately there is a very similar plant, a seedling from *C. × dallimorei* called *C. × osbornei*, which has not this disadvantage.

In comparatively recent times a new broom was introduced which was confidently expected to sweep clean through our gardens—*C. battandierii*; and if it has since proved something of a disappointment, it is not so much through doubtful hardiness as because it is variable, and some forms do not flower well. (Seed received at Kew from two different sources produced free-blooming plants with orange-yellow flowers, and sparse-blooming plants with paler flowers.) The type was found by M. P. de Peyerimhoff in the Atlas mountains of Morocco, at a height of 5,000-6,000 feet, and was named in 1915 in honor of Professor Jules Aimé Battandier (1848–1922), contributor to *Flore d'Algerie, Catalogue des Plantes du Maroc*, and other botanical works. Bean says that this species was introduced about 1922, but gives no details. One of the first specimens to be planted out-of-doors (the Kew plants were cautiously started under glass) must have been that grown in Hyde Park by Mr. T. Hay, which was six years old when it flowered for the first time in 1930; branches cut from it gained an Award of Garden Merit in 1931. This is another Cytisus that does not quite conform to the botanist's requirements, having no strophiole on its seeds; to the gardner's eye it resembles a small silky laburnum, with unusually large and velvety leaves. The flowers, which grow in upright, sausage-shaped racemes, have a fragrance which Bowles said changed every few minutes; it is usually compared to the scent of pineapple, strawberries or other fruits.

The name Cytisus is indirectly derived from Cythnus (now Thermia), "an extremely fruitful island of the Greek archipelago."[1] Here the first species was supposed to have been found—a plant much esteemed by the ancients and called by them *Kytisos*, but which modern botany identifies as a species of Medicago. One would not have thought the simple English name of "Broom" capable of much variation; but it occurs in every possible form in different parts of the country, among them breeam, breen, browme and brum, not to mention besom, bisom, basom or bizzom. In Dutch it is *brem*, but elsewhere on the Continent the plant is known only as *Genista* or its local equivalent. Apparently the common broom has a Welsh name that means "Goldfinch of the Meadows;" it is curious therefore to find its Chinese name translated as "Gold-Sparrow-Flowers."

NOTES: These Old World plants have found a home all across the U.S. but perhaps do best in the milder Pacific Northwest. But they are generally useful for poor, dry soils. Recently, a number of improved cultivars have been introduced from Europe, such as *C. × praecox* 'Moonlight', a selection with masses of pale yellow flowers. For rock gardens and draping over stone walls, *C. × kewensis* has long been a favorite. The Scotch Broom, *C. scoparious*, is available in several shades of yellow-red and the type has long escaped from cultivation along roadsides and open woods in the southeast.

The brooms are easy to grow, as evidenced by their range of cultivation across the country. They do best in a dry, slightly acid soil, such as that found on open sunny banks. Their brilliant yellow flowers in late spring and summer and the green branches make them desirable garden plants all year.

Daboecia

D. cantabrica (syn. *D. polifolia*), St. Dabeoc's Heath, Irish Whorts, Cantabrian Heath. It is not known when this large-scale heath was brought into cultivation, partly because, as Robinson gently complains, its name has been so often changed by botanists that it is difficult to trace it in their books. Linnaeus (who took the name from Ray) called it *Erica Daboecia*—getting the spelling wrong, as he frequently did, but nobody may alter it now; in Miller's *Gardener's Dictionary*[1] it is classified as an Andromeda; elsewhere it appears as Menziesia, or in even more unfamiliar guise, as Boretta. "It grows in the squalid and boggy Mountains of *Mayo*, and throughout all *Higher Connaught* and in *Galloway*,"[2] and was first discovered by the Welsh archaeologist, natural historian and botanist Edward Lhuyd[3] (1660–1709) after whom the Lloydia is named. He sent specimens to Ray in 1700, and to Dr. Richardson in the same year, with the information that in Galway and Mayo it was so common "that the people have given it the name of Frŷch Dabeôg . . . and sometimes the women carry sprigs of it about them as a Preservative against Incontinency." (Ray described and named the plant in the third volume of his *Historia Plantarum*, 1704.) As a garden plant it seems to have been a comparative latecomer; it is not known to have been in cultivation before about 1800. The beautiful white variety was discovered, according to Loudon, in "Cunnemara," about 1820. The Daboecia also grows in parts of southwest Europe, principally the Pyrenees—hence its second name, derived from the region of Spain formerly inhabited by the tribe of the Cantabrici. Later

[1]Maund, *The Botanic Garden*, 1825–42

[1]Martyn's edition   [2]Threlkeld   [3]His own chosen version of the name

it was also found in the Azores, together with the only other species of the small genus that is in cultivation, *D. azorica*—deeper in color, more compact in growth, but only half hardy—which was introduced in 1929.

I believe this to be the only genus that is named after a saint—with the possible exception of Veronica, the attribution of which is uncertain, and a dreary little cress called after St. Barbara. Little is known about Dabeoc, who was nevertheless one of Ireland's most venerated saints. He is said to have been a descendant of Dichu, St. Patrick's first Irish convert, and to have been at one time Abbot of Bangor. His chief fame comes from his association with the island in Lough Derg in Donegal, on which was situated the cave known as St. Patrick's Purgatory, where the courageous and penitent might gain direct access to the nether regions.[1] It is thought that St. Patrick, who visited the Lough about A.D. 550, left Dabeoc in charge of the church there, and that he founded a monastery on the island, which was named after him the Isle of Mabeoch, and was in subsequent centuries considered so sacred a place "that nobody dares injure another within a day's journey from the water."[2] The only personal fact recorded of the saint himself is that he loved to dip his head into a pool or pit of water—whether for cleanliness or godliness is not made clear.

NOTES: The heaths are notoriously difficult to cultivate here.

Danae, see under *Ruscus*

Daphne

It should at once be made clear that this family has no right to the name it bears. All authorities agree that the plant into which the daughter of Peneus was transformed in order to evade the pursuit of Apollo, was the Bay Laurel (*Laurus nobilis*), and the present genus usurped the name only because of some superficial resemblances, which caused some of its members to be given such names as "Dwarf Bay" and "Laureola" before the days of systematic botany. If this had indeed been the original Daphne, Apollo might have been congratulated on his escape; for besides being all sweetness on the surface and perfectly poisonous within, this nymph of our gardens displays a waywardness which suggests

[1] An early description of this famous cave is said to have inspired Dante [2] Latin MS.*c.* 1230; quoted by Shane Leslie, *St. Patrick's Purgatory,* 1932

an intellect permanently deranged by Apollo's attentions and the shock of her metamorphosis. Even the hardy mezereon is temperamental, and has a habit of dying suddenly when apparently in full health and vigor; and some of the rock-garden species are notoriously difficult and unaccountable.

It is uncertain whether *D. mezereum* is a native of Britain, or a naturalized alien; it was cultivated before 1561, but Gerard got his plants from Poland, and evidently did not know of it as a wildflower. Nor was it found by later botanists—not even the "indefatigable Ray"; though Martyn suggests that it may have been overlooked "on account of its early flowering, before herbarists sallied forth on their vernal excursions." It was first recorded in the wild in 1759, when Miller said there had been "a Discovery made of its growing in some Woods near Andover, from whence a great number of Plants have been taken in late Years." Not so very far away, on the other side of the same county, Gilbert White found it in 1778 at the southeast end of Selborne Hanger, "above the cottages"—where it may well have been a garden escape. Elizabeth Kent (1823) said it was very common in woods in Buckinghamshire; but Anne Pratt (1855) called it a rare plant, possibly native in Hampshire.

In cultivation the mezereon shows noticeably democratic tendencies, thriving best, like the Madonna lily, in cottage gardens. The white variety, then a rarity, was mentioned by Sir Thomas Hanmer in 1659, and various shades of red, together with double-flowered and autumn-blooming sorts, were known before the end of the eighteenth century. Miller in 1759 recorded variegated forms both of *D. mezereum* and *D. laureola*, "which some Persons are fond to have in their Gardens, but the plain is much more beautiful." The decorative value of the fruit was also appreciated: "When their Blossoms are gone, they still appear no less beautiful, being adorn'd with their Vermillion Berries, mix'd among their beautiful Leaves, which makes a most delightful Composition."[1]

Even if it were less ornamental in fruit and flower—and one severe critic has called the red variety harsh, and the white chilly—the mezereon would be worth growing for the sake of its delicious scent, warm, exotic and spicy, and on mild January days "discoverable at an extraordinary distance"[2] from a well-flowered plant. "The sense of smelling" is indeed, as Hanbury said, "peculiarly regaled by the flowers," but do not be deluded by this sweetness; the whole plant is acrid and violently poisonous. The berries are the worst, but a single flower, if chewed, will cause burning and irritation in the throat, and Phillips says that the branches "should

[1] Langley [2] Wood, *A Plain Guide to Good Gardening,* 1891

never be suffered to be cut for nosegays, as young people may be injured by putting the sprigs into their mouths." (In any case the temperamental Daphne takes pruning much amiss, and is apt to resent any but the most restricted cutting.) Linnaeus said that six of the berries would kill a wolf, and that he saw a girl die after taking twelve "to check an ague." "If a drunkard do eat one graine or berrie of this plant," Gerard informs us, "he cannot be allured to drinke any drinke at that time; such will be the heat of his mouth, and choking in the throte." Russian and Tartar women used to rub the berries on their cheeks, to produce by irritation a bright red color—to which the fruit itself may have contributed, as it was said to furnish a "fine lake-color" for painters. The bark (particulary the root-bark) was used externally to raise blisters, and the dried root had a great reputation at one time as an alleviant for toothache; and in spite of its poisonous properties the plant was an ingredient in a popular remedy called the Lisbon Diet Drink. Abercrombie, in 1778, said that gardeners near London used to "cultivate many of the Mezereons to stand several years to acquire large roots, then dig them up occasionally, cut off the root, and sell them for three or four shillings per pound to the druggists and physic-herb shops."

The berries that are so harmful to humans and animals are extremely popular with birds. Marshall makes the suggestion that daphne bushes should be planted on purpose to attract them: "Whoever is delighted with these songsters, should have a quantity planted all over the outsides of his wilderness-quarters." This was well enough when the birds contented themselves with eating the berries when they were ripe, and so helped to disseminate the species; but in recent years the greenfinch has formed a new and destructive habit of attacking the berries while they are still green, discarding the seedcoat and eating with great avidity the immature kernel. A most interesting study of the new trend has been made by Dr. Max Petterson. He found that before 1900 this habit was recorded only between Selkirk and south Lancashire, but by 1930 it extended from Perth to London, and by 1955 over the whole country, from Inverness to Deal and Devonshire, with the exception of Wales, the maximum incidence being in London and other urban areas. (It is quite unknown on the Continent.) Hitherto the mezereon has propagated itself in many gardens by self-sown seedlings, but if these depredations continue— and they seem unlikely to stop—it may become an increasingly rare plant.

There is no doubt that another Daphne, the Spurge Laurel (*D. laureola*) is a British native. It was well known to Turner in 1548—"Daphnoides called of the commune sort Laureola, in englishe Lauriel, Lorel or Lowry, groweth plëtuously in hedges in England, and some abuse the seed of it for coccognidio"—correctly the seed of another species, *D. gnidium*, then used in medicine. It received its common name because its evergreen leaves slightly resemble those of the laurel, and its small green flowers those of the spurge; as in several of the Daphnes, the leaves are borne close together at the top of a bare stem, and the plant has been described as looking like a grove of palms in miniature. It was popular in the eighteenth century, and Miller says that "of late Years, there are poor People, who get the young Plants out of the Woods, and carry them about the Town to sell in the Winter and Spring." Abercrombie called it a hardy and delightful little evergreen, and it was extolled by Hanbury for the sweet scent of its flowers, which "will often perfume the whole end of a garden; and when this happens early, before many flowers appear, the unskillful in flowers, perceiving an uncommon fragrancy, are at once struck with surprise, and immediately begin enquiring from when it can proceed."[1] Planted near windows its scent would "enter parlours, and ascend even into bedchambers, to the great comfort of the possessor and surprise of every fresh visitor."[2] Yet this is the plant whose flowers were said by Bean to have neither fragrance nor bright color to recommend them. He must have sampled them at the wrong time of day; for this Daphne sends out its fragrance only in the evening. (Nowadays we are advised to plant *D. pontica* instead—a similar but somewhat handsomer species discovered by Tournefort near the coast of the Black Sea, and introduced to Britain before 1752.) Hibberd instructs the gardener to have a few plants of *D. laureola* always at hand, "in case he should be at any time afflicted with a passion for daphnes in general," for this species was used as a stock on which to graft the finer evergreen sorts, while the mezereon served the same purpose for deciduous ones.

One of the species often grafted on the Spurge Laurel was the well-named *D. odora*, the penetrating sweetness of whose scent is outstanding even in this fragrant family. In its native China this shrub has been cultivated since the Sung Dynasty (A.D. 960–1279), and Mr. Li, in *The Garden Flowers of China* (1959) tells the story of its discovery by a monk of Lu Shan, who, having fallen asleep under a cliff, was conscious in his dreams of a delicious fragrance. When he awoke, he went to seek its source, and found it in a flower which he thereupon named *Shui Hsiang*, Sleeping Scent (later changed to Good Augury Perfume). Linnaeus called it *D. indica*,

[1]Quoted by Marshall [2]Ib.

although he knew of it only from China; apparently he did not clearly distinguish the two countries, for he named several Indian plants *"sinensis"* with equally lofty indifference. It was introduced to England in 1771 by Benjamin Torrin, Esq., and like most of the early importations from the Far East, was grown first in the "stove",

Daboecia cantabrica 'alba'
(IRISH HEATH)   F. W. Smith (1835)

then transferred to the cool greenhouse, and finally to sheltered places out-of-doors. It is a border-line shrub, hardy in the south and west but elsewhere requiring "a corner outside by the kitchen chimney, the sort of place where the cat goes to lie and where bluebottles sit on the wall in late autumn."[1] The kind with variegated leaves, contrary to the usual rule, is hardier and more floriferous than the type.

The popular Garland Flower, *D. cneorum*—in the sixteenth century so abundant in the Austrian mountains that bunches of the flowers were sold in the Viennese markets—was brought to England before 1739; and though normally "a very humble shrub," can be grafted on the mezereon, "whereby it acquires an elevation superior to what it has naturally."[2] It is recommended as an

[1]Bowles, *Journal of the Royal Horticultural Society,* 1953
[2]*The Botanical Magazine,* 1795

edging for beds of choice low shrubs, and Hibberd says that "for the dressed grounds this is a foreground gem, and hardly enough for any good garden south of the Trent"; but it must be admitted that it has its share of the Daphne temperament, and will sulk implacably in some gardens, while positively rioting next door. At one

Daphne collina
(DWARF BAY)   Ferdinand Bauer (1823)

time it was grown by the London County Council (L.C.C.) on Plumstead Common in Kent, where it received "treatment similar to that meted out to privets and such like"[1] and thrived on it.

These are the most familiar of the Daphnes; but there are between thirty and forty species in the family, all of them shrubs, many of them ornamental and at least reasonably hardy, and not all of them difficult. *D. retusa,* which will attain a height of 3 feet or so if both plant and gardener live long enough, has the reputation of being one of the most accommodating; it was sent by Wilson from China to the firm of Veitch in 1901, and its flowers, large for a Daphne, are "exquisitely fragrant, like lilac".[2] *D. genkwa* (also a Chinese plant) resembles a slender Persian lilac in appearance as well as in scent; it was procured for the same firm by Charles Maries in

[1]Bean   [2]Ib.

1878, a previous introduction by Fortune (1843) having apparently died out—for this is one of the difficult Daphnes, which finds it hard, as Dr. Stoker puts it, to resign itself to the will of the British gardener. (No Apollo, he!) Its name is the Japanese version of the Chinese *Yuan Hua*.

There is a handful of Daphne hybrids; two of them, *D. × neapolitana (D. collina × D. cneorum?)* and *D. × houtteana (D. mezereum × D. laureola)* occur in the wild, and are hardy but not particularly ornamental; whereas *D. × hybrida (D. collina × D. odora)* is ornamental and almost perpetual flowering, but not completely hardy. Best for general garden purposes are two vigorous Daphnes raised by the brothers Albert and Alfred Burkwood, in the present century. Working independently, each crossed *D. caucasica* with *D. cneorum,* and selected the best of the resultant seedlings; Alfred's plant was distributed by Messrs. Scott and Company under the name of Daphne 'Somerset', and Albert's by the firm of Burkwood and Skipwith as *D. × burkwoodii,* a name sometimes used to cover both plants, their parentage being the same.

Gerard gives Germaine Olive Spurge, Spurge Flax or Dwarffe Bay as alternatives for the name of mezereon, which is derived from the Arabic *māzaryūn,* meaning Destroyer of Life, on account of the plant's poisonous properties. (In Somerset, it is called Paradise Plant.) It is a Tuesday[1] plant, anciently dedicated to the Scandinavian god Tyr. The family belongs to the order of the Thymelaeceae, which might be taken as another instance of the general Daphne perversity, as it has nothing to do with the thymes, which are Labiates. Many of the species are characterized by very tough bark, which in some countries is used for the making of paper.

NOTES: Noted for their rich fragrance, the daphnes seem to rise and fall in interest and several interesting ones of the last century have been lost. But new ones come along, for example, *Daphne × burkwoodii* 'Carol Mackie', discovered in a New Jersey garden has variegated foliage as well as a profusion of pale pink flowers. In the south, *D. odorata* is a charming, useful, evergreen shrub, especially for the walled garden where the strong perfumed flowers in February are a garden highlight. The unopened flower buds are a spectacular ruby color and the calyx lobes become whitish when open. The variegated form appears to be somewhat hardier than the green type. Two species have recently been introduced from Japan—*D. kiusiana,* an evergreen with white flowers and *D. kamtschatica,* a deciduous shrub with yellow flowers, native to Hokkaido and other islands of the North Pacific

[1]Tyr's Day

ocean. Neither is in commerce at present.

The daphnes are somewhat difficult to transplant, particularly *D. odorata.* Once a site has been selected and the plants are performing well, it is best not to move them. They prefer a moist, acid soil and are improved by a mulch.

Deutzia scabra
(FUZZY DEUTZIA)   M. Drake (1834)

Dasiphora, see under *Potentilla*

Deutzia

The Deutzias are comparatively new to our gardens; with so many of our shrubs dating back to the Renaissance, a family that made its first appearance at the time of the Industrial Revolution must be regarded as something of a parvenu. As befits the period of its introduction, it is a large family containing some fifty members, and will probably become of increasing garden importance as we get to know it better; for two-thirds of the species described in the R.H.S. *Dictionary* were introduced only in the present century. The principal drawback so far has been that many of these species are too hardy for English gardens—a commoner, and more serious, defect than might be supposed. They come from a climate where a bitter winter, when it finally withdraws, is gone for good, permitting a glorious burst of bloom in spring; whereas milder, intermittent winters encourage early growths which then get caught by envious sneaping frosts. For this reason many of the deutzias do better in places like Edinburgh than they do in London.

D. scabra (syn. *D. crenata*) the senior and type of the genus, is one of the best in this respect—as Bean says, it is "not so restless in the early spring months, nor beguiled so easily into growth by a spell of warm weather," and flowering in late June, its buds usually escape the worst our springs can do. Although a native of China, it was first known as a cultivated plant in Japan, where it was seen and described by Kaempfer in 1712 and Thunberg in 1784. Kaempfer says that the wood, being hard, tough and smooth, was used for cabinet making and for "very fine bodkins"; and Thunberg, that the leaves, which are very harsh,[1] were used for polishing, including the polishing of articles made from its own wood—which one dimly feels ought to be prohibited by Mosaic law, like muzzling the ox or seething the kid. The introduction of this deutzia to Britain is usually attributed to John Reeves of Canton, in 1822, but Bretschneider says that it was his son and successor, John Russell Reeves, who sent it in 1833; possibly there were two consignments. At any rate, it seems to have been little known in 1838, for Loudon, in his *Arboretum et Fruticetum Britannicum* gives it only a brief mention in a "half-hardy" appendix to the related genus of Philadelphus. A double pink or pinkish variety was found by Robert Fortune in 1861 in the courtyard of a temple near Yokohama, and sent by him to the firm of Standish and Noble at Bagshot, who distributed it under the name of *D. crenata flore pleno*; and a pure white double named *candidissima* was in cultivation before the end of the century. Another white double, Pride of Rochester, was raised by the firm of Ellwanger and Barry of Rochester, New York, before 1881. By 1900 forms of *D. scabra* were to be found "in almost every shrubbery" and is now planted even in public parks—which, however, "often indicates toughness rather than garden merit."[2]

D. gracilis, on the other hand, must be given a carefully chosen lodging on the cold, cold ground if it is not to be excited into fatally early growth; but this very precocity makes it an ideal plant for early forcing under glass, and at one time it was extensively used for this purpose. Hibberd says that the nurserymen of his day

[1] Hence the name: *scaber* meaning rough [2] H. G. Hillier, R.H.S. *Tree and Shrub Conference Report*, 1938

(1898) got most of their plants of this deutzia from Holland, where large numbers were annually grown as a catch-crop between the beds of bulbs, and that they were sold in enormous numbers in the flower markets in the spring. It was introduced in the 1840s and was awarded a silver medal when first exhibited in 1851. In its native Japan it is called Snowflower, and is used for low hedges.

Even today, when we have so many others, the foregoing are still the best-known deutzias; but another species, *D. purpurascens*, though seldom seen in nurserymen's catalogues, is of considerable importance as a parent of garden hybrids. It was raised from seed sent from China by the Abbé Delavay to the French firm of Maurice de Vilmorin, in or before 1888. Vilmorin also raised another species, this time from seed collected by Père Fargés in 1897, which was named *D. vilmorinae* after Mme. de Vilmorin. It was another French nursery firm, however, who really put the deutzias on the map—the Lemoines of Nancy, father and son, who did a great deal of work on the genus, and created a number of fine hybrids, many of which are still in cultivation. They include *D.* × *rosea* (about 1896); *D.* × *magnifica* (about 1906) and its varieties *erecta* and *latiflora*; *D.* × *elegantissima*; the popular 'Mont Rose' and 'Contraste', and many others—all garden shrubs of great merit.

Early in the present century valuable new deutzias from China came in rapid succession. *D. discolor* var. *major* (1901) and *D. longifolia* var. *veitchii* (1905) were sent by E. H. Wilson to the firm of Veitch; *D. hypoglauca*, though discovered by Wilson, was actually introduced by William Purdom in 1910; *D. monbeigii* was sent by Forrest from Yunnan in 1913, and a tall form of *D. longifolia*, known as var. *farreri* (syn. *albida*) was found by Farrer the following year. But many think that the best of all is *D. pulchra*, found by Wilson in Formosa about 1918. Wilson sent seeds to Professor Sargent at the Arnold Arboretum, who, thinking the plant might be tender, distributed to gardeners in more favorable climates than that of Boston. Some went to Sir Frederick Moore, formerly of the Glasnevin Botanic Garden, Dublin, who in turn passed them on to the Marquis of Headfort, in whose garden at Kells in County Meath they went at last to ground, and plants were raised which flowered for the first time in 1922.

The deutzias are members of the Saxifrage order, closely related to Philadelphus; except for two which occur in Mexico, all the species are natives of China and other parts of Asia, and all those in cultivation are deciduous shrubs with white, rose or purplish flowers, sometimes fragrant but not on the whole remarkable for scent. The backs of the leaves are covered with minute stellate hairs of various forms, which afford a means of identifying the different species to those who care to take the trouble to examine them through a microscope. The genus was named by Thunberg after his friend and patron Johann van der Deutz (1743–84) a lawyer and town-councillor of Amsterdam, "who promoted and contributed to the expenses of his botanical expedition to Japan."[1] Deutz was a keen amateur botanist and corresponded with Sir Joseph Banks; a letter written in 1777, when Thunberg was on the way home, promised Banks a share of the plants he was expected to bring with him, and later in the same year some of Thunberg's seeds were sent. It is quite possible that we owe some of the plant introductions attributed to Banks, to this "respectable citizen of Amsterdam."

NOTES: Despite about 50 species, the hybrids developed by the Lemoine nursery in France over the period of forty years have captured the market. The Deutzias found in gardens are the forms and hybrids of the Japanese species, *D. gracilis* and *D. scabra*. Most famous cultivar of the latter species is the double flowered 'Pride of Rochester', introduced before 1900 by the famous Ellwinger & Barry Nursery. A few new cultivars have arrived. Of these, *Deutzia gracilis* 'Nikko', collected in Japan by S. March and myself for the National Arboretum, is a splendid dwarf shrub with slender arching branches. It is especially useful in the rock garden.

Deutzias are not demanding plants, requiring only good garden soil, routine fertilizing with a shrub fertilizer, and a sunny position. They have an excellent hardiness range.

Diervilla

A small family, allied to the honeysuckles, and falling into two sections, which follow a pattern already familiar: a group from North America, not of great garden value, from which the earliest introductions were made, and a much more showy group from China and Japan, the members of which did not reach the West until more than a century later, and which are more familiar to us under the name of Weigela.

D. lonicera was the first to be discovered; it was brought to France from "Acadie"—which we now call Nova Scotia—by a surgeon of the name of Dierville, who traveled in Canada from 1699–1700; and was named in his honor by the great botanist Tournefort. In Britain it was grown by the ever-inquisitive Miller before 1739,

[1]*Rees' Cyclopaedia*, 1819

and has kept a rather precarious foothold in cultivation ever since. Marshall, in 1785, says that it "forms an agreeable variety amongst other shrubs of its own growth, though the flowers make no great figure"; and this is as near to enthusiastic praise as its apologists ever get. It is rarely seen today except in botanic gardens, though it is still obtainable.

The Chinese *D. florida* (syn. *Weigela rosea*) is much the most important species in the family, the forerunner and chief progenitor of the garden forms we grow today. It was sent home by Robert Fortune in 1845 from the garden of a mandarin in Chusan. "The garden" wrote Fortune "... was often visited by the officers of the regiments who were quartered at Tinghae, and was generally called the Grotto, on account of the pretty rockwork with which it was ornamented. Everyone saw and admired the beautiful Weigela, which was also a great favorite with the old gentleman to whom the place belonged. I immediately marked it as one of the finest plants of Northern China, and determined to send plants of it home in every ship until I should hear of its safe arrival ... I have every reason to suppose it will prove hardy, or nearly so, in England."[1] His optimism was justified; three years later it was described in the *Botanical Magazine* as "A charming *hardy shrub,* for such it has proved to be, in the Royal Gardens at Kew, where it flowers in the open air in May, even as a standard. It is also very ornamental trained against a wall." In 1850 Fortune revisited some of the nursery gardens where he had previously obtained plants, and was able to inform the proprietors, much to their gratification, that their products had arrived safely in England and had been much admired, "and that the beautiful *Weigela rosea* had even attracted the notice of Her Majesty the Queen."[2]

In the following years a number of other Asiatic species were introduced—*D. coraeensis* (syn. *D. amabilis*) date unknown but before 1856, *D. floribunda* 1860, *D. hortensis* var. *nivea*, 1864, and *D. hortensis* itself, 1870 all natives of Japan; then, after a pause, *D. japonica*, 1892, *D. praecox* from Korea, 1894, and *D. venusta*, also from Korea, in 1905. Meantime the continental hybridists had busied themselves with the earlier discoveries with such celerity and zeal, that before the end of the century Robinson could remark that the species were "rarely found pure," and was able to select nineteen garden varieties of special merit from the large numbers available. Some of the names he mentions are still to be found on trade-lists today, e.g., 'Abel Carrière' and 'Van Houttei', and the yellow-leaved *looymansii aurea*. Most of them are mongrels

[1]*Journal of the Horticultural Society*, I, 1846 [2]*The Gardeners' Chronicle*, 1950

of mixed and largely unknown parentage, the darker varieties getting their richness of color from *D.
floribunda.* The lighter forms show more clearly the family characteristic of the flowers deepening in color as they
age, so that a variety of shades is found in the same
cluster.

The hybrid diervillas are for the most part of greater
garden value than their parents, and the originals are
now rarely seen; but one species deserves cultivation on
its own quiet merits—*D. middendorffiana*, which was introduced as long ago as 1850, but has never become com-

northern Germany, and author of *Flora Pomerano-Rugica*,
1769. Botanists now join the family with Diervilla, but
many nurserymen and horticulturists still keep them separate. A good English name is badly needed; those who
lack confidence in the correctness or pronunciation of
either of the Latin ones may be reduced to calling the
shrub "that pink thing." There are several Chinese
names, though. One means "Embroidered Belt Flowers,"
one "Silk Ribbon Flowers," and a third "Hair on the
Delicate Side-Temples."

NOTES: These native American shrubs are closely related

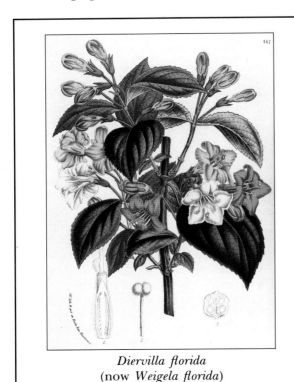

Diervilla florida
(now *Weigela florida*)
(OLD-FASHIONED WEIGELA) *Flore des Serres* (1857)

Drimys winteri
(WINTER'S BARK)   W. H. Fitch (1854)

mon. It is the only yellow Asiatic species, and is supposed
to be slightly tender. Its sulphur-colored, foxglove-
shaped bells, unlike the leopard, can change their spots;
for the markings on the lip, golden-yellow when the
flower opens, darken through orange and red to a port-
wine purple before it fades. It was named after Alexander Theodore von Middendorff, who traveled in north
and eastern Siberia in 1843–4.

Thunberg named the first Asiatic species of what was
then thought to be a new genus, after Christian Ehren-
fried von Weigel (1748–1831), Professor at Greifswald,

to Weigela and sometimes the names are used synony-
mously. But the true Diervillas are relatively limited in
ornamental value and rarely appear in nursery
catalogues.

Drimys

When Drake sailed round the world in 1577–80,
he was accompanied at the outset by four
other vessels; but by the time he had rounded

Cape Horn only one remained—the *Elizabeth,* under Captain Winter. A storm separated them; Drake in the *Golden Hind* sailed on, but Winter returned to the Straits of Magellan, and set a course for home. Presumably he had sickness on board, and at some point sent a boat's crew ashore in search of medicinal herbs; for he brought home with him the bitter, aromatic bark of a tree he had discovered, and for centuries afterwards Winter's Bark was highly esteemed in the medical world as a remedy for scurvy. Captain Cook was among those who used it (after removing the acrid taste by steeping it in honey), and the shrub received its official name of *Drimys winteri* from the naturalist who accompanied Cook's second voyage of discovery in 1772–5, John Reinhold Forster. The living plant, however, was not introduced to England until 1827.

Two members of the family are grown in British gardens. The true Winter's Bark, *D. winteri,* is a very stately plant—it may attain tree-height—with oval, evergreen leaves, red shoots, and fragrant ivory flowers. In the wild it extends, with little variation, the whole length of South America from Tierra del Fuego to Colombia and even Mexico; and plants originating in the extreme south of its range are much hardier than those whose parents came from the warmer regions. These hardier strains have been successfully grown in bleak places such as Harrogate and Perthshire. The other species, *D. aromatica,* only half the height, comes from Tasmania and is more consistently frost-tender; it thrives on the west coast of Britain as far north as Inverewe, where it seeds itself freely and is used for windbreaks. It was introduced in 1843.

Drimys in Greek means acrid. The family is related to the Magnolias.

NOTES: These two species have not had much success here.

Elaeagnus

Here we are on Tom Tiddler's Ground
Picking up gold and silver.
Children's Game

These are the shrubs to transform a garden into a Tom Tiddler's ground of silver and gold; yet gardeners are so unworldly that they do not plant them nearly so much as their merits deserve. Their flowers, though inconspicuous, are deliciously scented, their foliage is ornamental, the fruits of some species are both edible and decorative, and the occasional spines are

excusable where there is so much precious metal to be defended. In most members of the family all parts—shoots, fruits, leaves and flowers—are as scaly as a mermaid.

The first known species was the Oleaster, Wild Olive, or Jerusalem Willow, *E. angustifolia,* from south Europe and western Asia, which hovers uncertainly between the large shrub and the small tree. It is said to have been cultivated in England in the sixteenth century,[1] and was certainly grown by Parkinson before 1633; he said it flowered late here, and rarely perfected its fruit. (He

Elaeagnus angustifolia
(OLEASTER)   M. Hart (1828)

thought the flowers, described later by Hanmer as "extreame sweet, of the shape of little short hyacinthes and of a faded yellowish-green colour," might be used in perfumery.) Evelyn classed it among the plants that would not "kindly breathe our air . . . without the indulgent winter-house take them in"; but actually it is cut back only in very severe winters. Miller thought that the silver leaves, which are "as soft as sattin to the touch" made "a pretty Diversity" among other trees, and Loudon refers to their value in landscape gardening, for

[1] Gerard described it under the name of *Zizyphus cappadocica,* 'The Beade Tree of Cappadocia'

drawing attention to a focal point. He also says that it was called Tree of Paradise in Portugal, on account of the fragrance of its flowers, and the silvery whiteness of its foliage.

The Oleaster is a variable plant; Miller had two forms of it, which he distinguished as a separate species—a smooth one which he called *E. inermis,* and a prickly one, *E. aculeatus,* which he first saw "in the curious garden of the famous Boerhaave near Leyden." There is also the closely allied *E. orientalis,* which was sometimes classed as a variety of the Oleaster, and whose flowers are said to be even more fragrant than those of the type. In Persia it was called Zungeed or Zinzeyd, and the scent was reputed to have so strong an effect upon female emotions, "that twelve fursungs north of Teheraun, the men lock up their women when the flower is in blow."[1] Both species produce edible fruit; that of *E. angustifolia* is sweet, but dry and mealy, and in this country only eaten by birds (though some forms have larger fruits than others); that of *E. orientalis* is said to be superior and was probably the kind "constantly sold in the Constantinople markets under the name of Igidhé agághi"[2] and used for the making of sherbet. Several attempts have been made to establish this seductive shrub; Miller had it in 1739; in 1783 it was reintroduced by John Graeffer, and in 1828 it was again being hopefully raised in the London neighborhood from seed sent from Persia—but it is not to be found in trade lists today.

From the West in 1813 came *E. argentea,* the Silver-Berry—the only North American species; a still more whitely-shining shrub, whose leaves, silvery on both surfaces, have a metallic luster. At home in the Upper Missouri valley it grows to a height of 8 to 10 feet and suckers freely; its fragrant flowers are profuse, its silvery fruits edible; but in Britain, though hardy, it is smaller, and flowers and fruits less freely. It is worth growing, however, for its ornamental foliage.

From the Far East, in 1873, arrived the Japanese *E. multiflora* (syn. *E. edulis*)—not so metallic as the former two, the leaves being silvered on the underside only, but decorative from its abundant fruits, which in this case are juicy and a soft deep red or orange color with minute white spots. Professor Sargent of the Arnold Arboretum thought their pleasant acid flavor "far preferable to that of the Currant and Gooseberry" and that it might be further improved by selection and cultivation, to become "a highly-esteemed dessert and culinary fruit."[3] It has been used for jelly, and to make a spirit similar to Kirsch, but Sir Joseph Dalton Hooker "could find no notice of

the fruit being used for Sherbert"[1] in its native Japan. Birds will strip the bushes unless they are netted.

All these are deciduous species, but there are also evergreen ones, and it is here that the gold comes in; for the yellow-variegated forms of the Wood Olive, *E. pungens,* are among the best gold-leaved plants, and their bright colors, particularly valuable in winter, have made this shrub the most popular of the genus. The type was introduced from Japan in 1830; plants of the variety *variegata* were brought to St. Petersburg, also from Japan, by Carl Maximowicz in 1864; and we subsequently acquired vars. *aurea, aureo-variegata* (syn. *maculata) fredericii, dicksonii* and *tricolor,* all variously marked with pale or dark yellow. Another Japanese evergreen, the silver-lined *E. macrophylla,* is valuable for a different reason; it is highly resistant to damage by wind (though it will not stand much frost), and it makes a useful and beautiful hedging-plant for exposed seaside localities. It was described by Thunberg in 1784 and introduced for the firm of Veitch by their collector Charles Maries in 1879. The fragrant flowers of both *E. macrophylla* and *E. pungens* appear in October and November, but the fruit rarely appears in Britain. *E. glabra* (1880), another evergreen from Japan, has a rambling or climbing habit, and a hybrid between it and *E. pungens* called *E. × reflexa* is praised as a graceful evergreen for cutting; its dark leaves are reversed with shining brown—a copper coin among its silvery relations.

The Greek name was used by Theophrastus for some shrub that combined the qualities of the Olive (*elaia*) with those of the Chaste Tree (*agnos*); it cannot have been our Elaeagnus, for he said it bore no fruit—it was probably a willow. "Oleaster," from the Latin, also means olive-like.

NOTES: *Elaeagnus angustifolius,* the Russian Olive, is still the best in the north because of its gray foliage and exceptional hardiness. However, as a conservation plant and for bird gardens, *E. umbellatus,* the Autumn Olive produces the most abundant red fruits, especially the cultivar 'Cardinal', which was released by the U.S.D.A. In the South, *E. pungens,* native to Japan, is an outstanding evergreen hedge plant. There are several forms with conspicuous golden marking on the leaves. In the Soviet Union—where there is a dearth of vitamin C fruits—the American plant explorer Frank Meyer observed that the fruits of the Russian Olive are eaten raw or boiled in milk and sugar.

[1]*The Gardener's Magazine,* II, 1827   [2]Robinson
[3]Quoted by Robinson

[1]*The Botanical Magazine,* 1894

Embothrium

This is not a monotypic genus, but there is only one species that really matters—*E. coccineum*, the Chilean Fire-bush. Its conflagration has spread gradually northwards, with the discovery that one at least of its varieties is hardier than was at first supposed, and tongues of flame are beginning to glow and flicker in lime-free coastal woodlands in Yorkshire and Scotland.

The Fire-bush grows ultimately to a good height, but begins to bloom while still quite young and small. The collector William Lobb sent it from Valparaiso to the firm of Veitch in 1846; later, another variety was collected, also for Veitch, by Richard Pierce. The first plants flowered in 1853, and created a sensation when they were exhibited at a flower-show in 1855. The hardy variety is *lanceolatum* "Norquinco Valley form," which was collected in the Andes by H. C. Comber in 1925; it was awarded an F.C.C. in 1948 for the incandescent, honeysuckle-shaped flowers that smother every spray. The variety *longifolium* was first grown by Sir John Ross at Rostrevor, County Down, but it is definitely more tender. Chilean botanists, it is said, do not separate the varieties that seem to us so distinct; in the wild, leaves of all three types are often found on the same bush.

Dr. Stoker once called attention to the number of red-flowered evergreen shrubs that are natives of Chile; he mentioned *Asteranthera ovata*, *Desfontainea spinosa*, *Lapageria rosea*, *Mitraria coccinea*, *Philesia buxifolia* and *Tricuspidaria lanceolata*, to which might be added *Berberidopsis corallina*, *Escallonia macrantha* and the present subject— not to mention the deciduous fuchsias. Most are adapted for fertilization by birds, and protect their pollen from the damp climate by enclosing it in long tubes, or by hanging the flower downwards. The Fire-bush paradoxically demands a good deal of water; it likes a moist though not stagnant situation, and must never get dry at the root. The name is derived from *en*, in, and *bothrion*, a little pit, from the position of the anthers.

NOTES: This Chilean plant follows the pattern for most plants from that country. Despite their attractiveness, they are useful only in southern California.

Enkianthus

These are Oriental shrubs, and look it; *E. campanulatus*, for example, has a delicacy of line and color that is peculiarly Chinese (though it happens to come from Japan) and all the kinds bedeck themselves in autumn with all the splendors of the gorgeous East.

In China, *E. quinqueflorus* (the type-species of the genus) was held "in a kind of veneration, and the flowers deemed an acceptable offering to the gods."[1] Fortune described how the branches with their opening buds were brought down in boatloads from the hills in the neighborhood of Canton, in time for the New Year festivities, and of recent years it has been so much picked near Hong Kong that it has had to be protected. It was introduced to England about 1810 by one Robert Jenkinson, and was shortly afterwards purchased "at the sale of his collection" by the nurseryman Joseph Knight of Little Chelsea, with whom it flowered for the first time in 1814. Unfortunately this beautiful semi-evergreen plant is tender; Loudon said it was difficult in cultivation and grown with success only by a few, and it is not to be trusted out-of-doors even in southern gardens.

E. campanulatus, on the other hand, has proved a first-rate hardy garden shrub, of quiet charm in spring and flaunting display in autumn. It was introduced from Japan by Charles Maries for the firm of Veitch in 1880, and two specimens planted in Robinson's garden at Gravetye Manor, said in 1958 to be fifty years old, had grown to a height of 15 feet—unusually large for this species. The new century brought two other desirable hardy kinds—*E. chinensis*, introduced from west Hupeh by Wilson, about 1901, and later by Forrest from the Salween-Shweli divide, and the Japanese *E. cernuus*, in cultivation by 1900, but now rare, if not extinct, except in its fine variety *rubens*, much praised by the cognoscenti.

The genus was named by the Portuguese botanist Ioannes de Loureiro (1715–94/6), who called *E. quinqueflorus* Gravid Flower (*enkuos*, swollen, *anthos*, flower) "because its flowers are pregnant with others."[2] This was due to a misapprehension; what he took as the flower and described as being "solitary, of a beautiful red, crowned with a white fringe, and bearing five florets within,"[3] was actually an assemblage of bracts enclosing the true flowers. This and other errors in Loureiro's *Flora Cochinchinensis* (1790) was kindly excused by J. E. Smith as being "in consequence of his having studied almost entirely without communion with other botanists"[4] during the forty-seven years he spent as a missionary in Cambodia, Cochin-China and Canton. The Chinese name of *Tsiau-Tsung-hoa* means "Suspended Bells."

NOTES: The handsome bell-like flowers and the orange tints to the foliage in autumn make these modest plants a must for the shrub garden. *E. campanulatus* remains the most commonly grown and new cultivars are appearing. The red-flowered *E. cernuus* is still uncommon in our gardens. More often seen is *E. perulatus*, the

[1]*The Botanical Magazine*, 1814 [2]Rees' *Cyclopaedia*, 1819 [3]Ib. [4]Ib.

Embothrium coccineum
(CHILEAN FIRE-BUSH)   W. H. Fitch (1855)

Enkianthus quinqueflorus
(GRAVID FLOWER)   S. Edwards (1814)

smallest of the Japanese species. It has been in cultivation since time immemorial because of its flexibility in the garden—it can be planted in groups and trimmed into shapes or extensive ground covering masses. All of the species are admired in Japan for their flowers and autumn color. As with most plants in Japanese horticulture, there are forms with variegated leaves. In addition, Enkianthus is a popular bonsai subject. Their cultural requirements are similar to those for azaleas and they are used in combination with broad-leaved evergreens because of their neat habit and bright red autumn foliage. They can survive where evergreen azaleas and Pieris can be grown.

Erica

Of the five hundred or more species in this genus, some four hundred and seventy are concentrated in the Cape of Good Hope area of South Africa; the rest are strung out through tropical and north Africa and south Europe to the coasts of the Atlantic and to Norway. This longitudinal distribution, going up and down the world instead of around it, is very unusual in plant families, the only other example being that of the cactuses, which extend from top to bottom of the New World in much the same way as the heaths do in the Old.

Four or five ericas are natives of Britain, but up until the mid-eighteenth century they were not grown in gardens. Philip Miller was among the first to suggest that they might be worthy of cultivation: "notwithstanding their Commonness, yet they deserve a place in small quarters of humble flowering shrubs, where by the beauty and long continuance of their flowers, together with the diversity of their leaves, they make an agreeable variety." (He listed only three of the native species, and the Mediterranean *E. arborea*.) Most of the European kinds were introduced or reintroduced between 1760 and 1770, and shortly after this the family began its sudden and spectacular rise to fashion, with the coming of the tender heaths from South Africa. Seeds of two harbingers, *E. tubiflora* and *E. concinna*, arrived in 1772, but the real impetus was given by the dispatch of twenty Cape species to Kew in 1774 by Francis Masson, "who made two voyages to that spot, so abundant in rare plants, by the command and at the expense of the King of Great Britain."[1] After that, "so many different collectors produced new species, that it would only be a list of names to enumerate them,"[2] and it became the fashion

[1]Miller-Martyn   [2]*The Heathery*, 6 vols., 1804–12: also by H. C. Andrews

to assemble "collections" of these beautiful but difficult plants. By 1823 some four hundred ericas were in cultivation; nearly three hundred were illustrated in H. C. Andrews magnificent monograph on the subject, *Colored Engravings of Heaths,* published in four folio volumes between 1802 and 1830. Meantime, the infection had spread to the Continent; the Empress Josephine had a collection (imported, despite the Napoleonic Wars, from England) which was one of the glories of Malmaison, and shared with the English nurseryman John Kennedy the expense of maintaining a collector at the Cape.

It has been incorrectly stated that all the South African species were discovered and introduced "subsequent to the claim made on Cape Colony by the British Government in the year 1795";[1] but by that date so many had already been introduced that it looks rather as if the claim had been influenced by the demands of British gardeners.

The interest taken in the family as a whole stimulated the cultivation of the hardy heaths of Europe, and heath gardens began to be laid out, though in the earlier examples provision was still made for bedding out exotic heaths from the greenhouses during summer. Loudon (1838) suggests the formation either of an *Ericetum,* consisting of heathers only, or an *Ericacetum,* containing ericaceous plants and shrubs in general. He offers a choice of designs ranging from the formal terrace, with elaborate symmetrical plots, seats and fountains, to the so-called "informal", whose kidney- and sausage-shaped beds would "by their irregularity of outline show that no particular form is necessary for an ericacetum." (The alternative was "a favourable spot . . . selected in the shrubbery."[2]) By the end of the nineteenth century the heaths seem to have gone out of favor, for we find Robinson lamenting their neglect. Now there has been another swing of the pendulum; Cape heaths have almost disappeared from cultivation, and informal outdoor heather gardens are extremely popular, owing to their long flowering season and ease of maintenance. In Miller's day heaths were not to be had from nurseries, and those who wished to grow them were advised to dig up for themselves seedlings from the moors; and although trade lists now offer large numbers of varieties, a great many of them were originally "found" plants, the result of accident or patient search in wild habitats, rather than "raised" plants from gardens.

As far as the hardy species are concerned, the family divides conveniently into two groups—the low-growing or ground-covering kinds, and the tall or "tree" heaths. In the first group, four of the five main species are

[1]Lady Wilkinson, *Weeds and Wild Flowers,* 1858   [2]Phillips

natives of Britain, while the fifth, and perhaps the most important, is not. This regrettable oversight on the part of providence was remedied by George William, Earl of Coventry,[1] who introduced *E. carnea,* the Austrian Heath, from its home in central and southern Europe, in 1763. (This does not necessarily imply that the noble Earl brought it over in person, in the eighteenth-century equivalent of a sponge bag; seeds or plants were probably sent to him by a correspondent.) It was among the first plants to be illustrated in Curtis' new *Botanical Magazine,* in 1789, where we are told that it was usually "kept

Erica sebena   (syn. *E. petiveri*)
(HEATH)   Francis Bauer (1796)

in a greenhouse, or in a common hotbed frame"—trying treatment for "a jewel among mountain Heaths, and hardy as a rock Lichen."[2] Curtis thought its name "not very characteristic . . . but in genera where the species are very numerous, it is no easy matter to give names to all of them that shall be perfectly expressive." Actually Linnaeus was deceived by the pale green flower buds which appear in autumn, into thinking that he was dealing with two different species, an autumn-blooming one which he named *"herbacea"*—"herbaceous coloured" meaning greenish—and a rosy or flesh-pink spring

[1]See under *Chimonanthus*   [2]Robinson

bloomer which he called *"carnea."* There are now approximately thirty garden varieties of this popular heath, many of them originally raised and distributed by the firm of Backhouse of York; but the best one—and, some think, the best of all the two hundred or so hardy heaths in cultivation—was a foundling. It was a white variety, found growing on Monte Corregio, Italy, by Mrs. Ralph Walker of Springwood, Stirling; it was called 'Springwood' at first, but later on, when it became the parent of a colored offspring, the name was changed to 'Springwood White' to distinguish it from 'Springwood Pink.' It received an A.M. in 1930 and an A.G.M. ten years later.

One of the virtues of *E. carnea* and its varieties is that of "not having hysterics at a touch of lime in the soil, as most of the heathers do"[1]—the only other lime-tolerant kind being *E. mediterranea.* The lime causes no actual damage to the plant itself, but it kills the root bacteria on which most of the species depend to convey their nourishment from the soil, so that the plant perishes from starvation. Heaths also have another phobia—like other evergreens, they are sensitive to the effects of smoke and pollution in the atmosphere, and few will thrive in towns.

The four British heaths are among the lime haters, but are otherwise among the best for clean air gardens; two are common and two local, though in their own districts not rare. There is the Bell Heather, *E. cinerea,*[2] the most abundant of the heathers except the ling. It flowers before the ling, so it is sometimes possible to obtain honey from it undiluted by that of other flowers, and this honey is said to be of a reddish port wine color and a distinct flavor. No less than eleven varieties of this species were in cultivation before 1836, and now there are over thirty. The Cross-Leaved Heath, *E. tetralix,* is common in the wetter parts of the moors. Lyte (1578) called it "a little base plant, with many little twigs, or small slender shuts[3] comming from the root ... the floures grow at the tops of the strigs or twigs, five or six in a companie together, hanging downwards, of colour carnation and red, of making long and round, hollow within, and open at the end like a little tonnel ..." It has about fourteen garden varieties. The Dorset Heath, *E. ciliaris,* is thought to be the handsomest British native, but suffered a curious eclipse at the end of the eighteenth century; for though described as an English plant by the Swiss botanist Caspar Bauhin (1560–1624), Ray disclaimed all knowledge of it, and Professor Martyn in the 1807 edition of Miller's *Gardener's Dictionary* says that

"no one ... has heard of it being found in our island," and is very scornful of Miller for calling it a British plant.[1] (He dates its introduction to 1773.) This Heath is abundant in parts of Dorset and Cornwall, and also in southwest Europe; and the most celebrated of its nine or ten varieties, var. *maweana,* was found in Portugal in 1872 by George Mawe, the authority on the crocus and introducer of the chionodoxa. *E. vagans,* the Cornish Heath, is the tallest and bulkiest of our native kinds, and the only one to have bell-shaped instead of urn-shaped flowers; Ray found this "Juniper or Firr-leaved Heath" before 1677, "By the wayside going from Helston to the Lezard Point in Cornwall." Johns says that purple pink and white varieties grew together in the greatest profusion on the Goonhilly Downs in Cornwall, "covering many thousands of acres," but it is not hardy everywhere in Britain, and it has a habit that some find objectionable of blooming serially over a long period, and retaining the unsightly remains of the earlier flowers to spoil the appearance of the later ones. Again, two of the best of its fourteen varieties were discovered in the wild: 'St. Keverne', found in that locality before 1914 by Mr. P. D. Williams of Lanarth, and 'Lyonesse', found by Messrs. Maxwell and Beale in the Lizard area, and distributed by them in 1925.

These last two species (*E. ciliaris* and *E. vagans*) are members of the "Atlantic" or "Lusitanian" flora, and are supposed to have survived from a time when the English climate was warmer than it is today. So also has *E. mediterranea,* a western European heath, a form of which still lingers in one locality in Connemara. It may thus be regarded as a link between the foreign and the native species, and between the low bush heathers and the tree-heaths, for it is intermediate in size, and also considerably more hardy than most of the taller kinds. It was cultivated in the Oxford Botanic Garden in 1643, but was not mentioned by Miller in 1759, and had probably died out as a garden plant until reintroduced by one Joshua Brooks about 1765. This species has the additional charm of a delicious honey scent, most apparent very early in the morning. There are some eight or ten garden varieties, some of which flower a month or six weeks earlier than the type; the dwarfer and less floriferous Irish form has been thought by some botanists to be a distinct species. Best of all its progeny is the hybrid *E. × darleyensis,* a cross between *E. mediterranea* and *E. carnea,* which seems to have inherited all the virtues of both parents; it is hardy, fragrant and flowers from November to May. The original plant was a chance seed-

---

[1]Marion Cran, *Joy of the Ground,* 1929   [2]So named for its ash-colored bark   [3]i.e., "shoots"

[1]Gerard described it as "Chalice Heath," but did not say where it grew

ling that appeared in the great heath nursery of Messrs. James Smith and Son, of Darley Dale in Derbyshire, before 1898. It is variable, and there are three or four named forms.

E. arborea, the doyen and epitome of the tall tree-heaths, also has a very remote claim to be considered a one time native of Britain—one pollen-grain deposit having been recorded from Roman times; though I suppose it might have been the remains of a single plant, or even of a single spray, imported by a homesick centurion. The earliest known date of introduction in recorded times is 1658. In today's climate the Tree Heath is not hardy, and can only be grown near the sea with its friendly warmth; in the Isle of Wight it has attained its full height of about 20 feet. As the individual whitish blooms with which it smothers itself are smaller than a grain of (uncooked) barley, the numbers produced on a well-grown plant of this size defeat the imagination. It is said that if a juniper bush is planted by the door of a house, no witch can enter, for she is obliged to stop and count all the tiny leaves before she can cross the threshold; if the same were true of the flowers of this heath, one good specimen would keep a whole coven busy. Although the type-plant is too tender for most of us, there is fortunately a much hardier form, var. *alpina,* seeds of which were collected in the Cuenca mountains of Spain and sent to Kew in 1899. One of the original plants raised from these seeds was still thriving at Kew in 1945, and had never been affected by cold. It does not grow so tall as the type—a mere 10 feet or so, at most—and its upright plumes, which have been compared to Gothic architecture, have a rich green velvet effect when out of bloom.

Two more tree-heaths of only doubtful hardiness are *E. lusitanica* (syn. *E. codonodes*) and *E. australis.* The first, the Portugal Heath, was a comparatively late introduction; it is thought to have been in cultivation here from about 1800, but it is rather similar to *E. arborea,* and was not separated and named as a distinct species till 1835. Robinson calls it "most precious" with "fine long Fox-brush-like shoots," but admits that it is cut down by frost one year in five; the fact that it is a winter bloomer makes it particularly vulnerable. Bean says that at Lytchett Heath near Poole in Dorset the descendants of a single plant of this heath "have naturalized themselves in thousands and now cover one and a half acres of ground," and it has also "gone native" in Cornwall—so the traffic between moor and garden is not all one way. *E. australis,* the Spanish Heath, was another of the introductions of the Earl of Coventry, this time in 1769; it is universally praised as the loveliest of all the tree-heaths,

and worthy of wall space where it cannot be grown in the open. Its slightly hardier white variety, 'Mr. Robert', was found by Lieut. Robert Williams, son of J. C. Williams of Caerhays and "Mr. Robert" to all the staff and neighborhood, who in 1912 spent the whole of one ten-day leave in an arduous and ultimately successful search for a white form in one of the plant's natural habitats in the mountains near Algeciras, South Spain. In October 1915 he was killed in action at Loos, and never saw his plant in flower in England; it was named at Kew in his memory.

There is one tree-heath that is completely hardy—surprisingly enough, the Corsican Heath (what a geographical lot they are!)—*E. terminalis* (syn. *E. stricta*). But as so often happens, it is less ornamental in bloom than the tenderer kinds, though its bushy, upright growth of up to 8 feet makes it valuable for windbreaks and hedges. It was introduced about 1765, but was at first thought to be a Cape species and given greenhouse treatment. No varieties of this heath are recorded.

E. arborea in French is *la bruyère,* and it is from the wood of its knotty roots that the bruyère or "briar" tobacco pipes are made; their preparation has long been a considerable industry in parts of the Mediterranean, including France, Italy, Sicily and Corsica. Dr. Fairchild has described the process, as carried out at Reggio in Calabria in the early years of the present century. The great gnarled roots were dug out of the boggy ground and kept alive in moist sheds until required for use; they were cut up while still green, and the pieces boiled from eight to ten hours to prevent cracking. They were then dried, and "blocked out" by mechanical saws into smaller pieces ready for the finishing processes in England. Some five hundred factory hands were employed, those working on the "dangerous" saws receiving higher wages. The district was desperately poor, and many of the men had mutilated hands; the Italian Government paid a fixed rate of compensation for every damaged finger, and it was not unusual for a workman in debt intentionally to saw off a finger-tip, to get the promised bounty.

Apart from supplying the wood for this unique manufacture, the heaths have no uses, though they are said to have received the name, derived from *ereike,* to break, from their supposed medicinal property of dissolving the stone. (Others say it is because the branches of some kinds are noticeably brittle.) It is useless for heath lovers to attempt to call their daughters after their favorite flower—Erica as a girl's name has quite a different derivation; it is merely the feminine form of the ancient northern Eric or Eirik, meaning Ever King. Our name "heath" comes from an Old English word for unculti-

vated ground, and is native only to the midlands and the south; it was not originally applied to the Yorkshire or Scottish moors. Petulengro's wind blew over a Norfolk heath: "There's the wind on the heath, brother; if I could only feel that, I would gladly live for ever . . ."
NOTES: Miss Coats, writing for the English gardener, gives considerable attention to the various heaths. But here in America, at least in the East, they require so much coddling that even the most expert gardeners have eventually given up on them. In the Pacific Northwest, they are fine rock garden plants and several species are sold. *Erica carnea* and *E. cinerea* are worth attempting in the Northeast as *E. cinerea* has naturalized along the coastal sand dunes. Otherwise view them at their best in Scotland and England.

Escallonia

The Escallonias are not very old inhabitants of our gardens, dating only from the nineteenth century; they are nearly all evergreen shrubs from Chile or Brazil. The best are those that expose the smallest leaf area to the inclemencies of winter; the narrow or small-leaved species being hardier than the broad-leaved kinds, and the deciduous *E. virgata* the hardiest of all. (In this they resemble the Ceanothus family, where the deciduous *C. americanus* is also the hardiest species; in both families garden hybrids have been raised that are tougher than their parents.) Many escallonias are sticky and resinous, and at least two—*illinita*, and still more, *viscosa*—are said to smell of the pigsty; but to many people a good hearty smell of pig is not unpleasant, though perhaps out of place in a plant.

Among the earliest arrivals was the Chilean *E. rubra*, grown in the Liverpool Botanic Garden about 1827. It is rarely seen today except in its dwarf rock garden form *pygmaea*, but it is one of the hardier kinds and has been useful as a parent of garden hybrids. Next came the pig-pungent *E. illinita* in 1830, and afterwards a group of species sent to the firm of Veitch by their collector William Lobb in the 1840s—*E.E. organensis, pterocladon* and *macrantha*, from Brazil, Patagonia and the island of Chiloe, respectively. The last is one of the most important members of the family, for although tender inland, it was found to make an excellent seaside hedge, and has been extensively planted in the warmer coastal districts of the south of Britain. Living plants successfully weathered the long journey and flowered for the first time in Veitch's nursery in 1848; by the early 1900s it was abundant in Cornwall and its flowers were "sold in

the streets of watering places."[1] In the Isles of Scilly it is much used for windbreaks, and produces its red flowers freely in the spring, though on the mainland they do not appear until June or later.

It was the firm of Veitch, too, who introduced the one deciduous species in cultivation, *E. virgata* (syn. *E. philippiana*), this time by means of a later collector, Richard Pierce, who sent it home from Valdivia about 1866; it flowered for the first time in 1873 and was awarded an F.C.C. in 1888. It has white flowers, and has been described as resembling a spiraea when in bloom; but although in itself a "first rate shrub" (according to Robinson) its greatest value has been as a parent, imparting some of its hardiness to its many hybrid descendants.

Unlike the varieties of deutzia, diervilla and lilac, the majority of which were raised on the Continent, the garden forms of the escallonia are mostly products of Britain, though our nurserymen have hitherto omitted to raise a blue variety to join the red and white ones and complete the colors of the Union Jack. Here again Messrs. Veitch were among the pioneers; their first hybrid, raised at Exeter and named *E. × exoniensis* (*E. pterocladon × E. rubra*) received an Award of Merit in 1891, and was followed very closely by *E.× langleyensis* (*E. macrantha* var. *sanguinea × E. virgata*) raised in their Langley nurseries by the celebrated hybridist John Seden about 1893. This has proved one of the very best escallonias and is deservedly popular; it received an A.M. in 1897 and an A.G.M. in 1926. Scotland contributed two hybrids, *E. × edinensis* (similar to *langleyensis*, but paler in color) and *E. × balfourii*, both raised in the Royal Botanic Gardens, Edinburgh (then under the direction of Sir Isaac Bailey Balfour) and in cultivation by the early 1890s. Ireland's quota followed in the 1920s, with the famous group put on the market by the Donard Nursery Co. of Newcastle, County Down, and named Donard Beauty, Donard Brilliance, Donard Gem, Donard Seedling and so forth, like so many pedigree pups. Pedigree they are not, for their parentage is very mixed, and although all are the results of definite and intentional crosses, records of the particulars have not been kept. Most of them are so hardy as to have greatly increased the area in which escallonias may be grown. At the other end of the scale *E. × iveyi* is definitely tender; at Kew it can be grown only on a wall. It was an accidental cross between *E. montevidensis*, a tall-growing white-flowered species, and *E. × exoniensis*, which occurred in the garden of J. C. Williams at Caerhays in Cornwall. The seedling appeared close to the parent plants and was brought to notice by a gardener called

[1] J.H. Veitch

Escallonia macrantha
(now *E. rubra* var.)
(RED ESCALLONIA)   W. H. Fitch (1849)

Eucryphia pinnatifolia
(now *E. glutinosa*)
(BRUSH-BUSH)   M. Smith (1889)

Mr. Ivey. It is said to be the most spectacular member of the family, and received an A.M. in 1926.

The genus is named after Antonio José Escallon y Flores, a Spanish botanist about whom little or nothing is known, except that he was for a time the pupil and "indefatigable companion" of the much more famous Spanish botanist José Celestino Mutis (1732–1808); and his life can only be surmised in relation to that of his better documented preceptor. Mutis went to "New Granada" (Columbia) in 1760 as physician to the newly-appointed Viceroy, the Marques de la Vega Armijo, and we first hear of Escallon as the Viceroy's page. Mutis became Professor of Philosophy, Mathematics and Natural History at the University of Santa Fé at Bogota; and Escallon was his pupil in all these subjects. In 1776 Mutis left the capital and went to a mountain retreat, near Maraquito (a region with a very rich flora) where he presently became the head of "a botanical school, as it were, and the superintendent of a tribe of botanical adventurers, employed by the Spanish Government to investigate the plants of America,"[1] and with some of his

[1]Rees' *'Cyclopaedia*

pupils made a tour of the colony in 1783. (Escallon is said to have made "several journeys" through New Granada.) Mutis conducted a copious correspondence with Linnaeus till the latter's death in 1778, and afterwards with his son, and sent him specimens of the new plants found by himself or his pupils. (Escallon too corresponded with the younger Linnaeus.) Mutis was very anxious that a particularly handsome plant should be named after his favorite pupil; but the specimen he chose unfortunately was found on examination to belong to a genus already named, so Linnaeus gave the new name to the plant now called *Escallonia myrtilloides*, the first of the present genus. Mutis returned to Spain in 1797, and thereafter Escallon is no more heard of; quite possibly he died relatively young, since he disappears so completely from the annals of botany, and if it were not for these shrubs, he might now be completely forgotten.

NOTES: These South American bushy shrubs are especially popular in England. But they are rarely found in U.S. gardens except in California and similar climates. The hardiest species, *E. virgata*, is a deciduous shrub that grows to about 3 feet with small white flowers in June.

Eucryphia

This is an extremely small family, and only one of its four members can fairly be regarded as either hardy or a shrub—but *E. glutinosa* is beautiful enough for six, and no genus can be considered insignificant that has so lovely a member. Two of the kinds come from Chile; the others, less ornamental, from southeastern Australia and Tasmania.

E. glutinosa (syn. *E. pinnata*), a tall shrub or small tree with deciduous pinnate leaves, was sent to the firm of Veitch by their collector Richard Pierce in 1859, and was awarded an F.C.C. when exhibited by the same firm in 1880. It is a native of the Cordilleras in the neighborhood of Concepcion (Chile), where it is known as *Nirrhé*. This is a plant that breaks all the rules; it is the hardiest of the family, and therefore ought to be the plainest—but on the contrary, it is the most beautiful. It has never become common, because it is slow to establish and difficult to propagate except by seed; and the seedlings always include a large proportion of unwelcome doubles, which—again contrary to garden custom—are regarded as markedly inferior to the four-petaled, single bloom. The bushy stamens, persisting after petal-fall, "are in themselves very ornamental,"[1] and the leaves turn a fine orange in autumn.

E. cordifolia, the other Chilean species, is neither hardy nor a shrub; in its native land, where it is called *Muermo* or *Ulmo,* it grows up to 80 feet in height, and its timber is used for railway sleepers. It was on this species that the genus was founded by Cavanilles in 1797; but the date of its introduction seems uncertain. It may have been cultivated as early as 1831, but it attracted little notice until it flowered in Veitch's Coombe Wood nursery in 1897. It can only be grown in the warmest parts of Britain, but is important as one of the parents of the well-known hybrid, *E.* × *nymansensis* (*E. glutinosa* × *E. cordifolia*). Normally *E. cordifolia* flowers later than *E. glutinosa,* but occasionally their seasons overlap. This happened in 1913 at Nymans in Surrey, the garden of the late Colonel L. C. R. Messel, where both sorts were grown close together; and hybrid forms appeared when seedlings of the two species were raised the following year. In due course the best form among the seedlings was selected and named, and received an Award of Merit in 1924. It is intermediate between the parents, with both simple and compound evergreen leaves, and is hardier than *cordifolia,* but not so hardy (nor quite so beautiful) as *glutinosa*. Its special merits are its vigor and early maturity, and its tolerance of lime,

[1]*The Botanical Magazine,* 1889

which *E. glutinosa* detests.

The eucryphias hybridize freely in cultivation, and some other hopeful crosses have been raised, which have hardly yet had time to prove their garden worth. They include *E.* × *intermedia* 'Rostrevor' (*E. glutinosa* × the Tasmanian *E. lucida*; A.M., 1936) and *E.* × *hillieri* 'Winton' (*E. lucida* × the Australian *E. moorei*) which originated in Messrs. Hillier's Hampshire nursery about 1945.

The name, meaning "well-covered," refers to the neat little brown caps that protect the flower buds before they expand. For a common name, Robinson offers "the Brush-Bush"—surely a very prosaic title for what he himself called one of the most beautiful shrubs of recent introduction. We should do better than that.

NOTES: As with most Chilean natives, these shrubs do not succeed in American soil.

Euonymus

The spindlewoods are sharply divided between the deciduous species, all bearing ornamental fruits of the characteristic "biretta" shape, and the evergreen ones, many of which rarely flower or fruit in Britain, so that they are often grown without our knowing it—without, that is, realizing to which family they belong. *E. japonicus* is one of the most ubiquitous of evergreens but as it has no established common name, and nobody knows the Latin one, it remains completely anonymous, and often passes unnoticed, in spite of the "insistent glossiness" that some gardeners find wearisome. It was introduced from Japan (where it was called *Iso Curoggi,* black shore-tree), in 1804, and was described by Lindley some forty years later as "a *hardy Evergreen Shrub,* with much the appearance of a small-leaved Orange."[1] He noted that the gold-striped variety was apt to revert to plain green, while the silver-striped nearly always kept true. At the end of the nineteenth century this shrub achieved great popularity, and five of its varieties were awarded First Class Certificates; a number of different kinds are still to be had.

The other familiar, though frequently unrecognized, evergreen species is *E. fortunei* (syn. *E. radicans*) introduced by Fortune from Japan in 1860—a shrub which assumes as many different forms as Proteus, but is most frequently seen in the shape of a low bush with white-edged leaves that sometimes blush pinkly in autumn. In its basic form *E. fortunei* has much in common with ivy;

[1]In the *Botanic Register,* 1844. He said it was introduced from Belgium under the name of "Chinese Box"

it will creep along the ground, or climb a wall to a height of 20 feet, but will not flower until it reaches a certain height and maturity, when it changes character and becomes upright and bushy; and as with ivy, cuttings taken at this stage retain the bushy habit and can be used, like box, as an edging for borders. The type is said to be hardier even than ivy, and to grow on houses in New England where the latter will not survive;[1] but some of the variegated sorts are slightly tender. Both fruiting and sterile forms are grown.

As different from both of the foregoing as cheese from the chalk on which it delights to grow, is the British native wild spindlewood, *E. europaeus*. It is an undistinguished shrub or small tree, except in autumn, when its seed-vessels display their outrageous color scheme of orange and pink; the leaves also color well. Its autumn beauty brought it early into cultivation, and five varieties were grown by 1785. The gay, bright berries can be used for dyeing, the capsules yielding a red color, the seeds boiled with alum a green, and without alum a yellow— the latter known in Gerard's time. They have been used medicinally, but in general all parts of the plant have a bad reputation for poisonous properties. Gerard sweepingly remarks that it is "hurtful to all things," but Parkinson, in his painstaking way, notes that although Theophrastus and Dalechamps say that the leaves are poisonous to goats, Clusius asserts the contrary, saying that in Hungary it is called the Goat's Rose-tree—"so divers be the writings and opinions of men." The spindle-tree is popular, unfortunately, with the caterpillars of the moth named after it *Yponomeuta Euonymella* (which seem to find the leaves palatable and harmless), and still more lamentably, with the female black bean-aphis, which resorts to it to lay her eggs and pass a cosy winter. For that reason it has been banished from some gardens; but the broad-beans get the fly just as inevitably when no spindlewood is anywhere near, so it seems unfair to penalize it on this account.

Some of the foreign kinds are said to be less prone to insect attack, and may be safer garden denizens. The European *E. latifolius* is a handsome shrub which has been compared to a fuchsia when loaded with its red fruit, and is very ornamental if trained into a small tree. It was known to Gerard, by description at least, but does not seem to have been grown in Britain until the early eighteenth century. Two American species are notable in their native environs for their abundant brilliant berries, but fruit less freely in Britain; *E. americanus*, the Strawberry Bush or Burning Bush, which was introduced by Bishop Compton in 1683, and *E. atropurpureus*

(Lee and Kennedy, 1756), also known as Burning Bush or by the Indian name of Wahoo.[1] "Brer Fox look like he dead, yit he don't do like he dead. Dead fokes hists der behime leg en holler *Wahoo!* w'en a man come to see 'um," muses Brer Rabbit, and Brer Fox, shamming dead, reacts as expected. The name of any other shrub would doubtless have done equally well; but it was wahoo-bark that Uncle Remus used for weaving into horse collars.

Captain Kingdon-Ward tells alluring stories of euonymus species in Assam and Upper Burma, one of which had fruit as big as crab apples, and another a capsule that looked "something like a purple starfish . . . with several brilliant scarlet orange-capped cowries dangling from its edges. This marine curiosity"[2] as he calls it, would perhaps not be hardy here; but he suggests that much might be done by hybridization within the genus, which has not yet been attempted.

Several different interpretations have been given of the name *Euonymus,* which was used for the Common Spindlewood by the Greeks. Turner (1548) says it is because "the yonge braūches which come streyghte from the roote are al foure square, wherfore some cal it Euonomum."[3] (Elsewhere he says these shoots are "very faire grene, and so well foursquared as if it had been done with a plane.") Others derive it from the words meaning "of good name"—in a satirical sense, since its reputation is not of the best—or from Eunonyme, mother of the Furies. "Euonymos" was also the name of one of the Lipari islands, off the coast of Sicily. No English name was known to Turner, and he took that of Spindle-tree from the Dutch; the plant has since acquired a number of more or less uncomplimentary local names— Louseberry, Dogwood, Prickwood, Cat-rush, Gatteridge- or Gaitre-tree, and Death Alder—some of which it shares with the wild cornel and the guelder-rose. Spindle-tree of course refers to the use of its wood for spindles, skewers and suchlike articles—"of which" says Evelyn "they sometimes make bows for viols, and the Inlayer uses it for its colour, and the Instrument-makers for the toothing of organs, and virginal-keys, toothpickers etc." (It was still in use for piano and organ keys in the nineteenth century.) For skewers it was cut when in flower, the wood being then at its toughest. Turner also mentioned the Italian names of *Fusago* and *Fusaria*; in France it is called *Fusain,* and the same word is used for the artist's stick of charcoal, the best sketching-charcoal being made from the wood of this shrub. "But why they were wont to scourge parricides with rods made of this

---

[1] In New England it is called "Wintercreeper"

[1] According to Fernie; but the R.H.S. *Dictionary* gives the name to *Ulmus alata* [2] *Berried Treasure*, 1954 [3] *Eunomus* = orderly

shrub, before they put them into the sack, see Mod-estinus . . . cited by Mr. Ray," says Evelyn, and leaves us verbless and wondering.

NOTES: Two species of Euonymus, important to gardens in America, are not mentioned above: *E. kiautschovicus* is an evergreen shrub with a broad, loose habit and its cultivar 'Manhattan' (developed at Kansas State University) is a superior form, and *E. alatus,* the Burning Bush, particularly the selection 'Compacta', is outstanding for scarlet fall color and corky wings on the stems.

Mountain folk in Appalachia call *E. americanus* "hearts-a-bustin'," one of my favorite common names. The beautiful crimson pods burst open to display glossy

honor of Longwood Gardens. But it is so slow-growing and tiny that it is not offered by nurseries. A specimen at Brookside Gardens in Rockville, Maryland, has developed into a delightful evergreen foil on a tree trunk; it may eventually arouse interest.

Fatsia

When other shrubs are at their worst, the fatsia or Fig-leaf Palm comes into its own; in November, when the last rags of deciduous

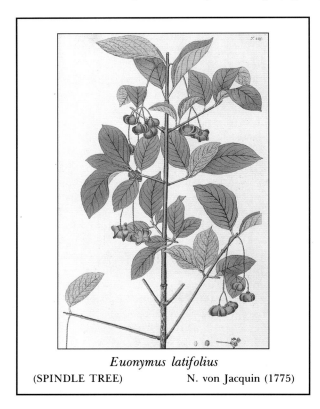

Euonymus latifolius

(SPINDLE TREE)   N. von Jacquin (1775)

Fatsia japonica

(FIG-LEAF PALM)   *Revue Horticole* (1896)

red seeds hanging from a thin thread, like drops of blood falling from a broken heart!

E. fortunei is among the most popular of the many Euonymus species in cultivation. This evergreen vine plant is used on fences, over low walls and rocks and as a groundcover. There are several forms with variegated leaves that turn purplish in winter. It is very hardy and its range of cultivation includes southern Canada. I once collected from a rock face in northern Japan an especially diminutive form, that I named 'Longwood' in

foliage have gone and the berries are going, it spreads shining, palmate leaves and rears above them its balls of greenish, ivy-like flowers, in a bold and formal design. Although not uncommon in gardens, this shrub (*F. japonica*) has been rather neglected in gardening literature. It was found by Thunberg in Japan, and named by him *Aralia japonica*; later, when it was decided to remove it from the Aralia family and establish it in a genus of its own, a name was chosen to imitate as closely as possible the native Japanese one, which was tran-

Forsythia viridissima
(GOLDEN BELLS)   M. Drake (1867)

scribed as *Fatsi*.[1] The shrub was brought to England in 1838, and for a long time was grown principally in conservatories; at Kew some specimens were planted out-of-doors for the first time, in 1891. Actually it is perfectly hardy, but since it has what are perhaps the largest leaves of any hardy evergreen it needs shelter from wind; it tolerates, and indeed prefers, partial shade. The variety *moseri* is rather more compact than the type; there is also a variegated sort.

An accidental, bigeneric hybrid occurred between this plant (*F. japonica* var. *moseri*) and the Irish Ivy (*Hedera hibernica*) in the nurseries of Messrs. Lizé Frères of Nantes, about 1910, and was given the unfortunate name of *Fatshedera lizei*, inevitably anglicized as Fat-headed Lizzy. It is said that the cross has never been repeated, and that all subsequent plants have been propagated (by cuttings) from the one original parent, which is sterile and produces no seed. It is more often listed by nurserymen than *Fatsia japonica*, but is a weak-kneed creature requiring the support of a stake or wall—an inheritance from its ivy parent—whereas the fatsia stands erect on its own sturdy legs.

The Fig-leaf Palm is sometimes confused with the Castor Oil Plant (*Ricinus communis*) whose leaf is somewhat similar, but which is not hardy and is usually grown only as pot-plant or a bedding-annual. (It belongs to quite a different order, the Euphorbiaceae.) Fatsia was believed for a time to be a monotypic genus, containing only this unique species; but *Tetrapanax papyrifera*, the plant from the pith of which Japanese "rice-paper" is made, has been transferred to the Fatsia family as *F. papyrifera*, to keep it company.

NOTES: This handsome shrub is native to the oceanside forests of warmer Japan. The northernmost locality is along the Pacific coast of Honshu and there remains an opportunity to acquire possibly hardier types. It is tolerant of salt spray, smog and other city conditions. The Japanese believe the hand-like leaves have the power to drive away devils. Several variegated forms are cultivated in Japan.

In milder localities such as the Southeast, Fatsia is a useful plant because it creates a tropical-like effect and perhaps should be used more widely. There is the possibility of improving its hardiness by selecting populations from its northern limit along the coast of Honshu, Japan. I once did this and the plants survived for several years at the National Arboretum.

Forsythia

Although the forsythias are so familiar in our gardens, they are not really very old inhabitants; their horticultural history goes back little more than a century. The first to be described was *F. suspensa*, a Chinese species also cultivated in the gardens of Japan. There it was noticed by Thunberg, who took it to be a kind of Lilac,[1] and named it *Syringa suspensa* in his *Flora Japonica*, 1784. It was another species, however, that was the first to reach this country—*F. viridissima*, found by Fortune in a mandarin's garden in Chusan and, later, growing wild in the mountains of Che Kiang, where, he said, it was even more ornamental. It was among the plants he sent home to the Horticultural Society in 1844. *F. suspensa* was introduced to Holland (by a Mynheer Verkerk Pistorius) in 1833, but did not arrive in England till about 1850, when it was grown by Messrs Veitch of Exeter. In 1857 it was considered a "rare and handsome" shrub, which would prove "equally hardy, no doubt, with the better-known *F. viridissima*".[2]

These two species are, as it were, the founder-members of the forsythia family. Both are variable. *F. viridissima* has produced a variegated form, a spreading, large-flowered form from Korea, and a dwarf form named *bronxensis* because it was raised in New York Botanic Garden, about 1939. The varieties of *F. suspensa* are much more numerous—seven or eight have been distinguished, ranging from the erect one found by Fortune near Pekin in 1861, and classed at first as a distinct species with the name of *F. fortunei*, to the dark-stemmed, early, primrose-colored variety *atrocaulis*, seed of which was collected in Kansu by Farrer (1914) from a plant which he had not seen in flower. But not content with all these natural variations, the hybridists set to work and made a cross between the two species, which was raised on the Continent about 1880 and named *F. × intermedia*. It was introduced to Britain some ten years later, and Robinson admits: "Though at first sight very little disposed in its favor, I have recently seen it in a better light." There are many forms of this hybrid, and until lately the best of them was thought to be the variety *spectabilis;* but this has now been surpassed by a bud-sport of its own, which appeared about a dozen years ago in a garden in Northern Ireland and was propagated and distributed by the Slieve Donard Nursery Co. Ltd., of County Down. Its full name is *F. × intermedia* var. *spectabilis* 'Lynwood.'

Meantime two more species had followed the others from Asia—*F. giraldiana* from Kansu in 1910, and *F.*

---

[1]More correctly, *Iats' de*

[1]It belongs to the same Order, the Oleaceae　[2]*The Botanical Magazine*, 1857

ovata, seed of which was collected by Wilson in Korea (again from a plant that had not been seen in flower) in 1917. They are not so familiar in gardens as the older varieties, but have been used in breeding, especially in America, where much interest has recently been taken in the genus. Some fine hybrids were raised from 1939 onwards by Dr. Sax of the Arnold Arboretum, of which we may hear more as time goes on. There is also a European member of the family—*F. europaea*, found in Albania in 1897, and raised from seed at Kew in 1899; but it is of little account.

The brassy, rather spurious gold of the forsythias aptly commemorates an able, pushing, somewhat unscrupulous Scot, William Forsyth (1737–1804). He became first the pupil, and later the successor, of Philip Miller at the Chelsea Physic Garden, and in 1784 was made Superintendent of the Royal Gardens at Kensington and St. James's, a post he held until his death. He is chiefly remembered in connection with an invention for which he was famous during his lifetime and infamous afterwards—Forsyth's Plaister. This was a "secret" concoction which would, it was claimed, if spread over the wounds of injured trees, heal the damage and cause them to grow into good sound timber; and sound timber was then so urgently required for the Navy that a specially-convened Parliamentary commission, after investigating Forsyth's experiments at Kensington, made him a Government grant of £1,500 (worth one hundred times that sum today) in return for the publication of the formula, "in order to diffuse the benefits of this discovery throughout the kingdom." (A further grant was promised when the results were substantiated, but there is no record that this was ever paid.) It is possible that Forsyth himself believed that his completely worthless compound, whose ingredients were cow-dung, lime-rubble, wood-ash, sand, urine and soapsuds, would perform all that was claimed for it; but he must have known that it was far from being his own invention, as the recipe differed very little from others put forward by his predecessors and contemporaries. This transaction has tended to overshadow Forsyth's genuine horticultural achievements; he was one of the seven founder-members of the Horticultural Society of London, now the R.H.S., and wrote a very popular book on the culture and management of fruit trees.

NOTES: We seem to regularly trade cultivars of the ever-popular Forsythia back and forth with Europe. New cultivars, often with only minor variations of yellow, still manage to attract buyers. 'Lynwood' remains among the best hybrids for large, intensely yellow flowers; another is Dr. Sax's 'Beatrix Farrand'.

Forsythias are among the easiest plants to grow and they are well-suited to urban conditions. If anything, their vigor can detract from their charms. The long branches may require regular pruning, but since they flower on branches produced the previous year, pruning should be done after flowering. The best method is to remove branches larger than your thumb at the base; always limit pruning to the oldest branches. Forsythias, because they bloom early in the spring, make good cuttings for arrangements and are always popular in flower shows.

Fremontia, see under *Carpenteria*

Fuchsia

Ninety-four of the hundred or so known species of fuchsia are natives of Central or South America, and the first to be recorded was found in San Domingo by Father Plumier in 1693. Charles Plumier (1646–1706) was one of the missionary-botanists in which France was so rich; he was a friend of Tournefort and became King's Botanist to Louis XIV. He made three voyages to the West Indies, and died when about to depart on a fourth; after his third expedition he published *Nova Plantarum Americanarum Genera* (1703), in which the first fuchsia was illustrated and described. He called it *Fuchsia triphylla flore coccinea,* and it was on his rather sketchy information that Linnaeus founded the genus. Seeds of this species (*F. triphylla*) were sent to Philip Miller from Cartagena in Colombia by Dr. William Houstoun[1]—this must have been before 1733, the year of Houstoun's death in Jamaica—but it soon disappeared from cultivation, and was not seen again for more than a hundred years.

The next kind to be recorded, *F. coccinea,* was also found by a French missionary, R. P. L. Feuillée, who described it in his *Histoire des Plantes de Chili* in 1714. The tale of its introduction and that of the very similar *F. magellanica* resembles the story of the Gondoliers, as related by the Duke of Plaza Toro. Transported (both of them, in this case) far from their homeland at about the same time, their identities soon became hopelessly confused. A plant of *F. coccinea* was presented to Kew by a Captain Firth[2] in 1788; and soon afterwards, the story goes, *F. magellanica* came to light in a very romantic manner. The famous nurseryman, James Lee, was showing a client around his establishment at the Vineyard,

[1] See under *Buddleia*   [2] Called "Frith" by Hooker

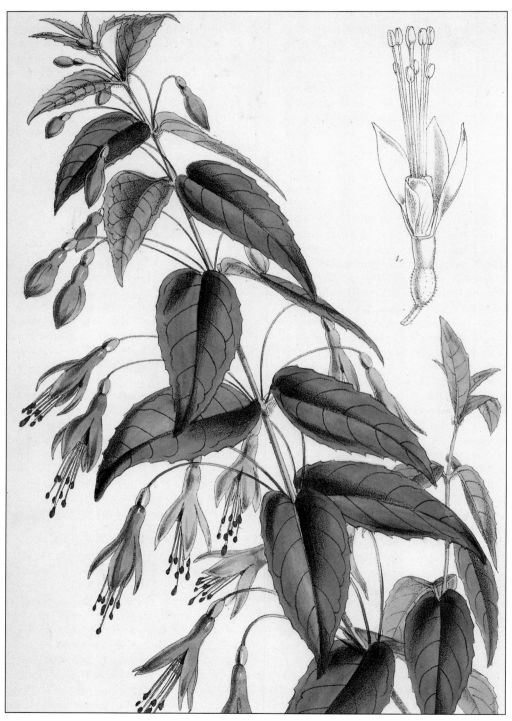

Fuchsia coccinea

(LADY'S EARDROPS)   W. H. Fitch (1868)

Hammersmith, when the visitor remarked that he had seen a plant that excelled anything in the collection, in the window of a poor house in Wapping. Off posted Lee in search of it, and found a fuchsia which was unlike anything then in general cultivation. The woman who owned it was unwilling to part with it, as it had been brought to her from the West Indies by her sailor husband, who was now again at sea; but when Lee emptied his pockets of all the cash he had about him (about eight guineas), she agreed to sell it, on condition that she should also have the first two plants to be propagated from it. Lee called a coach and bore off his prize to the nursery, where he stripped the plant of every bud and blossom, broke it up into cuttings, and propagated and repropagated with such assiduity in bark-beds and hot-beds that by next flowering season he had three hundred promising young plants for sale. As they began to bloom, he placed them two at a time in his show-house, replacing each as it was sold by another from stock. Society's horses "smoked off to the suburbs" as word of the new plant got around, "new chariots flew to the gates of old Lee's nursery grounds," and he had no difficulty in disposing of the whole three hundred at a guinea apiece. This plant was identified as *F. coccinea* and was distributed under that name, but was actually *F. magellanica.*

It must be admitted that doubts have been cast on this interesting story (which even attained the dignity of appearing in the Dictionary of National Biography), and both the source of the plant and its identity have been questioned. The version quoted above appeared nearly forty years after the event, in the *Lincoln Herald* of November 3, 1831, where it was credited to Mr. Shepherd, the respectable and well-informed conservator of the Botanic Gardens at Liverpool; but Loudon in 1838 says that Lee received the plant from one of his collectors, and in 1855 the *Floricultural Cabinet* made the suggestion that he actually "acquired" the plant from Kew, and made up the Wapping story to account for its possession. (It should also be noted that neither *F. coccinea* nor *F. magellanica* is a native of the West Indies). The only facts that emerge beyond "all possible, probable shadow of doubt" are that in the year 1793 Lee did sell a large number of fuchsia plants at a handsome profit— all accounts agree about this—and that for many years afterwards the plant in commerce under the name of *coccinea* was actually *magellanica.* The original *coccinea* disappeared from cultivation until 1867, when it was rediscovered by Sir Joseph Dalton Hooker in the greenhouse of the Oxford Botanic Garden—the only place in which it had survived; but some fuchsia-experts think that it is not a distinct species, but only another form of the ubiq-uitous *magellanica.*

All this time the true heir to the throne, Plumier's original *F. triphylla*, had been completely lost to sight— so much so that botanists had begun to doubt its very existence; but in 1873, Thomas Hogg, who then had a nursery in New York, raised a fuchsia from seed received from San Domingo, and sent some of the resulting plants to the fuchsia specialists, Messrs. E. G. Henderson of St. John's Wood, under the name of *F. racemosa.* Henderson sent specimens to Kew in 1882, and they were identified as the long-lost *F. triphylla.* It has always been a rare species, however, and has played little part in the general development of the family.

After this comic-opera start at the end of the eighteenth century, there was a pause in the flow of fuchsia introductions until about 1822, when they began to come thick and fast, fourteen or more species or near-species arriving between 1824 and 1844. The most important was *F. fulgens,* a glorious firework of a plant, sent from Mexico to the Horticultural Society by their collector, Theodore Hartweg,[1] in 1837. This may not have been its first introduction; Johnson gives the date of 1830, which is more in keeping with the report in the Botanical Magazine for 1841, that "Mr. Curtis,[2] in his beautiful nursery of Glazenwood . . . has succeeded in producing a great number of hybrids, by means of (this and) other species, and flowers of all kinds are the result." These were the first of the florists' fuchsias, the other parent-in-chief being *F. magellanica*; and the rapidity of their rise to popularity can be gauged from the fact that when the first monograph on the flower was written—Félix Porcher's *Histoire et Culture du Fuchsia,* in 1848—the author was able to list 520 species and varieties. Later in the century this number had risen to about seven hundred, and one author has estimated that in the 1880s there were some fifteen hundred named sorts. That was when fuchsia plants were sold at Covent Garden at the rate of ten thousand daily during the season; when an amateur named Edward Banks raised some 5,000-6,000 seedlings annually; when the pillars of the Crystal Palace were clothed with fuchsias from the bottom to the top; when specimens 8, 10 or more feet high and 4 through were exhibited at suburban shows, those too big to be carried by road in covered carts being transported by canal. Then the tide of fashion began to ebb; by the turn of the century only a few varieties were still in cultivation, and the big collections were no more. The First World War administered the *coup de grâce*; up until that

[1]Who also brought *F. splendens* from Mount Tontonpec in 1842
[2]Samuel Curtis, cousin of the founder of the *Botanical Magazine.* See under *Camellia*

time large numbers of fuchsia cuttings were grown in vinehouses before the vines were in leaf, but during the war the grapes were displaced in favor of tomatoes, and the fuchsias went too. Within more recent times there has been a considerable revival; the American Fuchsia Society was founded in 1929, and the British Fuchsia Society ten years later; many of the old varieties have been recovered and new ones raised, and the list of kinds has again reached fifteen hundred.

These florists' fuchsias, however, are too tender to stand the winter out-of-doors, though they may be used with fine effect in the garden in the summer. Only one species or group of species is hardy enough to be of service as a general garden shrub, and that is *F. magellanica*. In the wild it extends from Tierra del Fuego to Peru, and, as might be expected, varies considerably from one part to another of its vast range; several fuchsias introduced as distinct species are now recognized as forms of *magellanica*. These include (besides, possibly, *F. coccinea*) *F. m. gracilis*, seed of which was sent from Chile by Francis Place to the Horticultural Society in 1822; *F. m. conica* (1824) from the same source, which bloomed in the Society's Chiswick Garden in 1826; *F. m. discolor* (syn. *lowei*), discovered in 1828 at Port Famine in the Magellan Straits, and introduced in 1830 by the firm of Low of Clapton, and *F. m. globosa,* a descendant of *conica,* raised by Bonney of Stratford before 1832, but later found wild in Chile and Colombia. Garden forms and hybrids soon began to appear, such as *F. m. riccartonii*, whose Italian-sounding name is deceptive, as it was raised about 1830 by James Young, gardener to Sir Gibson Craig of Riccarton, near Edinburgh (probably from *conica* × *discolor*) and the dwarf *F. m. pumila,* raised about 1837 by Thompson,[1] gardener to Lord Gambier of Iver House, Buckinghamshire, who also bred the fastigiate fórm *F. m. thompsonii* (1840). There are a number of other named varieties, and at least two variegated kinds—not to mention the white *F. m. alba,* found by Mr. Clarence Elliott in southern Chile about 1930, and already well known. (It is not pure white, but a combination of ivory and pale pinky-lavender, and its exquisite delicate flowers are, like birds, better in the hand than on the bush.) Altogether sixteen or seventeen different forms of *magellanica* are to be found in trade lists today.

Although these relatively hardy fuchsias were among the first to be introduced, it is hard to say to what extent they were grown as outdoor plants. They were probably kept for a long time with the other species in the greenhouse, and we do not know what venturesome or careless gardener first risked leaving them outside during the winter. The different forms of *magellanica* vary considerably in point of hardiness, but the preference of them all is for a mild, moist atmosphere. (They thrive in Ireland, and in English seaside districts fuchsia hedges are common. At Kirkwall, away north in the Orkneys, Canon Ellacombe observed houses covered with fuchsias from the ground to the roof, with spaces cut out for the windows. The plant could hardly travel further; in the Orkneys it approaches as near to the North Pole as its progenitors on Cape Horn approached the South.)

For those who like their botany neat, the flowers of the fuchsia are "actinomorphic, epigynous and with a perigynous tube." In their native habitat they are pollinated by birds, but here bees visit them freely, and in Ireland they are said to yield considerable quantities of nectar, though the resultant honey is of poor quality. The fruits that may follow are edible, and at one time great hopes were entertained of them. "When gardeners discover the way to improve the size and flavor of *fruits*," says Johnson, "we cannot doubt but that those of the Fuchsia and Cactus will be among the first novelties of the dessert." Hibberd says they make good tarts if assisted with lemon-juice and sugar, their flavor otherwise being "somewhat flat, green and poverty-stricken." Loudon says that the fruit of the New Zealand tree-fuchsia, *F. excortica*, is so sweet "that the missionaries have been trying to introduce the species into Otaheite, as a sugar plant; but have been unable to procure seeds, as in New Zealand and the berries are eaten greedily by pigs, as soon as they appear." Some of the New Zealand species, it is said, have blue pollen, which is collected and used by Maori girls as a face-powder.

In his *Nova Plantarum Americanarum Genera* Plumier followed the practice of naming his new discoveries in honor of famous botanists; about fifty families were so designated, including the present one, which was called after Leonhardt Fuchs (1501–66). This Bavarian physician and botanist is remembered chiefly as the author of *De Historia Stirpium,* 1542, one of the most scholarly and beautifully illustrated of the early herbals; but in his own day Fuchs had a wide reputation as a practitioner and teacher of medicine, and was particularly renowned for his treatment of an epidemic of plague which swept Anspach in 1529. The greater part of his working life was passed at the University of Tubingen, where he died of "an acute disease attended with great watchfulness."[1]

Feuillée reported that in Brazil the native name of his fuchsia was *Thilco,* or *Molle cantu,* which signified Bush of Beauty. The only English name that is mentioned is

---

[1] Known as "The Father of the Heartsease" for his pioneer work in pansy breeding

[1] Rees' *Cyclopaedia*

Lady's Eardrops. In the Language of Flowers the fuchsia stands for "amiability" or "taste"—perhaps in the latter case because the petals were described on its first introduction, as resembling "a small roll of the richest purple-coloured ribband."[1] Farrer described the acolytes at a Tibetan lamasery as looking "exactly like old-fashioned fuchsias" in their scarlet and purple robes; but this typically ecclesiastical color-combination is not now considered the acme of sartorial taste, and what looks daring and delightful when worn by a fuchsia can look quite dreadful on a clergyman.

NOTES: *Fuchsia magellanica,* a dwarf shrub common in England and elsewhere in Europe, is the only species we can grow outdoors as a hardy shrub. Its hardiness is questionable along the eastern seacoast, and it is mostly of interest to those who cater to rare plants. I have struggled with one since I saw it used as a loose hedge in southern England and it has yet to survive out-of-doors in western North Carolina.

Garrya

G*. elliptica,* Quinine Bush, Silk Tassel Bush. David Douglas came upon this shrub in 1827, growing sparsely near the sea on the south side of the Columbia River. He sent it to England to bloom for the first time in the Horticultural Society's garden in 1834, where it created a great stir among the botanists, for it was found to represent not only a new genus, but a new Natural Order—now known as the Garryaceae, but containing only this one family—which connected other well-known families in an "unexpected and satisfactory manner."[2] It had been previously discovered by Archibald Menzies during his expedition with Captain Vancouver (1790–95) but does not seem to have been named by him, nor to have aroused any particular interest until Douglas's introduction. Since then a number of other Garrya species have come to light—about fourteen in all; but only the first is of value for gardens. It was the male plant, with its elegant, fragrant catkins, 6 inches to a foot in length, that was first introduced; and the female, with its shorter tassels followed by strings of purplish fruit, did not follow until several years later, but both sexes are now obtainable. Their beauty, austere rather than showy, has steadily won its way to our appreciation and respect; the male form receiving an Award of Merit in 1931, and an A.G.M., the highest horticultural honor, in 1960. W. J. Bean once pointed out how exactly the design of the catkins resemble that of the festoons and swags that so often appear in the work of the brothers Adam; but none of them could ever have seen the plant, as the last survivor of the Adam family died five years before the Garrya was brought to England and twelve before it flowered here. (Robert Adam traveled to Italy, but never to a place so remote as the Pacific Northwest.) The Garrya has become naturalized in the Avon Gorge in Somerset; and there is a famous old specimen of the male form in Cambridge Botanic Garden, more than 35 long paces in circumference.

Douglas named the new plant after Nicholas Garry, Deputy-Governor of the Hudson Bay Company from 1822–35 (the period during which Douglas made his American journeys) "to whose kindness and assistance he was much indebted."[1] It was only the help of the Company that made these journeys possible; Douglas traveled, sent back his collections, and received his supplies by means of its ships, lodged at its trading posts and made use of its guides and interpreters. Garry was also commemorated in the naming of Fort Garry near Winnipeg[2] and of Lake Garry in the Northwest Territories, and was the godfather of Spokane Garry, an Indian boy brought up by the Church Missionary Society, who on returning converted his own tribe to Christianity and was the means of establishing the first mission to the Indians of Oregon. But although his name is preserved by a lake, a fort, a botanical Order and a missionary, very little is known about Garry himself. His early years were overshadowed by the fact that he was illegitimate, and his later ones by mental instability. He emerges briefly into the light in 1817, when he became a director of the Hudson Bay Company. In 1821 he was sent to Canada to supervise the Company's amalgamation with a former rival, the North-West Company, and went on an adventurous tour of inspection with one of the North-West's directors. On his return he was made a Deputy-Governor of the Company, and in 1827–8 was in London, taking part in discussions on the Oregon boundary dispute. Douglas was called in consultation over the same question when he was in London in 1829, owing to his first-hand knowledge of the terrain; so the two men may have met.[3] Garry married in 1829, and had one son, who was only four years old when his father became deranged in 1835. A charming journal of Garry's Canadian journey of 1821 has been preserved, and one cannot but warm to a man who says that the sensation of a canoe grating on a hidden rock "can only be compared to your feeling when under the Hands of an unskilful

[1] *The Botanical Magazine,* 1789   [2] Lindley, quoted by Loudon

[1] Loudon   [2] Built in 1835, demolished 1852   [2] The shrub was not named until it bloomed, in 1834

Dentist." There is nothing to suggest that he had any particular interest in botany or horticulture.

NOTES: The garryas are western natives and thrive in the dry, sandy soils of that region. Unfortunately, in this country, their garden appeal is restricted to that area.

Gaultheria

The species in this family are numerous, but few are in general cultivation. Many are difficult, or of doubtful hardiness; some of the smaller kinds are grown

Garrya elliptica
(QUININE BUSH)   M. Drake (1834)

in rock gardens, but are apt to give trouble because of their invasive tendencies. They are grown chiefly for the sake of their fruits, which may be white, red, bright blue or purple; the late Captain Kingdon-Ward said that some of the numerous Chinese and Himalayan kinds grow to a height of 2 to 4 feet and have decorative blue berries, but that these are only lightly attached and come off "almost as easily as shirt-buttons in the laundry."[1]

The two really well-known species are *G. procumbens* and *G. shallon,* both from North America; the prostrate

[1]*Berried Treasure,* 1954

G. procumbens—Checkerberry, Boxberry, Mountain Tea or Creeping Wintergreen—being the first of the genus to be discovered and named. It was introduced into English horticulture by 1762, probably as a medicinal plant, for it is the source of that Oil of Wintergreen (methyl salicylate) whose scent is so familiar to sufferers from rheumatism and lumbago. (It was also used in perfumery, chiefly for soaps.) The leaves, if properly cured, were said to make a "most excellent tea," which was used "as a substitute for the Chinese vegetable."[1] Loudon found the wintergreen "difficult to preserve alive, except in a peat soil kept moist," but it is now generally recognized as a hardy

Gaultheria procumbens
(CHECKERBERRY)   *The Botanist's Repository* (1800)

and vigorous carpeting plant, whose chief fault is its modest way of concealing its attractive flowers and fruit among the leaves.

G. shallon was discovered by Archibald Menzies (1754–1842), named by the American, Friedrich Pursh (1774–1820)—in the belief that "shallon" was the native name of the shrub, but actually "salal" is more correct—and introduced by David Douglas, who seems to have regarded it with particular affection. It was the first thing he picked when he landed at Cape Disappointment

[1]*The Botanical Magazine,* 1818

on the Oregon–Washington border, on April 9, 1825, after a voyage which had lasted (with interruptions) eight and half months, and a perilous crossing of the river-bar. No wonder he wrote: "So pleased was I, that I could scarce see anything but it." Later he noted that "the fruit is abundant and very good, so I hope it will ere long find a place in the fruit garden as well as the ornamental border."[1] When he made his astonishing trek on foot and by canoe over the Rockies and across the continent to Hudson Bay (2,900 miles) he regretted that he had not included his favorite salal among the few plants he managed to preserve alive during the arduous four month journey. He did bring or send home a sample of the cake made by the Indians from the dried fruit for their winter provender, but Lindley, who tasted it, said that it "did not prove very palateable to an European."[2]

Arriving in Britain in 1826, the plant was hailed as "an object of great interests with botanists, as being likely to become an important addition to our Artificial Flora." A large number of seedlings were raised from Douglas's seeds, and by 1838 the salal had already been planted as game-covert in the North of England and in Scotland. It will grow on the poorest soils—provided they are lime-free—and in the most unfavorable situations, but under better conditions is liable to get out of control; it has sometimes been necessary to attack it with a powerful rotary scrub-cutter in order to prevent it from smoth-ering more valuable shrubs. Dr. Fred Stoker suggests that a sprig of it should be made the badge of the out-and-out imperialist, since "every method of land-appro-priation is known to it; peaceful penetration, aggression, underground insinuation and propaganda"—that is, dis-semination by seed. He admits, however, that it is an attractive shrub, and if it was as difficult to grow as some of the other members of the family "we would approach it with uncovered heads." Even as it is, we might doff a hat to it occasionally, for Douglas's sake.

The name of the family arose through the travels in North America of the Finnish botanist Peter Kalm. When Kalm visited Quebec in 1749, the Governor, the Marquis de la Galiffonnière, sent his personal physician, Dr. Gaulthier (1708–51?), to show him the local places of interest. Together they explored the lead mines near St. Paul, and together got drenched with blown spray from the waterfall at Montmorenci, and Gaulthier, who was keenly interested in all aspects of natural history, presented the visitor with a copy of two years' meteoro-logical and seasonal observations. Kalm returned the compliment by naming after the doctor the newly discov-ered Creeping Wintergreen (*G. procumbens*)—perhaps in-tentionally chosen because of its use in medicine. *G. shallon* has now become naturalized in many parts of Britain.

NOTES: These close relatives of the blueberry are pros-trate or low-growing, mostly evergreen shrubs. *G. pro-cumbens,* widely distributed in eastern America, is found in damp woods and bogs. The leaves are pleasantly aro-matic and were once used as a "tea." There is a counter-part in western America, *G. shallon* with similar garden uses. Our eastern species has bright scarlet fruits; the western species has black fruit. In Asia there are several

Genista lusitanica
(WOADWAXEN)   *The Botanist's Repository* (1805)

species still to make their way into our gardens. Some that I have collected in Japan may eventually make use-ful groundcovers.

[1]Quoted by Hooker, *Companion to the Botanical Magazine,* 1836
[2]*The Botanical Register,* 1828

Genista

You are all aware that
On our throne there once sat
A very great king who'd an Angevin hat
With a great sprig of broom, which he wore as a badge in it,
Named from this circumstance, Henry Plantagenet . . .
The Ingoldsby Legends (The Brothers of Birchington)
R. H. Barham, 1837

It will never be known whether the "Planta Genista" of the Plantagenets was the Petty Whin (*Genista anglica*) or the Common Broom (*Cytisus scoparius*) formerly also classified as a genista and so named in most European languages. Probably either plant served, and the warrior, stretching out a mailed fist to snatch a plume of yellow blossom, did not stop to consider whether or not the small protuberance known as a strophiole was present on the seed—the one sure method of distinguishing a Cytisus from a Genista. On the authority of the R.H.S. *Dictionary of Gardening*, we will take the genista, not the broom, to be the historic plant, and insert the story here.

Legend associates the broom with the Anjou family as far back as Fulk Nerra (Fulk the Black, *c.* 970–1040) and his son Charles Martel; but it is not until we reach Fulk's great-grandson, Geffroi le Bel (Geoffrey the Handsome, 1113–51) that we are on sure ground; a number of contemporary records attest that he took the name of Plantagenet, though there is no surviving confirmation of Camden's statement that this was because "he wore commonly a broom-stalk in his bonnet"[1] or of the legend of its first use as a badge of recognition on the battlefield. But this was only a cognomen or nickname, such as "still did die with the bearer and never descended to posteritie."[2] Surnames were not then in general use, especially in royal families; Geoffrey's son, who became our Henry II, bore the name of Curtmantel, and his descendants descriptive labels such as Lackland, Crouchback or Coeur de Lion, or were distinguished by their birthplaces, like Richard of Bordeaux. The name of Plantagenet was not used again, until it was assumed by Richard, Duke of York, when asserting his claim to the throne against that of the younger branch of Lancaster; and he possibly selected it only because it was the most respectable of the ancestral names from which he was able to choose. It appears first in the Parliamentary rolls of 1460, three centuries after the death of the original bearer. Richard of York never ascended the throne, so that although we use the name in a general way for all the descendants of Henry II, the

[1] *Remaines of a Greater Worke concerning Britiane*, 1605   [2] Ib.

only truly Plantagenet kings were Edward IV and V and Richard III.

Meantime, the broom, both in flower and seed, had become a favorite badge or emblem with the kings of France. *L'Ordre du Genest* was founded by St. Louis in 1234, and the plant made numerous appearances in heraldic ornament during the next two centuries. Richard II, although he did not take the name, frequently bore the emblem; it adorns his robes on his tomb in Westminster Abbey, and in the Wilton Diptych in the National Gallery he is shown wearing a gown embroidered with broom-pods and white harts, and a broom-pod collar; the eleven charming angels that accompany his devotions have all been presented with badges and collars to match, like bridesmaids at a wedding.

This historical plant (*G. anglica*) still grows wild on our heathlands, and its ancient name of Petty (or Petit) Whin suggests its French associations.[1] It is seldom grown in gardens, perhaps because it requires a mailed fist to deal with its prickliness; but several authors have considered it quite worthy of cultivation. "Why," asks Marshall, "should this, because it is common in some parts of England, be denied admittance into gardens . . . as it has many natural beauties to recommend it? . . . No garden is so small but it may be there planted, if the commonness of it be no objection to the owner."

Another pretty and interesting British native, *G. tinctoria*, the Dyer's Greenweed, was one of considerable commercial importance, as the yellow color it yielded was used in combination with woad to produce the celebrated "Kendal green"—so named from Kendal in Westmorland, where the process was first introduced by Flemish immigrants in the reign of Edward III (1327–77).[2] The plant does not seem to have been cultivated for the purpose, but was sufficiently abundant in a wild state, both in the Kendal neighborhood and in Norfolk and Suffolk. "Of this Pliny made mention; the Greenweeds saith he, do grow to die clothes with."[3] A double-flowered form was grown as a pot-plant in the late nineteenth century, and called "a remarkably pretty dwarf free-flowering plant,"[4] and several other varieties have been cultivated.

These are low-growing shrubs, and some of the other genistas, such as the native *pilosa* and the European *sagittalis* (introduced by 1588), are completely prostrate; but our gardens are also enriched by some tall and handsome foreign species, one of which, *G. aetnensis*, is almost

[1] Turner gives the name of "Petie Whine" to the Rest Harrow (*Ononis arvensis*), but Gerard applies it to the genista
[2] Robin Hood's "Lincoln Green" was probably dyed with weld, *Reseda luteola*   [3] Gerard   [4] Wood

tree-like, growing to a height of 15 feet or more. It comes from the slopes of Mount Etna, and anyone who has seen this shrub in golden eruption in late July must find it remarkable that it escaped the notice of botanists until 1813, when its description appeared in *Rare and Less Common Plants of Sicily* by the Baron Antonio Bivona-Bernardi. It was introduced to British gardens shortly after its discovery, seed being sent by Bivona-Bernardi to P. B. Webb, Esq., of Milford House, Surrey, where the plant flowered before 1826. Berenson, the art connoisseur, described in his diary the "rivers of hardened lava wriggling down from Etna, the recent ones sinister in their unrelieved blackness, the older ones covered with yellow broom, the flower chosen by Leopardi[1] as a symbol of the precarious condition of human life".[2]

G. virgata, the Madeira Broom, has been described as "twelve feet of splendor" and was collected by Francis Masson on his way home from the Cape of Good Hope in 1777; it is one of the few Madeira shrubs that have proved hardy. The white-flowered *G. monospermum*, on the other hand, is rather tender, although it only comes from Spain, where it grows on the seashore like a willow "as far as the flying sands reach"[3] and is invaluable for binding the shifting dunes and converting "the most barren spot into a fine odoriferous garden with its flowers, which continue a long time."[4] It also grows in North Africa, where it is known as Retam, and in the Near East; the bush under which Elijah sat in the wilderness (I Kings xix, 4–5), mistranslated as "juniper" is supposed to have been this shrub. It was among the plants brought to England about 1690 by William Bentinck, close friend and adviser to William of Orange and afterwards first Earl of Portland, from his garden at Soesdyke in Holland. As a conservatory plant, it has been praised as "one of the most deliciously fragrant shrubs in the world."[5]

The name Genista was used by Virgil, and some authorities have been puzzled as to why these shrubs should be given a name derived from *genu*, a knee; but others tell us that the true derivation is from the Celtic *gen*, a shrub. All the species must be established in their permanent places when in a seedling, or at least a very young, state, for "they shoot downright roots very deep, and if these be cut or broken . . . the plants frequently miscarry."[6] It may be some slight help in identification to know that if a plant has spines it is certainly a genista; all the cytisus species are smooth—as well as about two-thirds of the genistas.

NOTES: The dwarf brooms are not well known in the

[1]Giacomo Leopardi, Italian poet, 1798–1837 [2]*The Passionate Sightseer*, Berenson, 1960 [3]Martyn [4]Ib. [5]Ib. [6]*The Botanical Register*, 1836

U.S. except in the Pacific Northwest, although they are fine rock garden plants. For a start, one might try *Genista lyda*, a dwarf pendulous plant with a profusion of yellow flowers in June. They are difficult to transplant, which may account for their limited availability from the nursery trade.

I once placed a plant in a sunny location on top of a retaining wall. It grew better than expected and crowded out other small companions. I finally had to move it to a less conspicuous location. It is perfectly hardy here in the mountains of North Carolina.

Halimiocistus, Halimium, see under *Helianthemum*

Hamamelis

The half-dozen species which make up this small family are divided between North America and eastern Asia, the senior and serviceable American kind being now quite superseded in gardens by the late-coming but more decorative Orientals. This first arrival was *H. virginiana*, the American Witch-hazel, which was introduced by Peter Collinson in 1736; a few years later Catesby, too, procured a specimen, of which he wrote: "For this plant I am obliged to *Mr. Clayton,* who in the year 1743, sent it me in a case of earth from *Virginia*. It arrived in Christmas, and was then in full blossom." Unfortunately the flowers "make no show; but perhaps the time of their appearing . . . may make the plant desirable to some persons. Nothing farther need be said to the gardener concerning this shrub, which Nature seems to have designed for the stricter eyes of the botanist."[1] It is now used chiefly as an understock on which to graft the more desirable kinds. Long before the colonization of Virginia, its inner bark was used by the Indians for external application in cases of sore eyes, inflammations and tumors, and witch-hazel in various forms is still a useful and popular remedy; the list of uses on the label of a bottle of "Pond's Extract" rivals any of the claims put forward by the early herbalists.

The first of the oriental species to arrive was *H. japonica* in its tall variety *arborea*, which was introduced by the firm of Veitch in 1862. It is a very variable shrub—as one grower complained, "you sow seeds, and may get anything"[2]—and the better varieties, of which there are

[1]Marshall [2]Bean

several, must be propagated by grafting. But the most deservedly popular of all the witch-hazels is the Chinese *H. mollis.* It was discovered by Charles Maries in the district of Kiukiang in 1879, when he was collecting for Messrs. Veitch; but the plants he sent home stayed disregarded for over twenty years in the firm's Coombe Wood nursery, for they were thought to be only another form of the variable *H. japonica.* (One of the differences is that the petals of *H. mollis* are straight, while those of *H. japonica* are wavy "like unevenly shrunk ribbon.")[1] It was only in 1900 or 1901 that George Nicholson, Curator of

Hamamelis virginiana
(AMERICAN WITCH-HAZEL)   M. Catesby (1843)

the Royal Gardens at Kew, was making a tour of inspection of the grounds, and called attention to the fact that it was actually a rare and valuable species. Every available twig was promptly cut up and grafted for propagation, and the plant was put on the market the following year. (Kew received a specimen in 1902.) It was awarded an F.C.C. in 1918 and an A.G.M. in 1922, and its beautiful variety *pallida,* raised at Wisley before 1932, has also received the highest horticultural honors.

Hamamelis, from *hama* "together with" and *melis,* "the apple," was the classical name for some tree which has

[1]Stoker, *A Gardener's Progress,* 1938

never been identified, and is said to have been applied to *H. virginiana* because its fruit, which takes twelve months to ripen, is borne at the same time as its flowers. Bowles called *H. mollis* "the Epiphany-tree", because it is at its best about the Feast of the Epiphany, when it presents its offerings of gold, frankincense (in its perfume) and myrrh (in its astringent bark—strictly speaking, only applicable to *H. virginiana*). Bean thought the name of Witch-hazel was given to the latter because the early American settlers used its twigs, like those of the hazel at home, for the mysterious and occult practice of water-divining: but it is also possible that it is derived from the Anglo-Saxon *wice* or *wic,* meaning pliant or springy. This name was always applied to an elm or elm-like tree; in Tudor and Stuart England the wych- or wich-hazel meant the wych-elm, or occasionally the horn-beam. It seems that the name was first used for the Hamamelis by James Lee in 1760; Collinson in 1751 called the shrub the *White* Hazel.

NOTES: The witch-hazels are good illustrations of the relationship between the flora of Asia and eastern North America. They are admired in both regions for their early spring bloom at the first sign of warm weather. In Japan, villagers delight in this plant, known as mansaku, meaning "first flowering." The hybrids (*H. × intermedia*) have either yellow or bronze flowers and are becoming the most popular. As in America they have medicinal uses; in Japan the branches were also used to bind rafters.

The witch-hazels are easily grown in the garden, preferring an exposed, sunny location and acid soil. They are propagated by seed, cuttings or grafting, usually onto our common Virginia Witch-hazel. Grafted specimens need watching, as the rootstock may often sprout and displace the grafted variety.

Hebe, see under *Veronica*

Hedera

That headlong ivy! Not a leaf will grow
But thinking of a wreath . . .
Elizabeth Barrett Browning, *Aurora Leigh,* 1856

Ivy harbors as many legends and traditions as it does spiders, and its literature, like the plant itself, must be severely pruned to keep it within reasonable bounds. I will only mention that the ivy was closely associated with Bacchus—"his winter crowne, the vine is his crowne in summer"[1]—hence the use of a bush of ivy as the sign of an inn or tavern. "Good wine needs no Ivie Bush" was a Roman proverb before it was an English one, and many an English inn had no other sign, as late as the reign of James I. It therefore seems strange that the ancients should have believed the vine and ivy to be antipathetic, and that, for example, adulterated wine could be detected by means of a cup of ivy wood, which would retain the water while letting the wine soak through. A thin layer of the porous wood will filter liquids, especially when freshly cut, but more recent experiments seem to indicate that it is rather the water that seeps through, and the wine that is retained. A story is told of an ivy growing on the walls of Magdalen College, Oxford, whose roots penetrated into the wine-cellars, "and after branching about in the sawdust in which the bottles lay, found a cork through which some moisture was oozing, entered the bottle, drank up all the port, and then filled the bottle with a matted tangle of roots, all growing in search of more of the ambrosial liquor . . ."[2] This does not look as though any great antipathy was felt by the ivy.

To drink from a cup of ivy-wood was believed to be beneficial for several diseases, including whooping-cough; the numerous medical uses of the plant, so various indeed as to suggest experiment rather than experience, are mostly taken at a long remove from Dioscorides, who, however, warns us that the juice of the fruit causes sterility, and the berries "being taken in too great a quantitie do trouble the minde." Pliny, having begun by saying "the use of it in Physicke is doubtfull and daungerous," proceeds to give a long list of medical prescriptions; but one cannot avoid a suspicion that the ivy was so much used simply because it was always at hand. In the eighteenth century it began to be discredited; "that the Berries are a Secret against the Plague, as *Mr. Boyle* relates, is what can hardly gain Belief . . ."[3]

There is no need to attribute properties to *H. helix*, the Common Ivy; it has plenty of genuine points of interest. The broad evergreen leaf and winter-flowering habit show it to be a plant of relatively mild climates; it will not grow—and is consequently much admired and coveted—in places where the winters are very severe; and the prevailing climate of the various prehistoric periods can be gauged from the abundance or scarcity of ivy-pollen in the different deposits. On some autumn day 12,000 years ago a rhinoceros was grazing upon it near to where Clacton in Essex now stands; scrapings

[1]*Bullein's Bulwark of Defence, 1562* [2]Gunter, *Oxford Gardens*, 1912
[3]Threlkeld. Robert Turner (1687) said that ivy was planted for this purpose about Pest-houses

from a fossil tooth found in the locality contained 37 per cent of ivy pollen.

Whatever poets may say, the ivy does not "twine;" it does not need to clasp its support in order to climb, for its adventitious roots enable it to make straight for its objective—a place in the sun in which to flower and fruit—as soon as it meets with a vertical surface. That these fibrils are not true roots and convey no nourishment to the plant can be seen by the way an ivy withers if cut through at the root. They like a surface neither too rough nor too smooth, and do not take readily to

Hedera helix

(IVY)   *Flora Lordinensis* (1775)

evergreen trees, heavy shade being probably the deterrent factor; in the case of conifers, Hibberd suggests, the resinous bark may contain "something distasteful to its teeth." Opinion has always been divided as to the extent to which ivy is harmful to trees. "There is nothing more natural," says Andrew Marvell, "than for the ivy to be of the opinion that the Oak cannot stand without its support . . . whereas it is a sneaking insinuating imp,[1] scarce better than bindweed, that sucks the tree dry and moulders the building where it catches";[2] and Shake-

[1]Imp—a graft or offshoot; hence, later, an offshoot of Satan [2]*The Rehearsal Transprosed*, 1672

speare and Parkinson shared his belief that the ivy was a parasite. Evelyn and Humphrey Repton took the opposite view and thought that trees might even benefit from the association. Actually, a tree that is healthy and vigorous suffers little harm; it is only when it is on the decline that the ivy can get the upper hand, and produce its flowers and fruit.

Some of the earliest botanists, from Theophrastus onwards, realized that the ivy that crept along the ground was barren of fruit and flower "rather for want of age, and that in time it did beare, and turn into the other sort,"[1] yet many later botanists believed the creeping and fruiting forms of ivy to be quite different plants. The second stage, when the plant has reached a sufficient height and maturity to produce flowers and fruit, is of importance to gardeners, as cuttings taken from the shrubby form it then develops retain this quality and seldom revert to the climbing habit. The flowers, "many togither, like a round nosegay, of a pale colour,"[2] attract swarms of insects in late autumn—Hudson called their glittering concourse one of the most impressive sights of insect life—not only because there is little else in flower at that season, but because the nectar of the ivy is particularly concentrated, and so lavishly produced that it may even drip from the flower. Honey made from it is said to be greenish in color and with a pleasant aromatic flavor. The berries that follow are as valuable to the birds in the winter and spring as the flowers to the insects. Ray remarked that the seeds the berries contain resemble swollen grains of wheat, and passing through birds of the thrush family unaltered in shape, gave rise to stories of wheat raining down.

There is a divergence of opinion about the effect of ivy on buildings, as well as on trees. Linnaeus thought it did little or no harm, "but that can hardly be admitted (says Martyn) when we consider that it must harbour wet and filth, and that the branches will make their way into any fissure or defect in the wall, and enlarge it." The dirt must be conceded—Pope called it "the creeping, dirty, courtly ivy"—but actually the overlapping leaves shed off the rain and keep the wall warm and dry, and in the case of many old ruins the ivy preserves rather than destroys, to such an extent that it has been called the "vegetable keeper of historic records."[3] Hanbury said that the Common Ivy should be used to hide old and unsightly walls, but that the silver-variegated sort was an ornament worthy of the best and finest surfaces.

Ivies with yellow or white berries ("seldom seen in these Christian parts"[4]) were known in classical times, and Parkinson interprets a passage of Virgil, which he

translates "then Swanne more lovely, or then the white Ivie" as referring to a silver-variegated sort, also mentioned by Pliny, and believed to be the first variegated plant ever recorded; but on the whole there is little reference to garden cultivation till a comparatively late date. When formal gardens and "sheer'd evergreens" became fashionable in the late seventeenth and early eighteenth centuries, the ivy was valued for training into standards—which Evelyn said could be accomplished "with small industry"—and clipping into globes, cones and other shapes, being hardy, not particular as to soil, and vivid in color. "But since this taste has been exploded," explains Martyn, "the ivy is seldom admitted into gardens, unless to cover walls, to run over ruins, etc." This, however, was a function of increasing importance as public taste veered towards the romantic and the picturesque. Already, in 1778, Abercrombie was praising it as being "excellently adapted to overrun grottos, caves, cots, hermitages, root-houses, artificial ruins, etc., to give them a more rural appearance." Ivy then was just ivy, and only a few variegated sorts were distinguished; little is heard of it in detail during the next hundred years, and yet more and more varieties must have been creeping into cultivation, for when Shirley Hibberd published his delightful monograph on the plant in 1872, he had collected over 200 different sorts, and had sold examples of fifty of the best and most distinct to the nurseryman Charles Turner of Slough. This was the great heyday of the plant, when it was lavishly used for every conceivable purpose, including bedding-out—trained ivies in pots being plunged to furnish the empty borders in winter.

At one time all the kinds were thought to be varieties of *H. helix,* but now some half-dozen species are recognized as distinct. The origins of both species and varieties are remarkably obscure and few particulars of their introduction or discovery are to be found. Many of Hibberd's varieties were collected by himself, principally in Wales; others were obtained from nurserymen or from personal friends. The most important species are *H. chrysocarpa* or Italian Ivy, *H. colchica* and *H. hibernica.* The first is the yellow-berried ivy already mentioned, and very handsome in its fruiting form; its identification with the plant of classical references formerly earned it the name of *H. poetica* or *poeticarum,* the Poet's Ivy. It comes from southeast Europe. The large-leaved *H. colchica* extends from the Caucasus eastwards; Kaempfer found it in Japan, Fortune in China and Wallich in Nepal, where it was simply called *Sagooke* or *Gooke*—"the climber." It was introduced to cultivation before 1869 by Mr. Roegner, curator of the Botanic Gardens of Odessa, from the Caucasus coast of the Black Sea. It too has

[1]Parkinson, 1640 [2] Lyte [3]Hibberd, *The Ivy,* 1872 [4]Parkinson, 1640

yellow fruit, and its leaves have a pleasant fragrance when bruised. *H. hibernica,* the Irish Ivy, is rather a dark-leaved sort; Robinson says that in his day it was used for edgings in Continental public gardens to such an extent that the dark masses became wearisome, and helped "to obscure rather than demonstrate the value of Ivy as the best of all climbers in the northern world."

In Loudon's day, when vast numbers of ivies, trained to a height of 6 to 12 feet, were raised in pots for the market, it was possible, by placing such pots in the balconies, to cover the whole front of newly-built London house in a day, "as effectually as if it were an old building in a secluded rural situation." Loudon also said that "in very large drawing-rooms" ivy plants in boxes could be trained on wire parasols or espaliers "to form a rustic canopy for small groups of parties, who may seat themselves under its shade"; and during the next fifty years the ivy enjoyed an extraordinary vogue as an article of furniture. Portable ivy firescreens (which took three years to grow to perfection) and arbors over couches and sofas were praised and illustrated in Burbidge's *Domestic Floriculture* (1875); we are also advised to bring a few strands from outside the window to climb indoors over a "neat arch," to train ivy over the piano "so that the musician may sit before his instrument as in a little bower,"[1] or to grow it up the banisters of the staircase. It might also be planted in a wedge-shaped container concealed behind a mirror or picture, to trail around the frame; "and if the frame contains the portrait of some departed friend, Ivy is perhaps the most appropriate of all plants for the purpose here suggested."[2]

It must be admitted that many of the associations of the ivy, "one of the best of graveyard plants,"[3] are melancholy, if not downright macabre. Although not unblessed by the church (there are several examples of ivy leaves among the carved foliage of the thirteenth-century chapter-house at Southwell), in the numerous medieval Holly and Ivy carols the female ivy, as opposed to the male holly, usually has the worst of it: "Holly and his mery men They dancen and they sing; Ivy and her maidenys They wepyn and they wring"—or is allotted a faintly sinister personality: "I pray thee, gentle Ivy, Say me no villainy In landës where we go." In the nineteenth century it was constantly associated with death and decay: "For the stateliest building man can raise Is the ivy's food at last."[4] Anne Pratt repeats a story that when the coffin of Queen Catharine Parr was opened "a wreath of ivy was found entwining the temples of the royal corpse. A

berry which had fallen there and had taken root at the time of a previous exhumation, had silently, day after day, woven itself into a sepulchral coronal . . ." Richard Jeffries says that old country women told their daughters never to wear ivy leaves in hair or brooch, because "they puts it on the dead paupers in the unions and the lunatics in the 'sylums."[1]

Fortunately the plant has outgrown such morbid associations, as can be seen from its present popularity as a houseplant; many specially selected varieties, both variegated and plain, being propagated for cultivation in pots. Out-of-doors it has never quite regained its former position, though perusal of a handful of nurserymen's lists reveals the surprising fact that there are still over thirty different kinds available. Standards and bushes are now rare enough to excite surprise; and costs of maintenance as well as changes of fashion have probably contributed to the disappearance of the ivy edging or "garland." It is a pity that so few examples are seen of what Mr. Clarence Elliott calls a "fedge"—ivy trained up a light fence or posts and wires, which in few years will form a fine, glossy, self-supporting hedge, taking up little room and requiring little upkeep.

The derivation of "Ivy" (from the Old English *ifig*) is obscure: some trace it to Old English *iw*, meaning green, the root of the word "yew." In very old manuscripts the spellings of "jwy" or "yuye" give pause for thought; on the other hand, the Dutch name of Klimop is instantly comprehensible. The Latin name is said to be derived from the Celtic *haedra,* a cord.

NOTES: Cultivated and often escaped, ivy is one of our most common groundcovers and wall plants. Little needs to be added to what Miss Coats has written. They are so popular that there is a bewildering array of handsome varieties in cultivation. There is an American Ivy Society.

Helianthemum

The Cistus, Halimium and Helianthemum families are all very closely related; but whereas the Cistuses grow waist high and upwards and the Halimiums knee high and upwards, the Helianthemums seldom rise above the ankles, and on some Tree-and-Shrub lists are omitted altogether. Yet they are undoubtedly shrubs, and evergreen at that; and were cultivated in gardens long before even the most elementary rockwork was thought of. Numerically they are a large family, containing about a hundred species; fifty-three were described by Tournefort (who founded the genus) in 1700,

[1]Anne Pratt [2]Hassard, *Floral Decorations for the Dwelling House,* 1875 [3]Hibberd [4]Dickens, *The Pickwick Papers* ("The Ivy Green")

[1]*Field and Hedgerow,* 1889

and sixty-six by Loudon in 1838; but in the early days species and varieties were much confused, and every new form that appeared was given a new specific name, resulting in the complex legacy of synonyms that cumbers the family today. Four kinds are found in Britain, but three of them are rare: the annual *H. guttatum*,[1] the hoary *H. canum* and *H. apenninum* var. *polifolium*, found in a few Devon and Somerset localities, including Tor Hill near Torquay, "as we have been informed by Miss Southcote, an intelligent botanical lady of that place."[2]

enteenth and eighteenth centuries, though Curtis in 1777 mentions a number of different colors, and has heard of a double kind, "which, if it could be procured, would be a valuable acquisition to our gardens."[1] It was only in the early nineteenth century that the sun-roses began to share the attention that was then being given to the Cistus family in general. The European *H. glaucum* was introduced in 1815, *H. alpestre* in 1818, and Sweet's great work on the Order, *The Cistinae*, in which about seventy-six Helianthemums and Halimiums are

Helianthemum nummularium
(ROCK ROSE)   J. Hart (1823)

The fourth species, *H. nummularium* (syns. *H. vulgare*, *H. chamaecistus*) is common almost everywhere, especially on limestone or chalk; a curious variety of it with narrow, notched petals was found about 1920 near Croydon in Surrey and was named var. *surrejanum*.

Two or three Helianthemums or "Dwarf Cistuses" were known to Gerard, and Parkinson described some seven or eight in his *Theatrum Botanicum* (1640) but on the whole little notice was taken of them during the sev-

illustrated and described, was published serially between 1825 and 1830. In this book frequent mention is made of the fine collection of these plants then growing in the Chelsea Physic Garden, and the great variety obtainable from various nurserymen. A few more years, and their brief glory was over; Loudon, writing in 1842, said it was little use enumerating the varieties named by Sweet, "the greater part having been lost during the winter of 1837–8," which apparently was unusually severe. He

[1]Now renamed *Tuberaria guttata*   [2]Sweet

[1]*Flora Londinensis*

advises the would-be collector to get together what kinds he can, and endeavor by crossing them to produce new forms; "and considering their very great beauty as border and rock-work shrubs, we think they merit the attention of cultivators at least as much as many florist's flowers."

The modern garden types are the result of crosses between *H. nummularium, H. apenninum* and *H. glaucum,* and the principal twentieth-century specialist in the family was John Nicoll (d. 1926) of Monifieth, near Dundee, who raised a series of varieties named after Scottish bens—Ben Afflick, Ben Lawers, Ben Nevis and so on (including Ben Vorlich with a color described as "boiled lobster")—more than a dozen of which are still to be found in catalogues, and are the backbone of present-day collections, though there are many more recent additions. Most of them, however, have one habit that can be a serious drawback—they open only when the sun shines; and in bright weather their gay but fugaceous flowers are over for good, early in the afternoon. If the morning is cloudy, however, or the plants get morning shade, they will stay open later in the day; and Farrer tells a story of a lover of these flowers who, wishing to give a garden party in their honor, covered the blooms with umbrellas all the morning, in order to preserve them for his guests. Shortly before the Second World War Mr. William Christy was working to produce a race of Helianthemums that would stay open in the afternoon, and claimed to have produced at least four varieties that were still out at tea-time.

The Halimiums were separated from the Helianthemums in 1836; they are closely allied botanically, but in the garden the former are very distinct, being upright silvery shrubs 2 or 3 feet high, with yellow flowers (one has white) and, unfortunately, rather tender. The majority come from the Mediterranean region, Spain or Portugal. The first to be introduced was *H. halimifolium,* described by Parkinson in 1640, and grown in John Tradescant's Lambeth garden before 1656. Marshall (1785) described the "Sea-Purslane-leaved Cistus" as the most tender of all the sorts, but said it might be grown out-of-doors so long as a few specimens were kept under glass in case of accidents, so that "a fresh stock may be raised from the thus-preserved plants." Loudon's description of this plant is hardly calculated to attract, for he describes the shoots as "leprously white," and though producing "beautiful yellow flowers, spotless, or each marked with a small dark bloody spot at the base," both calyx and peduncle are again described as "leprous." The white-flowered *H. umbellatum* followed in 1731, and *H. lasianthum formosum* from Portugal in 1780; this is said

to be the handsomest of the genus, but authorities differ widely as to its hardiness. Sweet said it "will scarcely endure our winters in the open air without protection," while Bean called it perfectly hardy and capable of enduring up to 32 degrees of frost without permanent damage. *H. ocymoides* is quite definitely tender; but Sweet and Bean are again at variance, this time on the date of introduction, which Bean gives as 1880, while Sweet says his drawing was made from a plant in Colville's nursery in 1823.

The Halimiums form a link between the Helianthemum and Cistus families, and their closeness to the latter is shown by the fact that three bigeneric hybrids have arisen between them, which have been given the name of Halimiocistus. All are accidental crosses, two of which were found in the wild, and one in a nursery; and all have white flowers. *H. × ingwersenii,* discovered in Portugal in 1929, is thought to be a cross between *Halimium umbellatum* and *Cistus hirsutus; H. × sahucii,* found the same year in southern France, forms what one nurseryman calls "rapturous little bushes," and is probably *Halimium umbellatum × Cistus salvifolius;* and *H. × wintonensis* occurred in the nursery of Messrs. Hillier and Sons at Winchester, it is thought as the result of a liaison between *Halimium ocymoides* and *Cistus salvifolius.* In this case the pearly-white flowers are embellished with a feathered zone of maroon, and a yellow center; unfortunately it is the most tender of the three.

The Helianthemums have made very little attempt to be of use to man. Gerard quotes Pliny to the effect that the Kings of Persia were accustomed to anoint themselves with this plant "boiled with Lion's fat, a little Saffron, and Wine of Dates, that they may seem faire and beautiful, and therefore have they called it *Heliocaliden,* or the beauty of the Sun." In the seventeenth century the common Rock-rose was used as an astringent and vulnerary herb, and only sixty or seventy years ago a tincture of the flowers was still in use as a homeopathic remedy for shingles and nervous diseases. A very beautiful herbaceous species, *H. tuberaria,* received its name, according to Parkinson, from "those Spanish or outlandish puffs that are edible or fit to be eaten, because where that shrub groweth, they usually finde those puffes doe breede . . . ," an ancient and delightful way of detecting the elusive truffle.

"Helianthemum" comes from the Greek *helios,* sun and *anthemon,* a flower. In the eighteenth century, when the Cistus and Helianthemum families were united, they shared the English name of Rock-rose; when they were again separated, there seems to have been an attempt to retain "Rock-rose" as the name for the Cistus, and to rechristen the Helianthemum "Sun-rose"—a name which first appears, so far as I know, in Loudon's *Encyclopaedia of Gardening*, 1822. But to the English gardener it is the Helianthemum, not the Cistus, that has always been associated with rocks; and the plant soon slipped back to its older name. Helianthemums are the hardiest of all the Cistinae, and range from their center of distribution in the Mediterranean region right up to the Arctic Circle. Like certain of the Cistuses, some of the Helianthemum species have sensitive stamens, which, if lightly touched, lay themselves down against the petals with an engaging gesture, like a kitten rolling on its back to be caressed. NOTES: These dwarf, trailing members of the rose family are of Mediterranean origin. While hardy even as far north as southern New England, they are not common in our gardens.

Hibiscus

Although this is a large family containing some 150 species, only one is both hardy and a shrub—*H. syriacus*, the Syrian Mallow; and even this is not good in very cold situations, not because of any tenderness of constitution, but because of the lateness of its flowering season. "August is the month we may expect to be entertained by this bloom," says Marshall, "though in starved cold soils the flowers rarely ever appear before September;" and if delayed by a bad season, they may fail to do themselves justice before being overtaken by the frosts. But this very lateness makes them valuable, and after a hot summer they can be wonderful, looking, in one author's comparison, like "trees of carnations."[1] The shrub has a naturally upright growth which led Loudon to recommend its use for hedges, especially beautiful "when the different sorts are planted in harmonious order of succession, and when the plants are not clipped, but carefully pruned with a knife."

In 1597 Gerard reported that he had sown seeds of the Tree Mallow and was "expecting the success." We never hear whether his expectations were fulfilled, but Parkinson grew the plant in 1629, and found it rather tender—it "would not be suffered to be uncovered in the Winter time, or yet abroad in the Garden, but kept in a large pot or tubbe in the house or in a warme cellar, if you would have them to thrive." Yet today they are grown out-of-doors in east Perthshire without any trouble; and Gibson, describing Lord Devonshire's gardens at Arlington House in 1691, said there were six very large earthen pots, containing only "the tree holy-oak,

[1] Wood, *Good Gardening*

an indifferent plant, that grows well enough in the ground"—and presumably did not require to be housed in the winter. (Arlington House was pulled down in 1703, and a new house built close at hand by the Duke of Buckingham, which after some further vicissitudes became known as Buckingham Palace.) A much later Devonshire was also associated with the hibiscus; Cobbett in 1833 said there was a specimen "before the door of the farmhouse at the Duke of Devonshire's estate at Chiswick, that is full twelve feet high, and that blows regularly every year."

Hanmer in 1659 was the first to mention the raising of new garden varieties; he said that the two common sorts were a red and a white, from the seeds of which "other GRIDELINES[1] and PURPLES have been lately rais'd in England ... It is a pretty flowerbearing tree and not common." A red-and-white striped form made its first appearance in 1730, and Miller in 1759 described seven kinds—"the most common hath pale purple Flowers with dark Bottoms, another hath bright purple Flowers with black Bottoms, a third hath white Flowers with purple Bottoms, a fourth variegated Flowers with dark Bottoms, and a fifth pale yellow Flowers with dark Bottoms, but the last is very rare at present in the *English* Gardens; there are also two with variegated leaves which are by some much esteemed." Abercrombie in 1778 was a little more poetical in his descriptions; he called the plant "the greatest ornament of the autumn season, of almost any of the shrubby tribe" with flowers "having all dark middles, darting out each way in radiant directions towards the extremitys of each petal." Up to this time no mention had been made of any double form, but in 1838 Loudon said that doubles were common. In the 1900 edition of the *English Flower Garden* Robinson selects the twelve best of the varieties then available, three of which, 'Coeleste' (before 1879), 'Totus albus' (before 1898) and 'Duc de Brabant', are still favorites today. The fact that some seventeen named sorts are still on the market shows that the popularity of this shrub has not appreciably declined.

In China, where the plant has been cultivated for as long as records exist, the leaves were used as a substitute for tea, and the flowers as an article of food; they are said to have "a pleasing and rather slippery taste, and are regarded as a delicacy."[2] Other useful members of the family (not hardy enough for outdoor cultivation) are the nutritious *H. esculenta*, the fruits of which are eaten in the tropics under the names of Gombo or Okra; the handsome *H. rosa-sinensis*, which yields both face powder and boot blacking; and the exotic *H. abelmoschus*,

[1] From *gris-de-lin*, lint-grey—a greyish mauve
[2] Li, *The Garden Flowers of China*, 1959

whose seeds have been valued for centuries on account of their sweet musky scent. The still more valuable cotton-plant (*Gossypium herbaceum*) and the medicinal marshmallow (*Althaea officinalis*) are related to the family.

The specific name of *H. syriacus* is deceptive, for it is not a native of Syria as Linnaeus believed, but was introduced there at some forgotten date from its true home in India and China. In the early days it was known simply as *Althaea frutex*, the shrubby Hollyhock. *Hibiskos* was a Greek name, used by Dioscorides for another plant of the same Order, the Marsh-mallow; but the hibiscus of Pliny seems to have been an umbelliferous plant, and that of Virgil, one with pliant branches that was used for basket-making. (The botany of the classics bristles with difficulties.) According to Loudon, the name was "supposed by some to be derived from Ibis, a Stork, which is said to feed on some of the species." The Chinese name of another species, the variable *H. mutabilis,* whose flowers open white and change to deep rose, is obviously uncomplimentary in intention; it means "Civil Servant."
NOTES: Our technical skill at advancing garden performance is seen most clearly in the shrub althea (*H. syriaca*). The late Dr. D. Egolf, the leading geneticist for ornamental plants in the U.S., produced the first triploids at the National Arboretum. The cultivar 'Aphrodite' is pink with a white throat, 'Diana' is a pure white and 'Minerva' is lavender with a deep throat. Because the triploids set no seed, they flower over a longer period of time. The shrub hibiscus is used in the garden as a specimen plant, requiring little care. If pruned, last year's branches are cut back in early spring because the flowers will be on the new growth.

The popular *H. rosa-sinensis* is only useful in the subtropical parts of the country; in Florida and Hawaii they are exceptional garden plants. *H. abelmoschus*, a species from India, has edible leaves, and H. *cannabinus*, which for centuries has been used for its fiber, is now being investigated as a source of paper pulp.

Hippophae

H. *rhamnoides*, the Sea Buckthorn, is, as its name implies, a prickly shrub that prefers to grow near salt water; but it will thrive quite well in inland gardens, and but for one fault would probably be seen in them more often. The drawback is that although there is only one species in general cultivation, there are two sexes; and the shrub will not produce the abundant orange berries that are its chief attraction unless both kinds are planted, in the mormon-like proportion of one male to every six or eight females. This requires space; but where space can be afforded, no shrub is more rewarding. It grows wild on the eastern coasts of England, but neither Turner nor Gerard realized that it was a native plant. Its range extends to eastern Asia; Captain Kingdon-Ward found it abundant in gravelly river-flats in northwest Yunnan, where in September the silvery foliage "was illuminated as by neon-lights, every branch and twig encrusted with orange berries."[1]

Dioscorides, in the first century A.D., described it as "a sprigly shrub, thick, putting out on all sides,"[2] and Marshall praised it not for its shining leaves or glowing fruit but for its "singular appearance in winter; for the young shoots of the preceding summer are then found thickly set on all sides with large, turgid, uneven, scaly buds, of a darker brown, or rather chocolate colour, than the branches themselves: These give the tree such a particular look that it catches the attention, and occasions it to be enquired after as much as any shrub in the plantation."

The fruit is unpalatable to birds, and is usually left long on the tree; it is not poisonous to man, although it was thought to be so in parts of France and Switzerland, and Rousseau tells of a botanist at Grenoble, who saw him eating the fruit, but "was so polite, or regarded Rousseau with so much respect, that he durst not presume to warn him of his danger."[3] Pallas said that the berries were eaten by the Tartars, "who make a jelly or preserve of them, and serve them up with milk or cheese, as great dainties,"[4] and a similar jelly was used by fishermen in the Gulf of Bothnia "to impart a grateful flavour to fresh fish."[5] The leaves, roots and stems are said to produce a yellow dye, and it has been suggested that the name, from *hippos*, a horse, and *phao*, to brighten, was derived from the plant's employment in some equine tonic. (But as it is by no means certain that the plant so named by Theophrastus in the third century B.C. is the same as the one that bears the name today, experiments should be made with caution.) It is not actually a Buckthorn, but a relation of another silvery-leaved shrub, the Elaeagnus.
NOTES: Although hardy into Canada, *Hippophae rhamnoides* is too little cultivated in the U.S., perhaps because it is dioecious, so that both male and female plants need to be present for best results. The Sea Buckthorn, with its tolerance to salt spray, is particularly useful on coastal lands.

[1]*Berried Treasure* [2]Goodyer's translation, 1655 [3]*Rêverie du Promeneur Solitaire,* VII, quoted by Loudon [4]Loudon [5]Ib.

Hydrangea

Here is yet another of those American-Asian families of plants, whose primary species, introduced
from America, have been so greatly surpassed by
the later Asiatic importations that the first-comers are
now almost forgotten. It is a moderately large family,
and if we had not allowed our gardens to be dominated

is composed entirely or almost entirely of the showy
sterile flowers.

The first garden hydrangea to be introduced to the
garden was the white-flowered American *H. arborescens,*
which was obtained by Peter Collinson in 1736, though
it did not bloom until ten years later. Miller said it was
"preserved in Gardens for the sake of Variety, more
than for its Beauty"; but in 1823, when *the* hydrangea

Hibiscus syriacus

(ALTHAEA)   Sydenham Edwards (1789)

Hippophae rhamnoides

(SEA BUCKTHORN)   M. Smith (1905)

so exclusively by one kind only, we might realize it is a
very varied one. In South America, for example, there
are said to be a number of evergreen, red-flowered,
climbing species which have never been introduced in
Britain, and several fine hydrangeas still await importation from China. Apart from these unobtainable treasures, eighteen or more species are available today,
besides the very numerous varieties of the Common
Hydrangea. Nearly all of them bear variously-shaped
clusters consisting of many small fertile flowers and a
few large showy sterile blooms, so placed as to call attention to the whole inflorescence—a pattern which has been
independently evolved by some of the Virburnums,
although the families are not related. In both cases garden forms have been raised in which the inflorescence

had hardly yet got into its stride, this species is mentioned as being "one of the most common plants seen
on our balconies, windows, etc."[1] The sterile variety
grandiflora is said to have been found wild in the mountains of Pennsylvania about 1860; the date of its arrival
here is not recorded, but it received an Award of Merit
in 1907. It bears large rounded heads of flowers which
weigh down their slender stems and droop disconsolately
after rain. The name of *arborescens* is misleading, for it
rarely exceeds 3 or 4 feet in gardens; other species are
much more tree-like, notably the Japanese *H. paniculata,*
which in its native country reaches a height of 25 to 30
feet. This was introduced in 1861, and has now almost
ousted the American species from our gardens. It too

[1]Elizabeth Kent, *Flora Domestica*

has a variety (*paniculata*) *grandiflora*, established in England before 1881, whose large conical plumes are built up entirely of sterile florets. (In America it is called the Pee-Gee Hydrangea, which sounds interestingly Indian until one realizes that these are merely the plant's initials.) A superior variety of the original wild form, *H. paniculata* var. *floribunda*, was among plants brought from Japan to the Botanic Gardens of St. Petersburg by Carl

Hydrangea hortensis
(now *H. macrophylla* subsp. *macrophylla*)
(HORTENSIA)   P. J. Redouté (n.d.)

Maximowicz in 1864.

The diversity of the hydrangea is well illustrated by three species not uncommon in British gardens—*H. quercifolia*, *H. petiolaris* and *H. sargentiana*. The first, though the oldest of the three, is probably the rarest. It was discovered by William Bartram during his travels in Carolina, Georgia and Florida in 1773–8, and was described and illustrated in his book about the journey. Owing, probably, to the interest thus aroused, living plants were brought to the U.K. by John Lyon about 1805, and were dispersed at his sale shortly afterwards. The flowers of the Swamp Snowball or Oak-leaved Hydrangea, according to Bartram, "are truly permanent, remaining on the plant for years, till they dry or decay," and the

handsome lobed leaves color well in autumn; but it blooms more freely in America, in England is tender except in the south and west. *H. petiolaris* is very different—a vigorous Japanese climber, which will thrive on a north wall and ascend to a height of 60 to 80 feet by means of self-clinging rootlets like those of the ivy. It was twice introduced to Kew—the first time by Max Leichtlin of the Baden Botanic Garden, in 1878—each time under the name of *Schizophragma hydrangeoides*, an allied plant which is distinguished from the hydrangea by having only one large showy sepal to each of its sterile flowers, while the hydrangea has four. (The *Schizophragma*, however, has retaliated by appearing in the illustration in the R.H.S. *Dictionary*, under the name of its rival.) *H. sargentiana*, the third of these dissimilar hydrangeas, is a woodland shrub of rather awkward habit, with very large leaves, and shoots protected by a thick felt of rufous bristly fur. It was introduced by Wilson from China to the Arnold Arboretum in 1908, and was named in honor of the Director of the Arboretum, Professor Charles Sprague Sargent. At Kew it flowered for the first time in 1911.

But all this is beside the point. To the majority of gardeners there is only one hydrangea, in all its hydrocephalous variety—*H. hortensia*, or *macrophylla*, or *opuloides*, or whatever the botanists prefer to call it. Perhaps it would be as well to elucidate this complicated nomenclature before going further. The trouble arose because for many years the only examples known in Europe, whether of live plants or dried specimens, were of Chinese or Japanese cultivated varieties in which the heads consisted only of sterile florets; and lacking stamens to count and seed-vessels to examine, botanists found it hard to decide to which family the plant belonged. Kaempfer in 1712 called it an Elder (*Sambucus aquatica*); Thunberg brought home two dried specimens from Japan in 1784 which he labeled Virburnums (*V. macrophyllum* and *V. serratum*) and Loureiro classified it rather wildly as a primula. Others thought it a new genus, and the astronomer Legentil,[1] who brought some plants from the East Indies to the island of Mauritius in 1771, suggested that it should be called Lepautia, after Mme. Lepaute, wife of a celebrated clockmaker and herself an astronomer of merit. The French botanist, Commerson, who was at that time in Mauritius making a study of the natural history of the region, preferred however to give it the name of Hortensia; not after Mme. Lepaute (whose name was not Hortense) nor after Queen Hortense, the daughter of the Empress Josephine, (who was

[1]Guillaume Joseph Hyacinthe Jean-Baptiste Legentil de la Galaisière, 1725–92

not born until ten years after Commerson's death), but after Mlle. Hortense de Nassau, daughter of the Prince of Nassau, a distinguished botanist who, like Commerson, had accompanied de Bougainville on his voyage around the world in 1766. This name was taken up in France, where the plant became established as *Hortensia opuloides*. But this was not the end. In 1789 a plant from China was introduced by Sir Joseph Banks to Kew, and in 1792 the English botanist, J. E. Smith, at last identified it correctly as a species of hydrangea, but gave it the specific name *hortensis*, "of gardens," not realizing that a lady's name was in question. After this had been corrected to *H. hortensia* (1829), and again back to Thunberg's specific name, as *H. macrophylla* (1830), the matter was allowed to rest for more than a century; but it became more and more apparent that the hydrangeas cultivated in gardens under this name consisted of two or more quite distinct strains. Robinson remarked that the whole family was "in want of looking up by some enthusiastic admirers," and fortunately an admirer was found in Mr. Haworth-Booth, who has recently made a most careful and authoritative study of the genus. He suggests that Sir Joseph Bank's Chinese plant (the wild original of which was found by Wilson in 1917) should be separated, under the name of *H. maritima*, from Thunberg's Japanese plants, both garden forms of complex descent, for which he retains the names of *H.* × *macrophylla* and *H.* × *serrata*, while recognizing their hybrid origin. The distinction is important, because the species vary widely in their horticultural requirements. *H. maritima* is a strong-growing plant, making excellent seaside hedges, but tender inland, inclined to be sparse of bloom away from the coast, and flowering from terminal buds only. *H.* × *serrata* is probably descended from three Japanese woodland species: the white *H. japonica* and the pink *H. acuminata* (both introduced by Low of Clapton in 1844), and the deeper-colored *H. thunbergii* (Cripps of Tunbridge Wells, *c.* 1870). Hydrangeas of this type are better away from the sea; they flower from both terminal and axillary buds and require partial shade. *H.* × *macrophylla* is thought to be of similar descent to *H.* × *serrata*, plus an admixture of the blood of *H. maritima*. The name "Hortensia" has been retained to distinguish the varieties with globular heads of sterile flowers from those with flat corymbs of flowers of both sorts, to which Mr. Haworth-Booth has given the appropriate name of 'Lacecaps.'

The "Chinese Guelder-Rose" as it as called, never lacked notice from the start; the flowers, green when immature, excited curiosity, it is said, even among the Customs House officials. Banks is always given the credit

for its introduction, though Gilbert Slater procured a plant about the same time, which is believed to have been actually the first to flower. By the mid-nineteenth century it was already "much valued as a chamber plant," and by the 1880s and '90s thousands were being raised for market and amateurs were transporting large specimens to flower shows on wheelbarrows. Meanwhile, in 1879, the firm of Veitch received from their collector, Charles Maries, two Japanese lacecap forms (vars. *rosea* and *mariesii*) the first of the type to be seen in this country. They aroused great interest, being regarded as "the most curious of the many varieties of the Common Hydrangea," and later became much valued for breeding, especially in France. Some fine lacecaps were raised about 1903 by the firm of Lemoine of Nancy, and in 1905 the same enterprising nurseryman found a few fertile flowers among the sterile ones in a bloom of the Hortensia type, by means of cross-fertilization obtained a few seeds, and produced the first European-grown Hortensia hybrids in 1907–8. Since then, the majority of the new varieties have been raised on the Continent, though an English specialist named H. J. Jones of Hither Green grew some good sorts about 1925. The number of garden varieties is now estimated at over 370; Mr. Haworth-Booth describes some 327 of them in his book *The Hydrangeas* (1959) with names ranging from 'Gartenbaudirektor Kuhnert' to 'Petite Soeur Thérèse de l'Enfant Jésus.' ("Their names reveal their origins.")

The chameleon-like propensity of the Hydrangea to change the color of its flowers soon attracted attention; the *Botanical Magazine* describes a plant owned in 1796 by the Countess of Upper Ossory, which produced red flowers one year and blue the next, though growing in the same pot. At one time the plant acquired the reputation of taking the color of anything by which it was shaded, but it was soon realized that the change had something to do with the medium in which the plant was grown; experiments in the production of blue flowers were reported in the Horticultural Society's *Transactions* in 1818. Phillips tells a touching story of a magnificent blue-flowered specimen that grew in a cottage garden on a "dreary common" in Hampshire, at a time when hydrangeas of any sort were rare: the owner, although extremely poor, refused an offer of ten guineas for it, because it had been "reared by a child whom she had lost," but was eventually persuaded to part with a few cuttings—all of which produced pink flowers when transferred to a new environment. By 1875 it had been discovered that watering with alum or mixing iron filings with the soil would turn the flowers blue, but "the natural delicate rosy hue is preferable," says Burbidge primly, "and we do not recommend trickery in window-gardening." It is now known that hydrangeas will turn blue on an acid soil if "free" aluminium is also available and many recipes are given in horticultural works for producing the desired result; but different kinds vary widely in their capacity to change color, according to the part played by the wild species in their constitution. The white-flowered *H. japonica* will not turn blue in any circumstances, whereas *H. acuminata* blues readily; so pale shades bearing a high proportion of *japonica* blood do not change color so easily as darker ones.

Shirley Hibberd praised the hydrangea as a pet, because it responds so willingly to any treatment that its owner may think fit to give it; but Loudon drily remarks that it is "particularly suitable for persons who have little else to do than attend to their garden, or greenhouse," as it requires so much watering. A large specimen in warm weather will consume ten or twelve gallons of fluid daily. The name *Hydrangea*, though it means a water-vessel, has nothing to do with this excessive thirstiness; it was given to the first (American) species to be discovered, on account of its cup-shaped seed-vessels. The middle syllable, we are told, should be pronounced "ran" not "rain"; so Loudon's mnemonic "head-ranger" leads us astray. In the Language of Flowers the Hortensia signifies a boaster, because the flowers are not succeeded by fruit, but are like fraudulent company promoters offering shares in nonexistent gold-mines.

In Japan the dried leaves of *H. thunbergii* were used to make a very special beverage called "Celestial Tea" (*Ama-tsja*), with which the images of Buddha were bathed on his birthday.

NOTES: As Miss Coats points out, this family of important garden shrubs again illustrates the close alliance of Asiatic plants to those of the southeastern U.S. As is often the case the Asiatics have far overshadowed their American counterparts in our gardens. But recently, an outstanding selection of our native *H. quercifolia* 'Snow Queen' (Pl. Pat. 4458) was released by William Flemer III.

The other noteworthy change to the genus since the first edition of the book is that *H. petiolaris* has changed to *H. anomala* subs. *petiolaris*. I have seen this species in the woods of northern Japan with its relative *Schizophragma* in full bloom, gloriously climbing over tree stumps. This was so effectively placed by the hand of nature—no landscape architect could do as well.

Hypericum

With its universal signature of yellow flowers (common to almost every species of this large genus), its abundant radiating stamens and its summer season of flowering, the St. John's Wort naturally became associated with the sun and summer solstice, probably long before the Church took over the pagan festival and named it in honor of St. John; and it plays a prominent part in all sorts of midsummer magic, with protective powers against lightning, witches and the Evil Eye. It was considered invaluable as an amulet "to be put into one's Hat, to be laid under one's Pillow in the Night, to be often smell'd to, to be strewed up and down the House, to be hung up upon the Walls, etc.";[1] it would also put demons to flight, and in the early confusion between magic and medicine was called *Fuga daemonium* and used, as Burton testifies,[2] in the treatment of melancholia and insanity. Gerard gives a recipe for extracting a blood-red oil from the flowers which was "a most pretious remedie for deep wounds and those that are thorow the body, for the sinues that are prickt, or any wound made with a venomed weapon," and it was oil of *Aparicio* (=hypericum) that was used by the fair and frivolous Altisidora to anoint the face of Don Quixote when it had been scratched by a cat.

It was the common herbaceous St. John's Wort, *H. perforatum*, that was considered by the Devil to be the most valuable for the above purposes, for it was the leaves of this kind that he pricked over and over with a needle, in an attempt to destroy their usefulness. (The holes, visible when a leaf is held against the light, are actually translucent oil-glands.) But the shrubby *H. androsaemum* (a British native), was used with almost equal freedom—and probably with equally good effect; Turner explained that it was called Tutsan (from *Toutsain* or *Tota Sana*) "because it heleth all", and *Androsaemon* "because it hath iuce like mannis blood." (The early botanists classed it apart from the other hypericums, under this name of Androsaemum.) Other early names were Park Leaves, "because it is so familiar to Parks and Woods that it seldom grows anywhere else,"[3] Touch-and-Heal or Touchen-leaf, and Sweet Amber. Its garden virtues are hardly so great as its magic or medical ones, for its flowers are relatively small; but its habit is good, with its "two and two leves ever comminge oute one agyanste another, resembling a byrdes winges stretched furthe, as when the bird doeth flye,"[4] and its fruit is handsome in late summer. It is useful for planting in difficult places

[1]Tournefort, *The Compleat Herbal* (1731)   [2]*Anatomy of Melancholy*, Pt. II, Sec. 5   [3]Tournefort   [4]Turner, 1551

and under trees.

Another St. John's Wort, also a good carpeter for shady places, is *H. calycinum*, commonly known as Aaron's Beard or Rose of Sharon, and less familiarly as Terrestrial Sun. It was discovered by Sir George Wheler, who traveled in Greece and Turkey in 1675–6, growing at the village of Belgrade, a rural pleasure resort near Constantinople. He sent seed of it to Robert Morison of the Oxford Botanic Garden, who wrongly described the plant in his *Plantarum Historiae* as being of Mount Olympus. It has since become naturalized in the United King-

Hypericum androsaemum

(TUTSAN)   *Flora Lordinensis* (1777)

dom, causing sundry botanists to believe it indigenous to Scotland and Ireland. At Hulne Abbey in Northumberland there is a tradition that this plant was brought from the Holy Land by the Abbey's Crusader founder, Sir Ralph Fresborn, for the sake of its medicinal properties; but it does not grow in Palestine, unless as a garden plant, and there seems little foundation either for this legend or for the name of "Rose of Sharon." (Linnaeus thought the biblical Rose of Sharon to be a Cistus, but modern botanists identify it as a bulbous plant, probably a tulip.) The application of this name to the hypericum seems comparatively recent; to

Gilbert White in 1790 it was still "Sir George Wheler's Tutsan," nor is the name of Rose of Sharon mentioned by Loudon in 1838.

Both these St. John's Worts, *H. calycinum* and *H. androsaemum,* can become a nuisance in the garden; they spread by underground stolons, and as Johnson puts it, "They stole freely."[1] The garden plants derived from the Chinese species, *H. patulum,* have no such vices. The original type was sent to Kew by Richard Oldham in 1862,[2] but having come from a mild region it was not reliably hardy, and it was soon superseded by two much finer forms, *H. patulum* var. *henryi,* found by Augustine Henry near Ichang and sent by him to Kew in 1898, and *H. patulum* var. *forrestii,* collected by George Forrest in Yunnan in 1918 or 1920. Both are hardy, and have larger flowers than the type. The group also includes some valuable hybrids, the oldest being *H.* × *moserianum* (*H. patulum* × *calycinum*) raised about 1887 at Moser's nursery at Versailles. This was before the better and hardier varieties of *patulum* had been introduced, and the resulting hybrid is not satisfactory in cold gardens; the plant known as 'Sungold' (*H. patulum* var. *grandiflorum,* or H. × *penduliflorum*) is possibly the result of the same cross, using a hardier *patulum* parent. Finally, there is the famous 'Hidcote Gold', whose unknown descent is probably from *H. patulum* × an evergreen species, perhaps *H. leschenaultii.* Universally acclaimed as the finest hardy St. John's Wort, it was raised at Hidcote, Gloucestershire, in the 1940s by the late Major Johnston from seeds collected, some say in a garden in Kenya, others, in China. Shapely, robust, semi-evergreen, perpetualflowering, indifferent to 30 degrees of frost, it has all the hypericum virtues, and was given an Award of Garden Merit in 1954.

Those who garden in favored climates may grow some other St. John's Worts, which have large flowers but tender constitutions; they include *H. hookerianum,* its variety *rogersii,* and *H. leschenaultii.* The first was discovered by Sir Joseph Dalton Hooker, and introduced from Assam by Thomas Lobb when collecting for the firm of Veitch in 1848–51. Plants raised from a further consignment of seed collected by Reginald Farrer on his last expedition in Upper Burma in 1920 are perhaps a little hardier than the original importation. The taller and larger variety *rogersii* was found on the Victoria Mountains in Burma and named after an officer in the Indian Forest Service, and *H. leschenaultii,* said to have the

[1]Sir George Wheler noticed this, at the very beginning: "It spreads upon the Ground in heaps; so that seldom one shall find one Stalk alone." [2]This may not have been the first introduction; Loudon gives the date as 1823

largest flowers and the poorest habit of any of the hypericums, was discovered in 1805 by the French naturalist
Leschenault de la Tour in the mountains of Java, and
was in cultivation here before 1882. This group too has
produced a famous garden variety in 'Rowallane
Hybrid', an accidental cross between *H. hookerianum* var.
rogersii and *H. leschenaultii,* raised from a self-sown seedling by the late Mr. Armytage Moore at Rowallane Gardens, Saintfield, County Down, between 1935 and 1940.
There is a close rivalry between this and the Hidcote
plant, which is about the same age, for the title of King
of the Hypericums; Rowallane Hybrid has the larger
flowers, but is much the less hardy.

Yperikon was the name used for the family by Dioscorides; some etymologists derive it, most unconvincingly,
from *yper,* "on account of" and *ereike,* "heath," because it
grows in such places; others say it comes from *huper
eikon,* "over an image or apparition," from the plant's
power over evil spirits. Not all its properties are beneficial; *H. perforatum* is poisonous to cattle and sheep, and
having been introduced into Australian cottage gardens
about 1850 because of its reputation for salves, in seventy years spread over and ruined some 750,000 acres
of grazing country. It was eventually checked by means
of a species of beetle, imported for the purpose; but
undaunted, the plant has gone on to colonize California,
and to ruin the ranges in the cooler parts of the state.
The Devil with his needle would probably be warmly
welcomed in either locality.

NOTES: The hypericums are useful garden plants for
their bright golden flowers in mid-summer. *Hypericum
patulum* var. *henryi* is most familiar to gardeners and the
hybrid 'Hidcote' described by Miss Coats is the one usually grown. Although it dies to the ground in northern
winters, in the South, particularly on the coastal plain, it
remains semi-evergreen. The bright yellow flowers
appear from early summer until autumn. The St. John's-
wort, *H. calycinum* is a good groundcover with silver-gray
foliage, which produces golden yellow flowers all summer. It tolerates poor soil and sun or shade and can be
grown successfully from New York City southward. The
other species are not common garden shrubs in America.

Jasminum

An undecided family, the jasmines, unable to make
up their minds whether to be climbers or shrubs;
most of the so-called climbers can be grown as
rambling bushes, and many classed as shrubs are at their
best against a wall. (The only garden jasmine that is
unmistakably a shrub and not a climber or rambler is
the diminutive *J. parkeri,* found by Mr. R. N. Parker in
northwest India in 1919 and introduced in 1923, which
is less than a foot high.) There are about 200 species,
with yellow, white, or occasionally rosy flowers, of which
twelve or fourteen are in cultivation in Britain; nearly
all the white species are remarkable for the strength and
sweetness of their scent, but most of the yellows have
little or no perfume. The family belongs to the Natural
Order of the Oleaceae, which also includes the scented
lilacs and osmanthus.

Jasminum multiflorum
(STAR JASMINE)   *The Botanist's Repository* (1807)

J. officinale, the common white jasmine or jessamine,
is so old and so widespread in cultivation that its country
of origin is uncertain. It is believed to be indigenous to
Persia, northern India and perhaps China; but Mr. Li
says that in the third century *J. sambac* and *J. officinale*
were recorded in China as "foreign" plants, and in the
ninth century their habitats were given as Byzantium and
Persia, respectively. The Chinese name of *J. officinale* was
Yeh-hsi-ming, obviously a version of the Persian or Arabic
name. In 1825 it was reported in the *Botanic Register* that
"some late Russian naturalists" believed it to be truly wild
in the Black Sea area of the Caucasus. One of the theo-

ries about the manner of its distribution in Europe is that it was carried westward by the Turks after their capture of Constantinople in 1453. In southern Europe it has been so long naturalized that Linnaeus was deluded into thinking it a native of Switzerland; and it was common in London gardens before Turner wrote his *Names of Herbes* in 1548.

For a plant so long in cultivation and so wide in distribution, the white jasmine has remarkably few varieties—perhaps because it is seldom propagated by seed, which is not freely produced even in warmer countries than ours. Silver- and gold-variegated forms were grown in the eighteenth century, grafted on the common jasmine; and Miller noticed that "it often happens, that the buds do not take, but yet they have communicated their gilded Miasma to the plants." There was also a rare double, now apparently lost; but it is still possible to obtain *J. officinale* var. *affine,* with larger flowers pink-tinged in bud, seed of which was sent from India by Dr. Royle and raised by the Horticultural Society before 1845.

The common jasmine was also used as a stock on which to graft certain choice and tender species, such as *J. grandiflorum* and *J. sambac.* When John Aubrey wrote (at the time of the Restoration) that "Jessamines came into England with Mary, the Queen-Mother,"[1] he was probably referring to *J. grandiflorum,* the Great Spanish Jasmine, which was first described by Parkinson in 1629. It is a subtropical plant, but was much cultivated in Italy and Spain, especially Catalonia, and so acquired the name. *J. sambac,* the Arabian jasmine, is said to have been grown in the royal gardens at Hampton Court as early as 1665, "but being lost there, was known in Europe only in the garden of the Grand Duke of Tuscany, who would not suffer any cuttings or layers to be taken from his plants; until Mr. Miller received a plant of it from the Malabar Coast (in 1730) by Captain Quick."[2] This kind was described by Hanmer as "extraordinary sweet, surpassing any of the other Jasmyns, as partaking both of the Sent of the Orange and Jasmyn flowers"; unfortunately neither the Arabian nor the Spanish jasmine can be grown in England except in a greenhouse.

All three of the species so far mentioned have been used to produce jasmine perfume, which is of great antiquity. Dioscorides tells of an oil of "Jasme" which was made by the Persians by steeping what he called "the white flowers of the violett" in sesame oil, the use of which was "entertained amongst the Persians in their banquets, for the sweet smells' sake that it yields." Today the perfume is extracted by the process called *effleurage,* in which the florets are pressed into shallow layers of

fat, and replaced daily by a fresh supply, until the fat is saturated with sweetness; it is then dissolved in alcohol, and the floral essence distilled. The first part of this process was known to the ancient Greeks, and used to make unguents, but the distillation of essences came somewhat later. "There is no small curiosity and address in obtaining the oil, or essence, as we call it, of this delicate and evanid flower," says Evelyn, "which I leave to the Chemist and the Ladies, who are worthy the secret." "Jessamy gloves" were popular in England in the late seventeenth century, and jessamy butter, to rub on the leather and renew the perfume, could be bought at the glover's for a shilling an ounce. Some find the scent of the flowers so strong as to be disagreeable. To Matthew Arnold, "open, jasmine-muffled lattices" represented the epitome of summer, but Gilbert White wrote in his journal on 17 July, 1783, that "the jasmine is so sweet that I am obliged to leave my chamber." It is essentially a flower of the night, and Moore speaks of

> *. . . timid jasmine-buds that keep*
> *Their odours to themselves all day,*
> *But when the sunlight dies away,*
> *Let the delicious secret out*
> *To every breeze that roams about.*

White Jasmine is associated with oriental female beauty; but there is nothing feminine about the Chinese winter jasmine (*J. nudiflorum*), which has no fragrance, is indestructibly bold and hardy, and flaunts the imperial yellow, the color that only the most privileged might wear. The measure of its value is shown by the fact that, although introduced only in 1844, it is now as widespread as the much older white species, and is grown in almost every garden. It was found by Dr. Alexander von Bunge at Pekin in 1830–1, and was introduced to England by Robert Fortune from the nurseries and gardens of Shanghai. Fortune described it as "a very ornamental dwarf shrub, and I have no doubt of it being perfectly hardy in this country. It is deciduous; the leaves falling off in its native country early in autumn, and leaving a number of large prominent flower-buds, which expand in early spring, often when the snow is on the ground, and look like little primroses."[1] Like most new introductions it was treated as a greenhouse plant before being tried in the open ground; but it soon proved its hardiness by surviving a series of severe winters, including that of 1879–80, which culminated in a twelve-day fog in February. Few today will agree with Sir Joseph Hooker's comment, made when the plant was still a novelty, that "it is to be regretted that the foliage,

---

[1]Henrietta Maria, known for her love of flowers   [2]Martyn

[1]Quoted in *The Botanical Magazine,* 1852

scanty at best, does not appear at the same time with the leaves."[1] The dramatic beauty of blossoms on bare winter branches does not seem to have been appreciated in the early nineteenth century; yet no such criticism was ever made, so far as I know, of the summer-flowering Spanish Broom, which has exactly the same combination of bright yellow flowers on bare green rush-like stems.

These two invaluable hardy climbing jasmines (*officinalis* and *nudiflorum*) have two nearly hardy counterparts, also a yellow and a white, both relatively new, but likely to become of increasing importance and popularity—*J. primulinum* and *J. polyanthum*. Both can be grown outside on sheltered walls in the south and west, but elsewhere need the protection of a cool greenhouse. *J. primulinum*, the yellow one, may possibly be an escaped garden form of *nudicaule*, with larger, semi-double flowers; it was collected by E. H. Wilson, on the way back from a perilous trip to visit Augustine Henry at Szemao, in Yunnan, and was sent by him to the firm of Veitch, with whom it flowered in 1901. It is slightly hardier than *J. polyanthum*, which is, however, worth every effort to grow, with its many-branched racemes of intensely fragrant white flowers, which are pink on the outside when grown in the open, but tend to lose this coloring under glass. It was found in Yunnan by the Abbé Delavay in 1883, and was introduced to cultivation by Major Lawrence Johnston of Hidcote, after his expedition to China with George Forrest in 1931.

Two hardy jasmines commonly described as shrubs, but often grown on a wall, are *J. fruticans* and *J. humile*, both yellow-flowered evergreens, and both very old garden plants. The former is a native of southern France, and its triple leaf and shrubby habit induced the early botanists to classify it with the melilots and brooms. Gerard grew it under the name of Shrub Trefoil or Makebate (while admitting that it had flowers "not much unlike the yellow Iasmine") and it is the only member of the family to which medical properties were attributed. It has had its admirers, but, superseded by many better plants, it seems now to have gone out of cultivation. *J. humile* was grown by John Tradescant before 1656; it was known as the Italian Jasmine, because plants of it were "annually brought over from thence by those who come over with Orange trees."[2] Opinions differ as to its country of origin, but it has two close relatives in the Himalayas, which are sometimes classed as varieties—*J. revolutum* and *J. wallichianum*, both introduced in 1812. All three have rather handsome yellow flowers in late summer.

The lolloping habit of the white jasmine has pre-

[1]Sic—error for "flowers"? *The Botanical Magazine*, 1852 [2]Miller

sented a problem to the gardeners of every age. Gerard puts it among the plants "which have need to be supported or propped up, and yet notwithstanding of it selfe claspeth not or windeth his stalkes about such things as stand neere unto it, but onely leaneth and lieth upon those things that are prepared to sustain it, about arbors and banqueting houses in gardens, by which it is held up." Liger in 1706 complained that all the Jessamines grew in "a very irregular Form," which "the Gard'ner is not permitted to rectifie by pruning"; nevertheless, his translator London added, "we make Standards of them, either with round or Pyramidal Heads, and plant them out in the open Beds or Borders, in most Places of our Plantations"—but such standards were admitted (by Miller) to be "very difficult to keep in any handsome order." Abercrombie (1778) gave detailed instructions for training it neatly on a wall or fence in the manner of an espalier fruit tree, but Marshall (1785) pointed out that being deciduous, its "brown and dirty-looking bark" was unsightly in the winter, and suggested that it should be grown as a shrub, with the older branches frequently cut back to ground level. Miss Louisa Johnson (*Every Lady Her Own Flower Gardener*, 1839) admired "inventive little arbours, where the plant has been trained up behind them, and the branches allowed to fall over their front in the richest profusion, curtained back like the entrance of a tent." The same difficulty was experienced in finding the ideal situation for the yellow winter jasmine; Gertrude Jekyll suggested that it should be planted so as to trail over a large rock, and for those who happen to have a rock about the size of a small greenhouse conveniently to hand, it would doubtless be the perfect solution.

The name comes from the Persian *jásemin* and Arabic *ysmin* and has been variously anglicized during the centuries. Turner gives "Gethsamyne" as the English version; it has also appeared as Jessamine, Gelsemine, Gessemine, Jessimy and Gesse. In Turkey and Greece the young shoots of this so-called creeper, which can leap 10 to 20 feet in a year, were used when about two or three years old to make tobacco pipes—presumably for hookahs—"and they may be seen in Constantinople, eight feet or ten feet long, twisted in various ways." This was written by Loudon, before the days of rubber tubes; he also tells us that jasmine and potato are the food-plants of the larvae of "that very remarkable lepidopterous insect," the Death's-head Hawk-moth.

NOTES: Other than *Jasminum nudiflorum* these handsome shrubs are not common; they tend to be of limited hardiness or simply rare. But *J. nudiflorum*, with its green stems and bright yellow flowers in early spring, compensates for the loss of the others. It is shown to its best advantage at the top of a bank or above a stone wall, its long flexible branches drooping. The branches make excellent cuttings for flower arrangements because the flowers are easily forced. It does tend to root down when branches touch the soil and it can become invasive. It tolerates most soil conditions and dryness. The winter jasmine, hardy through the eastern states and in the southern region, is particularly delightful because it will have scattered flowers throughout the winter.

Kalmia latifolia
(MOUNTAIN LAUREL)   M. Catesby (n.d.)

Kalmia

In the year 1747 Peter Kalm, a Finnish botanist and a pupil of the great Linnaeus, was sent by the poor but enterprising Swedish government (with the aid of some university grants), on a botanizing expedition to America. It was thought that since much of the northern part of the continent lay within the same latitudes as Sweden, plants of economic importance might be found there which would thrive in the difficult Swedish climate. Kalm was away for three and a half years, making trips to New York, Canada, the Niagara Falls and elsewhere from his main base in Philadelphia; and during his trav-

els he greatly admired a plant which was variously known as Mountain Laurel, Spoonwood, or (simply and surprisingly) Ivy. It was an evergreen—"when all other trees have lost their ornaments and stand quite naked, these chear the woods with their green foliage"—and when in bloom, Kalm's description continues, "their beauty rivals that of most of the known trees in nature; the flowers are innumerable, and sit in great bunches. . . . Their shape is singular, for they resemble the crater of the ancients;[1] their scent however is none of the most agreeable." Blushing like a Kalmia with pride, he further reports: "Dr. Linnaeus, comfortable to the peculiar friendship and goodness which he has always honored me with, has been pleased to call this tree . . . *Kalmia latifolia.*"

Kalm, however, was not the first European naturalist to become acquainted with the shrub that bears his name. Mark Catesby saw it during his travels in Carolina, and after his return to England in 1726 imported both seeds and plants. He described it in his book (*The Natural History of Carolina,* with which Kalm was familiar, though he said it was too costly for a poor man to buy) under the name of *Chamaedaphne foliis tini,* and seriously weighed its claim to be considered the most beautiful of all flowering shrubs. In 1734 Peter Collinson wrote to his correspondent Colonel John Custis, a planter in Virginia: "There's two or three plants in your country I should be glad of, viz. the Dogwood tree and a sort of Laurell or Bay that bears bunches of Flowers not unlike the Laurus Tinus it is by some Improperly called Ivy and if the Sheep Eat it Kills them." Custis duly sent him plants, which he received safely in 1736, and these seem to have become established, for in a letter to Custis in 1741, he praises "the Beautiful plant called an Ivy which you formerly sent Mee. This flowers Annually but I Really think Exceeds the Laurus Tinus—tho' it must be allowed that the latter's flowering all the Winter gives it in that point the preheminence."[2] Kalm spent six months in England on his way to America, and visited Collinson's garden on 10 June, 1748; so he may very well have seen the Kalmia before ever setting foot on the American continent.

For a long time the shrub remained rare in British gardens, owing to the difficulty of its propagation. Catesby had no success with either the plants or seeds he imported, until he got some plants from Collinson which had come from a more northerly part of America, "which climate being nearer to that of *England* than from

[1] Ruskin compares the inside of the flower to "bosses in hollow silver, beaten out apparently in each petal by the stamens of a hammer" [2] Quoted by E. G. Swem, *Brothers of the Spade,* 1949

whence mine came, some bunches of blossoms were produced in July 1740 . . . in my garden at Fulham." Miller and Collinson both agree that the only man who could raise kalmias from seed was James Gordon, a highly skilled nurseryman of Mile End, who was also successful with rhododendrons and azaleas. (We still find it a capricious shrub in cultivation, and even now its requirements are imperfectly understood.) The same difficulty was experienced in growing the smaller *K. angustifolia,* which was imported by Collinson two years after the first species. Miller, in 1759, reported that the only place where he had seen it thrive was in the gardens of the Duke of Argyll at Whitton near Hounslow, Greater London, Collinson, however, had it in bloom at Peckham, London, in 1743, and proudly records that a plant in his Mill Hill garden in 1767 had forty-five flowers and was "a surprising beauty."

Unfortunately, "the noxious qualities of this elegant plant [*K. latifolia*] lessens that esteem which its beauty claims";[1] it is a poisonous shrub, and deadly to sheep, which are liable to eat the leaves in spring when they are the only green things visible above the snow. Kalm reported that some calves which had browsed on it were restored only by the use of "gunpowder and other medicines." It is said that partridges that have eaten the berries become poisonous to humans, and this is made the basis of an incident in one of Maria Edgeworth's *Popular Tales* (1804). Even the nectar secreted by the flowers is suspect, and in districts where the kalmia is abundant, beekeepers take the precaution of feeding a sample of their honey to the dog before daring to allow it to the children or the public. The smaller *K. angustifolia* is equally toxic, as can be seen from its names of lamb-kill and poison-berry. The name of spoon-wood or spoon-tree was given to *K. latifolia,* because the Indians used to make spoons and trowels from the wood of the root, which is soft and easily worked when newly dug, but very hard and smooth when dry; Kalm brought home one such spoon, made by an Indian who had "killed many stags and other animals on the very spot where Philadelphia afterwards was built." The stem-wood is almost as close-grained as box, and was used by the early settlers for a number of useful purposes. The name of Calico Bush, now generally used, which seems so appropriate to the bunchy pink-and-white flowers, does not appear in the records of Collinson and Catesby—it would have appealed to the former, who was a linen-draper—and according to a correspondence in the *Gardener's Chronicle* (1930) this name is unknown, or nearly so, in America.

[1]Catesby

Linnaeus was ill in bed in June 1751, but was quickly restored to health when he heard that his favorite pupil had arrived back from America with a load of new plants: "He rose from his bed and forgot his troubles." Kalm was made professor of botany at the university of Abo—one of those that had contributed to the costs of his journey—and published an account of his trip "by intervals" on account of the expense; he had spent all his own savings, as well as every penny his government could scrape together, on his American venture. The

Kerria corchorus
(now *K. japonica*)
(JAPANESE ROSE)   S. Edwards (1810)

first volume of *El Resa til Norra Americana* appeared in 1753, the second in 1756, the third in 1761, and the fourth and last, though complete, was not in print at the time of Kalm's death in 1769: the still unpublished manuscript was destroyed by a fire in 1827. An English translation by John Forster of the part of the work concerning America was published in 1770, but Kalm's account of his visit to England was not translated till 1892.

NOTES: Our native Mountain Laurel, *Kalmia latifolia,* is still referred to as "ivy" in the Appalachian Mountains while *Rhododendron maximum* is called "the great laurel" and *Rhododendron catawbiense* the "purple laurel." The largest Mountain Laurel on record is located in the North Carolina Arboretum at Bent Creek, N.C. It is 58 feet in circumference with a trunk diameter of 18 inches and is 25 feet high.

The efforts of an American plant breeder, Dr. Richard Janes, have made the greatest impact on *Kalmia latifolia*. Through selective breeding over many years, the improvement from pink or white flowers to clones that display vivid red clusters is exceptional. Dr. Janes's 'Red Bud Strain', which includes these flower qualities, particularly bright red buds, now includes several superb clones, of which 'Sarah', named for his wife, is one of the best selections. Because the Mountain Laurel is difficult to propagate, the success with micropropagation techniques is also an essential element in the availability of these new clones.

Kerria

This is a monotypic genus, containing only a single species—*K. japonica*. Although known to Europeans from its description by Kaempfer in 1712 (under its Japanese name of *Jamma (Yama) Buki*), and from Thunberg's *Flora Japonica* in 1784, it was not introduced until 1805, when the double form was sent to Kew from China by William Kerr. As in the case of the hydrangea, the lack of the parts of fructification in the double flower made the plant difficult to classify, and Thunberg assigned it to the family Corchorus (allied to the lindens, and the source of jute), one of whose members, the yellow-flowered *C. olitorius*, was known as far back as 1640 as the Jew's Mallow, because its leaves were "a favorite salad with these people, and they boil and eat them with their meat."[1] This name got transferred most unsuitably, to the Kerria, with which it has no connection whatsoever, but which still appears as "Jew's Mallow" in books and catalogues today. The plant was allotted to its correct Order by de Candolle in 1817, and named by him after its introducer; his diagnosis was confirmed when the single form arrived in 1835, and it could be seen that it belonged unmistakably to the Rosaceae and was akin to the spiraeas and the brambles.

William Kerr was appointed by Sir Joseph Banks at Kew to collect plants in China and the Far East, at a salary of £100 a year; he arrived at Canton in 1804, and thence made expeditions to Java and the Philippines. Most of the Chinese plants he sent home were purchased at the Fa-Te nursery-gardens, near Canton; and he was

[1] Bryant, *Flora Diaetetica*, 1783

not very successful in shipping them, many arriving in a dead or dying condition. Nevertheless we owe to him the introduction of *Lilium tigrinum, Lonicera japonica, Nandina domestica, Paeonia suffruticosa* var. *papaveracea,* and *Rosa banksiae,* to name only a few. It was said that he was at first very active, but after three or four years became greatly changed. "He was then unable to prosecute his work, in consequence of some evil habits he had contracted, as unfortunate as they were new to him."[1] Yet in 1810 Banks wrote to him that in consideration of his good conduct as a plant-collector, he had been appointed Superintendent of the new Botanic Garden which was about to be established in Ceylon. Kerr took up the post, but died only four years later. Looking at the uncombed orange-yellow flowers of his namesake plant, one is tempted·to speculate whether Kerr was a Scot of the sandy, shaggy-haired type; but of this there is no evidence.

Once introduced, the double kerria spread rapidly. By 1810, though only five years later and still cautiously being cultivated as a greenhouse or even a "stove" shrub, it was already "in most of the principal collections abut London,"[2] and by 1838 it was "so common as to be found in the gardens of even labourers' cottages."[3] Three years earlier the single form had been sent to the Horticultural Society by John Russell Reeves, the son of the John Reeves who did so much to facilitate the importation of Chinese plants during his long residence at Canton. It arrived in bloom, according to Loudon, in the spring of 1835, and was planted at once in the open air (without the usual probation period under glass), with the happiest results. Since then little attention has been paid to it, though its elegance and long-flowering season earned it an A.G.M. in 1928; and it is still rather undervalued in gardens. Forms with gold- and silver-variegated leaves, the latter less sturdy than the type, were grown towards the end of the century. There was a mild flutter of excitement among botanists in the late 1860s when it was thought that the newly introduced variegated kerria might contradict the theory just put forward by Professor Morren, that double flowers and variegated foliage never occurred together—the first condition arising from strength and the second from weakness; but in spite of a misleading picture in *L'Illustration Horticole* the variegated kerria proved on better acquaintance to have single flowers, and upheld the theory. Apart from these, there are no garden varieties. There is little chance of improvement in monotypic genus, where there are no other species to share in the making of new hybrids;

[1] *The Chinese Repository*, 1834; quoted by Bretschneider
[2] *The Botanical Magazine*, 1810 [3] Loudon

moreover, vegetative propagation of the kerria is easy, and in the case of the double and variegated forms, essential, so the plant is rarely raised from seed. A closely allied family, the white-flowered *Rhodotypos kerrioides* might possibly provide a bigeneric cross.

Although long cultivated in Japan, the kerria is a native of China; it was found wild in Hupeh by Wilson in 1900, and earlier by Augustine Henry. The pith is said to be used in Japan to make tiny replicas of buds and flowers, for floating in cups of saké.

NOTES: There have been few changes in the development of Kerria as a garden plant except in seeking to create larger flowers. It should be mentioned, however, that while Miss Coats described *Kerria japonica* only as a Chinese plant, we now know that it is listed as occurring from Kyushu to Hokkaido in Ohwi's *Flora of Japan*. Regardless of its origin it is a fine ornamental shrub.

Kolkwitzia

We are very lucky to have this beautiful shrub, *K. amabilis*; I believe it has only twice been found in the wild, once by an Italian missionary called Guiseppe Giraldi, who worked in Shensi from 1890 to 1895, and once by Wilson in the mountains of northern Hupeh, when he was collecting for Messrs. Veitch in 1901. In both cases the bush was in seed when it was discovered, and the flower was unknown until plants raised from Wilson's seed bloomed in Veitch's nursery at Coombe Wood in 1910. These flowers are such that the Americans have named the shrub "Beauty-Bush." It seems a pity that the German botanist who named Giraldi's specimens did not choose some other plant to commemorate Professor R. Kolkwitz of Berlin; a name might then have been found for the present subject more in keeping with its epithet *amabilis*, meaning lovable. Perhaps it is due to its first name that the shrub is not better known. It is perfectly hardy, but blooms best after a hot summer, and becomes more prolific as it becomes mature.

Kolkwitzia is a monotypic genus, allied to the Abelias.

NOTES: Introduced from China by the famous plant explorer, E. H. Wilson, this large deciduous shrub is commonly called Beautybush in America. It has pink flowers and is hardy in most northern states. It received immediate acclaim after its introduction in 1901 and has remained a reliable plant since. Because plants raised from seed often produce washed-out pink flowers, selected forms raised from cuttings are better. The selection 'Pink Cloud' has superior color.

Laurus

The Gods, *that mortal Beauty chase,*
Still in a Tree did end their race—
Apollo *hunted* Daphne *so*
Only that She might Laurel grow.
And Pan *did after* Syrinx *speed,*
Not as a Nymph, but for a Reed.
The Garden, Andrew Marvell, 1681

This is the true and only laurel, into which the daughter of Peneus was transformed—the Bay Laurel or Sweet Bay, surnamed *nobilis* by Linnaeus because of its many exalted uses; the type and original whose name has been usurped by so many inferior and unrelated plants, which resemble it only in the possession of an oval, evergreen leaf. Gardeners should look rather more carefully to their laurels; the botanists have already done so, for a great many species that were formerly classified under Laurus have been separated, and allotted to different, though closely allied families. These near-relations include the tropical plants that yield camphor, cinnamon, cassia and the avocado pear, and the American spice-bush and sassafras-tree. One can see where the bay gets its fragrance; the miracle is that it manages unfailingly to produce it, even in alien soils and climates.

After the famous metamorphosis, Apollo, whether contrite or merely frustrated (Lyte said he was "much astonied", and well he might be) made the scented bay his sacred tree; and it was thereafter associated in many ways with the Sun-God and his worship. Poets were entitled to wear it, because poetry is "a kinde of prophesie or soothsaying, which Apollo governeth and ruleth,"[1] and so were conquerors; but it would be interesting to know how victorious generals managed to send their dispatches to the Senate, as they were said to do, wrapped in its wiry and intractable leaves. Already in Roman times the plant was clustered with superstitions, and its untimely withering was regarded as a sure presage of disaster. It was believed to afford protection against thunder and lightning, and in a bad storm, the Emperor Tiberius was accustomed to cover his head with laurel-branches and retreat under his bed. Similar beliefs still lingered in seventeenth-century England: ". . . neither witch nor devil, thunder or lightning, will hurt a man in a place where a Bay tree is."[2]

The date of introduction of *L. nobilis* to this country is unknown; it probably arrived very early, although one cannot perhaps rely on Chaucer's statement that it was

[1]Lyte [2]Culpeper

used to garland the Knights of the Round Table. Turner, in 1548, said that it was common in gardens in the southern part of England; its leaves he later described as "blackishe grene, namely when they are olde. They are curled about the edges, they smell well. And when they are casten into the fyre, they crake wonderfully. The tre in England is no great tre, but it thryveth there many partes better and is lustier then in Germany . . ."

specimens he ever saw, some 28 feet high, were growing under tall Scotch firs in Windsor Great Park, "immediately adjoining that barrenest of all spots, Bagshot Heath." If given liberty to do so, the bay will both flower and fruit, even in relatively cold districts; but usually it is kept too closely pruned. Six varieties, with broader, narrower, curled or variegated leaves, were in cultivation before 1865.

Kolkwitzia amabilis

(BEAUTY-BUSH)   M. Smith (1914)

Laurus nobilis

(LAUREL)   Ferdinand Bauer (1823)

John Evelyn, a hundred years later, gave high praise to what he called case-standards, imported from Flanders, "with stems so even and upright, heads so round, full and flourishing . . . that one tree of them has been sold for more than twenty pounds . . . I wonder we plant not whole groves of them, and abroad; they being hardy enough, grow upright, and would make a noble Daphneon." (At such prices, one can hardly share his wonder.) Miller said that "headed plants" or standards, such as Evelyn described, were best raised from seed, as they were then more vigorous and less inclined to produce the side-shoots that convert most specimens into large bushes rather than small trees. Given some shelter, the bay does not object to a little shade, and actually prefers a poor sandy soil; William Cobbett said that the largest

These leaves, according to Parkinson, "serve both for pleasure and profit, both for ornament and for use, both for honest Civill uses and for Physicke, yea both for the sicke and for the sound, both for the living and the dead: And so much might be said of this one tree, that if it were all told, would as well weary the Reader, as the Relater: . . . from the cradle to the grave we have still use of it, we have still need of it." Almost every old writer gives a different medical use. Dioscorides, modestly enough, recommends the laurel for bee and wasp-stings; Pliny, for a cough; Turner, to "weish out frekles" and for medicines "which refreshe them that are werye or tyred"; Bulleine says that it is "good for a cold liver, dronk in strong wine," whereas Gerard prescribes it against drunkenness; Parkinson uses it for "cold griefes in the joyntes," Culpeper to "settle the pallate of the mouth into place" and Evelyn for ague. Sometimes the leaves and sometimes the berries were employed; Woodville said that the latter were imported from "The Streights" (Gibraltar)—though by that time (1793) they were no longer much in demand. Evelyn also mentions that "walking-staves, straight, strong and light, for old Gentlemen" were made from the wood.

All these uses are now forgotten; the only purpose for which the bay is still generally employed is for its flavor in cookery. In 1634 it was described as "a notable smell-feast, and is so good a fellow in them, that there is almost no feast without him";[1] and a bay-leaf is still an essential part of a *bouquet garni*, as all good cooks know. (It should not be confused with the Cherry Laurel, a leaf of which will impart an almond flavor, but which is poisonous if used in any quantity, whereas the true laurel is not only harmless, but beneficial.) In Cornwall it is known as "Fish-Tree" from its use in former days in the pickling of sardines and pilchards, and one of its French names is *laurier du jambon*, Ham-Laurel. Bay-trees in tubs are often to be seen at the entrances of smart hotels and restaurants, but, as Mr. Jason Hill points out, they might more usefully be placed beside the *kitchen* door. As to the use of the bay as a perfume, it seems doubtful whether either bay or rum is used in the production of Bay Rum—a popular hair-lotion first produced in the saloons of New York in the mid-nineteenth century; actually the plant employed is *Pimenta acris*, the West Indian "bayberry."

The Greek name of the bay-tree was *daphne*, the Latin *laurus*, which is variously interpreted as being derived from the Celtic *blaur*, green, or the Latin *laus*, praise. "Bay" is indirectly derived from the Latin *bacca*, a berry, and was originally applied only to the fruit—"The frute of Lauri tre ben clepid baies".[2] It was transferred to the tree itself before 1530—possibly another indication of the antiquity of its cultivation in Britain. In classical times, it was the custom to crown young doctors in physic who had succeeded in their examinations, with wreaths of this laurel in berry (as the plant of Apollo's son Aesculapius, the healer), and from this we get our baccalaureates, and thence—by an association of ideas not always in accordance with the facts—the word "bachelor." Some of the more reprehensible bachelors may indeed spread themselves like their namesake shrub, for the "green bay-tree" to which the wicked are compared in Psalm 37 (verses 35 and 36) is believed actually to be the evergreen

laurel, which occurs, though not abundantly, in Palestine. It is not often that we have two Christian names derived from, or associated with, a single tree. "Daphne" has only become popular during the present century; before that, according to Charlotte M. Young,[1] it was used exclusively for dogs. Laurence (and its feminine, Laura), comes from the Latin *Laurentius*, meaning "of Laurentium"—the town which got its name, and its reputation as a health-resort, from the abundance of its laurel-trees.

NOTES: Except in the warmest localities, this historic plant is mostly grown for its foliage in tubs in conservatories because it is so tender.

Leycesteria

L. *formosa*, the Flowering Nutmeg or Elisha's Tears, is a woodland shrub from the Nepal area of the Himalayas, where it is called *Nulkuroo*. Loudon says it was introduced in 1824 and flowered soon afterwards in the nursery of Allen and Rogers in Battersea; and it was grown about the same time in the gardens of the Horticultural Society from seed procured from India by Dr. J. F. Royle. It proved a disappointment to those whose expectations had been raised by a rather highly colored plate of it in Wallich's *Plantae Asiaticae Rariores* (1830–32). "Its leaves are a pale dull green," complained Lindley in the *Botanical Register* (1839), "it has a rambling inelegant mode of growth, and the color of the bracts is not at all brighter than what is represented in the accompanying plate" (a dull crimson). It was expected to be tender and to require "a conservative wall," but it was soon found to prefer a cool position, and Lindley makes the ambiguous statement that "if grown in the shade, it is most likely to be a beautiful object." Its hardiness and ease of propagation have led to it being regarded as a background-shrub, and its peculiar merits are rather overlooked. The design of its hanging racemes is unusual and striking; its berries have a pleasant taste which Mr. Jason Hill compares to "a soft raisin flavored with caramel"; and the blue-green bark of its shoots has a fine effect in winter. Birds, especially pheasants,[2] are fond of the berries, and the shrub is sometimes planted for game-coverts.

Only one other species is in cultivation—L. *crocothyrsos*, which has handsome yellow flowers, but unfortunately is tender, and cannot be grown out-of-doors except in the mildest districts. This is a species of great botanical inter-

[1] *A Strange Metamorphosis of Man*, quoted by Brand, *Popular Antiquities* [2] *De Proprietatibus Rerum*, Bartholomeus Anglicus, trans. Trevisa, 1398

[1] *A History of Christian Names*, 1884 [2] Some of the ancestors of our pheasants may have come from the same region as the shrub

est. Seed was collected by Kingdon-Ward in 1928 from a single plant which he found growing in the Delei Valley in Upper Assam—the only specimen that has ever been discovered; fortunately the seed germinated well, and the plant became established in cultivation, for it is quite possibly now extinct in the wild. It is the most primitive form of the genus, and may have been dying out when so dramatically given a new lease on life by the chance find of a plant collector. The genus itself is believed to be the basic or type-form of the Order to which it belongs, the Caprifoliaceae (which includes Lonicera, Diervilla and Symphoricarpos); "certain species of Leycesteria clearly foreshadow some of the lines along which other genera have developed."[1]

Dr. Nathaniel Wallich named the family after his "highly esteemed friend William Leycester Esq., Chief Judge of the principal Native Court, under the Bengal Presidency, who, during a long series of years, and in various parts of Hindustan, pursued every branch of horticulture with munificence, zeal and success which abundantly entitle him to that distinction."[2] Wallich was Danish by birth and perhaps could hardly be expected to foresee that within fifty years or so the name of the worthy Justice would be corrupted into "Elisha's tears"—which yet seems strangely to suit the plant, with its pendent white flowers and its persistent bracts which darken to a somber blood-red as the season advances.

NOTES: These two species are rare in cultivation and are not readily available from most American nurseries.

Ligustrina, see under *Syringa*

Ligustrum

"The meanest of all mean shrubs"—so Robinson called *L. vulgare,* the Common Privet; only popular, he said, because "its weed-like facility of increase" made it "dear to those to whom something growing with a fungus-like rapidity is a treasure . . . The commoner sorts have no beauty whatever, and they all have the same vile odour in summer days when they flower—a sickly smell." He continues for a column and a half to lament its universal employment in gardens and for hedges, to the exclusion of many better plants; and while admitting that it will stand any amount of ill treatment and will grow where nothing else would thrive, he insists that it is "not worth having anywhere."

At a time, however, when the choice of garden shrubs

was much more restricted than it is now, privet was highly regarded, especially for topiary work. Parkinson in 1629 said it was much used for hedges and arbors, "whereunto it is so apt, that no other can be like unto it, to bee cut, lead, and drawne into what forme one will, either of beasts, birds, or men armed, or otherwise; I could not forget it, although it be so well knowne unto all . . . ," and it was praised in much the same style by Culpeper in 1653. By 1675 it was beginning to lose ground—"Privet is a plant that has been in great request for adorning Walks and Arbours, till of late other new and more acceptable Plants by degrees begin to extirpate it out of the most modish Plantations,"[1] but it continued to have occasional apologists. Curtis noted its ability to withstand city smoke, "so that whoever has a little garden in such places, and is desirous of having a few shrubs that look green and healthy, may be gratified in the Privet";[2] Cobbett (1833) said that a low privet hedge "when white and red roses are planted with it, makes as pretty a fence as can be conceived," and Loudon (1838) suggested that trained as a standard, it would make an attractive small tree, whose black fruit would hang on all winter. By 1728, the common form was becoming superseded in gardens by the "Italian Green Privet" (var. *sempervirens*) which was more reliably evergreen; it was introduced, according to Batty Langley, by "the ingenious Mr. *Balle* of *Kingsington,*" and this in turn was superseded by the Japanese *L. ovalifolium,* introduced in 1842, and its variety *aureum,* the golden privet—both good hedge-plants, retaining their leaves well in winter, but inferior in flower and fruit to our own species.

Miller makes the surprising statement that the privet has only fibrous roots, and *therefore* robs the ground less than any other shrub; most of us find the contrary to be the case, and we are so repelled by the ubiquity, greed and dullness of the common privet, and still more of *L. ovalifolium,* that we neglect a number of species that have been introduced from time to time, and which are handsome both in flower and fruit. These include the deciduous *L. sinensis,* introduced from China by Robert Fortune in 1852; *L. quihoui* (1862) also from China, whose oriental-sounding name actually commemorates a superintendent of the Jardins d'Acclimation in Paris, and *L. japonicum* var. *rotundifolium* (syn. *L. coriaceum,* 1860) a dwarf evergreen species from Japan. Most impressive of all is *L. lucidum,* a tall-growing kind introduced by Sir Joseph Banks about 1794; it is said that a good specimen in bloom is "one of the most striking autumn garden pictures."[3] In its native China it grows into an evergreen tree sometimes as much as 60 feet in height, and is of

[1] Hill, *The Botanical Magazine,* 1934 [2] *The Botanical Magazine,* 1839

[1] Worlidge, *Systema Agriculturae* [2] *Flora Londinensis,* 1777 [3] Bean

considerable economic importance as one of the two host plants (the other being an ash, *Fraxinus chinensis*) of the wax-producing scale-insect, *coccus pela*, which is cultivated in China with almost as much assiduity as the silkworm. At first the function of this insect was not understood in Europe, and it was thought that in some way the tree itself produced the "vegetable" wax.

The English privet also has its associated insect; the leaves support the caterpillars of the Privet Hawk Moth (*sphinx ligustri*) an ornamental but not a serviceable creature. The shrub had several medical uses, but none of great importance; "The Moon is lady of this," said Culpeper, "It is little used in Physic with us in these times, more than in lotions, to wash sores and sore mouths, and to cool inflammations, and dry up fluxes." The berries could be used for dyeing,[1] for tinting maps and prints, or for coloring wines, "but for these several purposes," said Curtis, "there are much better materials in common use." On the whole the privet is more valuable to birds and insects than to man; bees work the flowers eagerly, and will sometimes even desert the limes if the privet[2] blooms before the lime is over—to the great disgust of the beekeeper, as privet-nectar is of a strong bitter taste and a dark color, and will spoil any honey with which it is mixed. The berries, as Gerard noted, are "a pleasant meate in winter for owsels, thrushes and divers other birdes"—particularly the bullfinch. Eighteenth-century bird-catchers used large branches of the privet in fruit, with the top twigs limed, and a call-bird underneath, to attract their bullfinch prey; and it is still said that the best way of stopping the depredations of these lovely wretches is to use a trap-cage baited with privet-berries.

One authority derives the shrub's old English name of Prim-print "from its regular appearance when cut;" Loudon, influenced perhaps by Tusser's instructions to "Set Privie or Prim," thought its name might have originated "from its being frequently planted in gardens to conceal privies." In the fifteenth century there seems to have been some confusion between the names of the privet and the primrose, and several vocabularies of the period have entries such as "hoc ligustrum, a Primerose." Canon Ellacombe took this to mean that the privet was called "primrose," and derived Gerard's Prim-print from *primé-printempts*; but as the privet is not particularly early in leaf and very late in flower, this interpretation seems unconvincing. Turner in his *Libellus*, 1538, takes particular pains to refute the error; "*Ligustrum arbor est, non herba ut literatoru vulgus credit; nihil que minus est quam*

[1] In the nineteenth century they were used for dyeing black kid gloves   [2] *L. ovalifolium—L. vulgare*, flowers earlier

a Prymerose."[1] Miller identified the *ligustrum* of Virgil's second Eclogue, used as a warning to youth not to trust to its beauty, with the Italian Privet, "for here are the white Flowers of the Privet appearing early in the Spring, which is an Allusion to Youth; but these are of short Duration, soon falling away; whereas the Berries, which may be applied to mature Age, are of long Continuance, and are gathered for Use." Unfortunately the privet, as already remarked, does not flower until June; nor would it have occurred to us to use the whiteness of the flowers as a standard of comparison, as did another Italian poet:

> Amaryllis, yet more fair
> More white than whitest privets are,
> But than the cruel aspic still
> More cruel, wild, and terrible.[2]

NOTES: The privets are everyone's utilitarian plant and there are several extremely hardy species not included by Miss Coats. For example, *L. amurense*, Amur River-North, and *L. obtusifolium*, the Regel Privet, are among the hardiest of hedge plants. In the South, *L. japonicum*, the Japanese privet, is a widely used evergreen. More and more the cultivar 'Rotundifolium' is seen because it forms a compact upright plant with curly leaves that are unusually densely arranged.

The privets are ironclad in hardiness and tolerate dryness as well as poor soils. They make excellent utilitarian hedges and respond well to shearing.

Lippia, see under *Aloysia*

Lonicera

The honeysuckles are a much more decided family than the jasmines; the shrubs among them are definitely shrubs, and the climbers climbers, so much so that some botanists divide the genus into two sections, the climbing Periclymenums and the bushy Xylosteums or Chamaecerasus. (A small third section shares some of the characteristics of both groups.) Five-sixths of the members of the large family are shrubs, and less well known to gardeners than the more handsome climbing species, all of which are said to be worthy of cultivation. Among the climbers, only *L. periclymenum*, the common honeysuckle, is regarded as being a true

[1] "Privet is a tree and not a herbaceous plant, as most writers suppose, and there is nothing which has less resemblance to a primrose."   [2] Guarini, *Il Pastor Fido*, 1590

Leycesteria formosa

(FLOWERING NUTMEG)   Gorachand (1831)

Lonicera periclymenum

(HONEYSUCKLE)   *Flora Lordinensis* (1775)

Ligustrum vulgare

(COMMON PRIVET)   J. Sowerby (1798)

native of Britain, and doubts have been cast even on this, though it is one of our most familiar wild-flowers. Richard Jeffries points out that its leaves always unfold far too early, and that whether brought over by "Richard Conqueror" (sic) or the Romans, "it still imagines itself ten degrees further south, so that some time seems necessary to teach a plant the alphabet."[1] If it was an introduction, it must have been a very early one; the Earl of Lincoln had honeysuckles in his Holborn garden (later to become Lincoln's Inn) before 1286; Chaucer refers several times to the woodbine, and in 1548 Turner said it was "commune in every wodde." Like most of the climbing honeysuckles, it was designed for pollination by the long-tongued hawk-moths, and the scent is therefore more freely produced at night or early in the morning, when the moths are on the wing. Not all the species are scented, however; unfortunately some of the showiest have no perfume.

One of the most delightful of the many descriptions of the common honeysuckle comes from *Bullein's Bulwarke of Defence*, 1562: "Oh how sweet and pleasuante is *Woodbinde* in Woodes or Arbours, after a tender soft rain; and how friendly doe this herbe if I maie so name it, imbrace the bodies, armes and branches of trees, with his long windyng stalkes, and tender leaves, openying or spreding forthe his swete Lillis, like ladies fingers emong the thornes or bushes . . ." Unfortunately the friendly embrace quickly becomes a stranglehold, and the pretty honeysuckle probably does a great deal more harm than the ivy, which is given such a bad name as the villain of the woodlands. A slender support is preferred, but Darwin says the plant can climb a young beech nearly 4½ inches in diameter. Young trees are sometimes so tightly clasped that the spiral of the honeysuckle becomes actually embedded in the wood, and "handsome twisted walking-sticks are thus formed."[2] Dr. Johnson once made a pithy remark about "honeysuckle wives" who "are but creepers at best, and commonly destroy the tree they so tenderly cling about."

The honeysuckle, like the jasmine, presents a troublesome problem to the gardener, as its native untidiness is difficult to curb, and it is also subject to infestation by aphides. "Although it be very sweete," said Parkinson, "yet doe I not bring it into my garden, but let it rest in his owne place, to serve their senses that travell by it, or have no garden." The Elizabethans used it to cover arbors, such as the "pleached bower" in *Much Ado About Nothing* (1600) "where honeysuckles, ripened by the sun, Forbid the sun to enter"; but later the plant was forced to adapt its rambling habits to suit the formal gardens

fashionable in the late seventeenth century, and was even trained into standards. Fortunately by that time the Early Dutch or Belgian honeysuckle (var. *belgica*) had been imported, which was of a relatively compact habit and more amenable to training than the native type. No details of its introduction are recorded; it may have accompanied the Late German or Dutch honeysuckle (var. *serotina*) which is said to have been introduced before 1715 by the Flemish florists. Both of these ancient varieties are still popular and easily obtainable; but others mentioned in the seventeenth and eighteenth centuries—Hanmer's "little Red Woodbind, called by some the Spanish Woodbind," Parkinson's Double, and strong-growing Red German; the Oak-leaved, the Midsummer, and the Long-blowing—seem to have disappeared from cultivation.

Many other climbing species are grown, some of them of considerable antiquity; the European *L. caprifolium* for example, now naturalized in several localities, and described by Gerard as having "faire, beautifull and well-smelling flowers, shining with a whitish purple colour, and somewhat dasht with yellow, by little and little stretched out like the nose of an Elephant . . . and when the flowers are in their flourishing, the leaves and flowers do resemble sawcers[1] filled with the flowers of woodbinde; many times it falleth out, that there is to be found three or four sawcers one above another filled with flowers as the first." This formation, with the heads of flowers gradually decreasing in size, earned it the delightful name (mentioned by Evelyn) of "Tops and Topgallants." *L.* × *italica* is thought to be a natural hybrid between *L. caprifolium* and the Mediterranean *L. etrusca*;[2] it was introduced about 1730, and at some early date was taken to America, where it escaped and became naturalized, so that some botanists, receiving it from there, were deluded into naming it *L. americana*. It is one of the handsomest of scented climbing honeysuckles, and was awarded a belated A.G.M. in 1955. Very different in appearance, but also very sweetly scented, is *L. japonica*, which belongs to the small third section of climbers that produce their flowers, like those of the shrubby kinds, in pairs in the leaf-axils. Although first discovered in Japan, this species was introduced from China, being sent from Canton by William Kerr to the Court of Directors of the East India Company (via the *Hope*, Captain Prendergass) in 1806. In China it was much esteemed for its beauty and fragrance, and was called "Gold and Silver Flowers" because of the way the pale blossoms deepen in color as they age—a characteristic of the hon-

---

[1] *Field and Hedgerow*, 1889   [2] Johns, *Flowers of the field*, 1853

[1] The paired leaves of *L. caprifolium* are joined at the bases, forming a single receptacle   [2] Introduced in 1750

eysuckle family, as of the related Diervillas. Almost better known than the type is its golden variegated form, (*L. japonica* var. *aureo-reticulata*) which was introduced from Japan by Robert Fortune (*c*.1860); it does not flower freely, but has been described as "exquisitely rich in its foliage, and well worthy of attentive observation."[1]

Gay and gaudy, but quite scentless, are two foreign species, widely separated in space and time. The North American Trumpet Honeysuckle, *L. sempervirens*, was grown by the younger Tradescant before 1656, and though rather tender, has kept a place in our gardens and greenhouses ever since, on account of its brilliant scarlet and yellow tubular flowers. The Chinese *L. tragophylla* is a relative newcomer, introduced by E. H. Wilson in 1900; it is hardy, and bears very large, bright yellow blossoms. Chinese plant and American plant met and were mated in the garden of the Royal Hungarian Horticultural Society in Budapest, to produce the hybrid *L. × tellmanniana*, which was distributed by the German firm of Spath in 1927. It is colorful and hardy, and so is another hybrid in the same group, *L. × brownii* (*L. sempervirens × L. hirsuta*), but none of these showy honeysuckles has any perfume, and a honeysuckle without a scent is like a man without a shadow.

Although few of the shrubby members of the family are in the first rank as garden plants, a hunt through nurserymen's catalogues reveals that a surprising number are in cultivation; nearly thirty species and varieties of bush-honeysuckles are to be had. Gerard grew two in 1597: *L. alpigena*, the Cherry Woodbine, and the rare native *L. xylosteum*, the Upright Fly Honeysuckle. Parkinson in 1640 had two more, *L. coerulea* and a kind with black berries, probably *L. nigra*—all European, and not particularly ornamental. More important are two species that produce their small, creamy, scented flowers in the winter—*L. fragrantissima*[2] and *L. standishii*, both introduced from Chinese gardens by Robert Fortune in 1845. They are much alike but *standishii*, named by Sir William Jackson Hooker "in honour of the active and intelligent nurseryman to whom many of Fortune's rich Chinese collections were consigned"[3] flowers a fortnight or three weeks earlier than *fragrantissima*, is deciduous instead of semi-evergreen and has a stiffer habit of growth; a hybrid between the two, *L. × purpusii*, which appeared as a chance seedling in the Darmstadt Botanic Gardens in the 1920s, is said to be superior to either plant.

Of the spring and summer-flowering bush-honeysuckles, the best are probably *L. tatarica* and *L. syringantha*. The former was described in 1815 as "one of the most desireable shrubs we know of," making a close round bush "feathering on all sides down to the ground"[1] and covered with flowers in mid-April. It is a native of Russia and Siberia, and was grown by Miller in 1752 from seeds sent from the Imperial Garden at St. Petersburg, "to which they were conveyed from Tartary." By 1838 it was very common in British gardens, and in 1914 Bean mentioned that it was running half-wild in some places. It is an extremely hardy plant, and a very variable one; Rehder describes twenty-five varieties in his monograph on the genus, and says that it also hybridizes very freely with other species. A deep red form known as *punicea*, and a white, *alba*, are among those still available. The Chinese *L. syringantha* (the Lilac Honeysuckle) also reached us by way of Russia; seeds of it were sent by the explorer Przewalski to the St. Petersburg Botanic Garden, where it bloomed in 1889, and a plant seems to have been sent to Kew the following year. Long before Farrer set foot on the Tibetan marches, he had seen a specimen of this shrub trained on a wall in Ireland, and was so charmed by it that he afterwards searched for it in all the leading nurseries in Europe, for many years in vain. (It was rare in cultivation, for it is difficult to propagate except from seed, which is seldom produced in captivity.) Great was his surprise and delight to discover it in 1914, growing freely on the cold mountain ranges of north-eastern Tibet, with its "innumerable bunches of little crystalline pearl-pink stars of the most ravishing fragrance, scenting all the air, and turning at last to knots of ruby-scarlet fruits."[2]

Another Chinese shrubby honeysuckle, *L. nitida*, has become familiar to us as an evergreen hedging plant, although it was introduced only in 1908, by E. H. Wilson from west Szechwan. It rapidly became popular—some think over-popular, for it has its defects as a hedge plant, requiring frequent clipping, and being miserably resentful of an industrial atmosphere. The original type, distributed by Messrs. Veitch of Exeter, rarely, if ever, flowers and never fruits; but plants raised from a later consignment of seed collected by Forrest in Yunnan flower and fruit freely. The flowers are inconspicuous, but attractive to bees; the berries are ornamental.

The fruits of the bush-honeysuckles are botanically peculiar, in that two flowers unite to produce one berry, or in a few cases a double berry joined like Siamese twins. Some, at least, of the species have edible fruit. Farrer mentioned a Tibetan kind whose berries, ignored by the natives, were eagerly gathered by American missionaries, who found them delicious in pies, jams and jellies, and he prophesied "a most important future for

---

[1] Hibberd   [2] Not known in the wild   [3] Bretschneider

[1] *The Botanical Register*, 1815   [2] *On the Eaves of the World*, 1917

this new fruit, which can hardly fail to be a treasure for temperate climates."[1] (Unfortunately he did not give the species a name, only a number, and I do not know whether it is now in cultivation.) Anne Pratt said that in Kamchatka the berries of *L. caerulea* formed a favorite article of food, "not inferior to the finest cherries"; those of *L. tatarica* were also said to be eaten by the local (Russian) inhabitants "although disgustingly bitter, and not entirely innoxious."[2] Duhamel, on the other hand, warns us that the fruits of the European bush-honeysuckles are purgative and vomitory, and says that children should be prevented from eating them.

L. tatarica and a few other species were also notable as host plants for the Cantharides or Spanish Blister Fly—a small iridescent beetle (*Meloe vesicatorius*), an extract of which was formerly much employed externally as a rubefacient or counter-irritant, and internally—and usually illegitimately—as an aphrodisiac, whose use was excessively dangerous and frequently fatal. Various parts of the common climbing honeysuckle were used in medicine in Tudor times—Turner recommends the seed drunk in wine for "ye hitchcoughe or yisking"—and by the mid-seventeenth century a decoction of leaves and flowers had a firmly established but quite unwarranted reputation as a mouthwash or gargle; Culpeper complained that "Dr. Tradition . . . hath so grounded it in the brains of the vulgar, that you cannot beat it out with a beetle." A conserve made from the flowers was employed both as a remedy and as a cosmetic. But most of the uses of the plant were long ago discarded; only the traditionally minded dormouse continues, whenever possible, to weave his winter nest of honeysuckle bark.

> *"Why* Lonicera *wilt thou name they child?"*
> *I asked the gardener's wife, in accent mild:*
> *"We have a right," replied the sturdy dame,*
> *And* Lonicera *was the infant's name . . .*[3]

and evidently with the accent correctly placed on the third syllable. The plant was christened in honor of Adam Lonicer or Lonitzer (1528–86), a German physician who practiced in Frankfurt, and who produced among other works a book on natural history, the second part of which, published in 1555, contained much curious information about plants. *Periclymenum* comes from two words meaning "to roll round about"; *caprifolium*, or Goat-leaf (*Chevrefeuille* in French) probably received its name rather because these animals find the foliage palatable, than because (as has been suggested) plant and animal share a propensity to climb. Both of

[1] *Journal of the Royal Horticultural Society,* October, 1916 [2] *The Botanical Register* 1815 [3] Crabbe, *The Parish Register,* 1807

the common English names were originally applied to other flowers; "hunnisuccles" in the thirteenth century meant clover or trefoil, and "uuidubindae" was used from the ninth century onwards for several climbing plants, principally convolvulus and ivy. It is probably convolvulus that is intended in the familiar Shakespeare passage—"So doth the woodbine the sweet honeysuckle Gently entwine . . ." At an early stage the plant seems to have been called by the one name and the flower by the other; Murray gives several references to "woodbyne, whiche beareth the honeysuckle." An old Latin name was *mater silvarum,* Mother of the Woods, and English local names include honey-bind, suckle-bush and lily-among-thorns.

NOTES: The honeysuckles—shrubs with bright red or orange fruit and several vine species, including the ubiquitous weed, *L. japonica*—are dependable garden plants that are particularly favored in our coldest regions. *L. fragrantissima* is one of the most common in the South while *L. tatarica,* with several selections, is popular in the North.

A rather uncommon evergreen species in the United States is *L. nitida,* the box honeysuckle. It is used in the Pacific Northwest for low-trimmed hedges and can serve equally well in the Southeast.

Miss Coats does not mention *L. caerulea,* a native of the northern Pacific region of Asia. In Japan and the Soviet Union the sweet, dark blue fruits are a favorite choice for making preserves. I collected seed in Japan in 1961 and plants were subsequently evaluated, but this species has not yet attracted much attention here.

Lycium

The Lyciums or Box-thorns are rather unimportant members of the sinister Solanum Order, and are not remarkable for use or beauty. Their botany seems to have been much confused. Four of the species in particular have names that were almost interchangeable—*barbarum, chinense, europaeum* and *halimifolium*; the horticultural writers seem to select one of the names almost at random for whichever plant they happen to be describing, and each of them has at one time or another been listed as a variety of one of the others. (All four in turn have been coupled with the English name of the Duke of Argyll's Tea-Tree.) There are actually four distinct plants, but from the horticultural point of view *L. barbarum* and *L. europaeum* can be dismissed, as the R.H.S. *Dictionary* says that the true species are probably not in cultivation, and both are said to be tender. Cambridge Botanic Garden has specimens of the remaining two, *L. chinense* and *L. halimifolium*, obligingly planted side by side, so that their differences may be seen; the former is more robust and leafy, and is said to have larger fruits, but the latter is a better garden plant with its neat greyish leaf and its profuse, if rather dingy, purple flowers. Both are frequently found naturalized, especially in southern coastal districts, appearing, probably

Beaufort—though none of the family really merits such exalted patronage. Only Loudon shows any enthusiasm for *L. halimifolium*. "Trained to a strong iron rod," he says, "to a height of twenty feet or thirty feet, and then allowed to spread over an umbrella head, it would make a splendid bower.[1] Its shoots would hang down to the ground, and form a complete screen on every side, ornamented from top to bottom with ripe fruit, which is

Lycium europaeum

(BOX THORN)   Ferdinand Bauer (1819)

Magnolia altissima (now *M. grandiflora*)

(MAGNOLIA)   G. D. Ehret (n.d.)

bird-borne, on the tops of old walls and on cliffs.

It should on the face of it be *L. chinense* which is entitled to the name of the Duke of Argyll's Tea-tree; Collinson noted that in 1752 "my honoured friend the Duke of Argyll, presented me with the curious trees and shrubs undermentioned, from his garden at Whitton, on Hounslow Heath," including "One China purple-flowered Lycium, sent from China to the Duke for the Tea-tree." Accounts vary as to whether this was an accident due to the mixing of the labels on the way, or an intentional subterfuge by the wily Chinese. Other sources of introduction are given, but in view of the confused nomenclature one can only say with certainty that one Lycium was grown in the Royal Gardens of St. James in 1698, and two in 1709 and 1712 by the Duchess of

bright scarlet or yellow, with unripe fruit which is of a lurid purple; or with blossoms, which are purple and white. . . . If it were required to open the sides of a bower covered with this plant, the shoots could be tied together so as to form columns, at regular distances all round; but they must be untied an hour or two afterwards, to prevent the shoots in the interior of the column being heated so as to cause them to drop their leaves and fruit."

Lycium, "an ancient name of no meaning,"[2] was used by Dioscorides for a thorny shrub, now thought to be *Berberis lycium*; it may have been derived from Lycia in Asia Minor. There is a large number of species in the genus, many of them tender plants from the Cape of

[1]Splendid indeed—as high as a house   [2] Johnson

Good Hope; but few are in general cultivation.

NOTES: Although *L. chinense* is hardy into New England and can be effective as a thorny groundcover for banks, it is not generally considered sufficiently ornamental to be cultivated as a garden shrub in America.

Magnolia

This is where we reap the benefit of our declaration that any woody plant of spreading growth that starts to bloom at a height of 5 feet or less may be regarded as a shrub, however tall it may afterwards become; for specimens of even the smaller magnolias will in time become high, wide and handsome. Like a child with big feet, they seem destined to grow to a size proportionate to their large and noble flowers. These flowers, which seem to the layman sufficiently elaborate, are regarded by botanists as of extreme simplicity; they consider the magnolia to be one of the most primitive of flowering plants, ranking in antiquity with the ginkgo or maidenhair-tree. Fossil magnolia remains of five million years ago are common over a wider range than that occupied by the family today, confined as it is to eastern Asia and northeastern America. About eighty species are so far known to science, more than half of them tropical evergreens unsuited to Britain. Of the thirty-two deciduous or moderately hardy evergreen kinds, twenty-seven are in cultivation there.

The earliest to be discovered were naturally the American magnolias; the first seen by a European being the Mexican *M. dealbata* (not in cultivation) which was described by Francisco Hernandez, physician to Philip I of Spain, who made a special expedition from 1570 to 1577 to study the plants of the New World. But the name-plant of the genus and the first to be introduced to gardens was the North American Swamp or Sweet Bay, *M. virginiana* (syn. *M. glauca*), which was among the plants sent by John Banister to Bishop Compton in 1688, and which has remained in cultivation ever since. Catesby mentioned that in his day (*c.* 1730) it was "growing at Mr. *Fairchild's* in *Hoxton* and at Mr. *Collinson's* at *Peckham*, where it has for some years produced its fragrant Blossoms, requiring no protection from the Cold of our severest Winters." About thirty years later, John Clarke, a butcher of Barnes, was noted for his success in raising this shrub from seed. By many writers it was very warmly commended, especially for its perfume; but its highest praises came from those who, like Catesby, Kalm and Cobbett, had seen it growing in its native land. To Cobbett it was a shrub excelling every other, the scent of whose flowers was "the most delightful that can be conceived, far exceeding that of the rose; in strength equaling the jasmine or tuberose, but far more delightful. . . . None of the other magnolias are nearly so odoriferous as this; all but this are somewhat tender; this might be in every shrubbery in England with the greatest ease, and I cannot help expressing my hope that it may be one day as common as the lilac." At the time that these words were written,[1] however, the Asiatic magnolias were practically unknown—the Yulan was still on probation in the conservatory—and once they became familiar, the chorus of praise for *M. virginiana* died away. Its habit of producing its cream-to-apricot flowers in succession over a long period is as much a drawback as an advantage, as there are not enough out at a time to make any great display.

The other well-known American species is *M. grandiflora*, the only hardy evergreen kind in general cultivation. It was not known to Miller in 1724, but shortly afterwards several importations seem to have been made. Catesby in 1738 says that it flowered first in the garden of "that worthy and curious Baronet, Sir *John Colliton*, of *Exmouth* in *Devonshire*, where, for these three years past, it has produced plenty of blossoms; since that, and in 1737, one of them blossomed at Parsons-Green, in the garden of the Right Honourable Sir *Charles Wager*." It was also grown about the same time in the nursery of Christopher Grey at Fulham. Seed-raising was already in train, for Miller in 1752 said that many young plants were killed by the severe winter of 1739, and that in consequence this magnolia was scarce in nurseries for some years afterwards. Grey's original Fulham specimen grew in time to a tree 20 feet high and as much wide, whose fragrance when in bloom perfumed the whole neighborhood; by 1822 it was dead, but its trunk, which had been preserved, measured 4 feet 10 inches across. This seems about the maximum size to which the "Big Laurel" can grow in Britain, but in its native land it reaches 70 or even 90 feet high, and specimens left standing on account of their beauty when land was cleared for cultivation were regarded by Michaux as some of the finest productions of the vegetable kingdom. It is said that the Indians would not sleep under such a tree when in bloom, because of the overpowering scent of its flowers—one of which, if kept in a bedroom, could cause death in a single night.

It has long been the practice in England to grow this plant against a wall, and Loudon describes a wall at White Knights, near Reading, 145 feet in length and 24 feet high, entirely covered with twenty-two magnolias of

[1] 1833

this kind, planted by the owner, the Marquess of Bland-ford, in 1800—a time, says Loudon, awed, when such plants cost £5 apiece. An evergreen magnolia on the wall of a fine house (bearing the flowers "as big as tay-kettles" of which Mrs. O'Dowd boasted in *Vanity Fair*) was then considered almost as essential as the wall itself; now, Dr. Stoker suggests, walls are not what they were, and it would rather be a case of the magnolia supporting the wall. Many garden varieties have been raised—nine, including *exoniensis* (syn. *M. grandifolia* var. *lanceolata*), were in cultivation before 1863. The Exmouth variety, with narrower leaves rusty-felted on the underside, has the advantage of starting to flower at an earlier age than the type; it is the best for general use, but in southern gardens is becoming superseded by the grand but rather tender variety 'Goliath', which received an A.M. in 1931 and an F.C.C. in 1951.

The first of the Asiatic magnolias to arrive was the Yulan, *M. denudata* (syn. *M. conspicua*) introduced from China by Sir Joseph Banks—in 1789 according to Loudon, or 1780 as stated in the *Botanical Magazine*. (Perhaps Banks had read a French report, published in 1778, in which this shrub was compared to "a naked Walnut Tree with a Lily at the end of every branch."[1]) In China it had been cultivated since the seventh century, both out-of-doors in the vicinity of temples and palaces, and dwarfed in pots and boxes for a succession of indoor blooms; so highly was it valued, says Loudon, "that a plant in flower, presented to the Emperor, is thought a handsome present, even from the governor of a province." For a number of years after its introduction to this country it was strangely neglected; it is not mentioned in Martyn's (1807) edition of Miller's *Dictionary*, and John Sims, writing in the *Botanical Magazine* in 1814, describes it as a conservatory plant, and only ventures a cautious opinion that it is "not improbable that it may be hardy." (The same author also deprecates the fact that it flowers before the leaves appear, and is therefore "far less agree-able" than *M. grandiflora*.) Loudon in 1838 said that the Yulan had only been found to be hardy within the last twelve years, and it was still listed among half-hardy spe-cies as late as 1863. In the early days another Asiatic species, the red-flowered *M. liliflora*, seems to have been better known than the Yulan; it was introduced to England from Japan by Thunberg in 1790,[2] and is the lowest in growth of the family, rarely exceeding 10 or 12 feet. Long cultivated in the gardens of both China

[1]Quoted by Salisbury in *Paradisus Londiniensis*, 1806

[2] According to Loudon. Thunberg was in Japan only in 1775–76; but the plant may have spent the intervening years growing and being propagated in the botanic garden at Upsala

and Japan, it is probably Chinese in origin, but is unknown in the wild. It has been closely associated with the Yulan ever since Kaempfer described them both in 1712 under the Japanese name of *Mokkwuren*. In 1791 Sir Joseph Banks published a series of plates from draw-ings made by Kaempfer, and unfortunately the descrip-tions belonging to the pictures of these two magnolias got transposed, so that the French botanist Desrous-seaux, allotting Latin names to the plants in the same year, believed the flowers of *denudata* to be red, and of *liliflora* to be white, and gave this species the name that was obviously intended for the Yulan or Lily-Tree. The first specimen of *M. liliflora* to flower in this country bloomed in the greenhouse of the Duke of Portland at Bulstrode; Curtis in 1798 was "favoured by the Countess of Coventry with a small plant of it, about a foot high, which flowered with her Ladyship in town," for illustra-tion in the *Botanical Magazine*. It is a species of question-able garden value, but important as one of the parents of the group of hybrids known as *M. × soulangeana*.

The first of these was raised from some seeds of a specimen of *M. denudata*, which stood near one of *M. liliflora* in front of the château of M. Soulange-Bodin at Fromont, near Paris; it flowered in 1826 and proved to be an accidental hybrid between the two species. (The Chevalier Soulange-Bodin was an army officer who retired after the peace of 1815, and devoted himself to gardening; he was the founder of the National Horticul-tural Society of France.) In this country the new magno-lia was distributed by the firm of Young at Epsom, who procured plants from M. Soulange-Bodin as early as 1828. It rapidly became the most popular of all the sorts for smaller gardens, though it blooms a little later than its *denudata* parent; the flowers, appearing with the leaves, lack the startling beauty of the great snowy blos-som on the bare branch, which we were so slow at first to appreciate. A number of named forms have since been raised from the same cross, nearly all stained with the wine-lees purple of the misnamed *liliflora*, and rang-ing from the white *alba* to the deep claret *nigra*, which was introduced from Japan by John Gould Veitch in 1861, and was thought to belong to this group, but which is now believed to be no hybrid, but a pure-blooded vari-ety of *M. liliflora*.

These are the only colored magnolias that can be grown in small or chilly gardens, for the magnificent pink kinds—*MM. campbelli, sargentiana* and *sprengeri diva*—are tree-size, and too tender for cultivation except in the south and west. But for those who prefer their magnolias white and small, there is plenty of choice; *M. salicifolia, M. stellata* and *M. sinensis* being particularly desirable. The first, an upright-growing shrub or small tree whose bark smells of lemon verbena and whose smallish white flowers are produced at an early age, is a native of Japan, and was introduced by way of the Arnold Arboretum in 1906. One can almost tell at a glance that the familiar *M. stellata* came from the same country; it has a Japanese prettiness about it, as compared to the noble Chinese beauty of the Yulan, and it is no surprise to hear that it grows naturally on the slopes of Fujiyama, where Richard Oldham found it in 1862. (It was not mentioned by Kaempfer, though he described another Japanese sort, *M. kobus.*) Plants of *M. stellata* brought from Japan bloomed in the Coombe Wood nursery of Messrs. Veitch in 1878—apparently its first appearance in the west. At the beginning it underwent a number of rather boring changes of name; fortunately the one now accepted as correct is aptly descriptive of its starry flowers, which have a greater number of "tepals"—twelve to eighteen—than have those of any other member of the family, as well as a sweet "bean-field" scent. Gardeners looking proudly down on a plant 2 feet high and a yard square, and covered with bloom, should ponder the fact that there is a specimen of this magnolia at Inverewe, Scotland, which some years ago was 28 feet high and 75 feet in circumference. There is a variety *rosea*, introduced by Maries for Veitch before 1880, whose pale-pink buds are apt to fade to white when the flower opens; but the color is said to improve as the plant ages, and to be stronger in dry seasons.

When these spring-flowering magnolias are over, there are still the summer ones to enjoy—almost more beautiful, with their solid porcelain-white flowers enhanced by crimson anthers. There are three of them, the first to be introduced being *M. sieboldii* (syn. *parviflora*) from eastern China, about 1865. Although the flowers are by no means small (at 4 to 5 inches across) they are less large than those of the other two species, *M. sinensis* and *M. wilsonii*. Both of these come from western China, and were introduced early in the present century by E. H. Wilson; they are much alike, but the more upright *wilsonii* will grow into a small tree, while *sinensis* flowers somewhat later, and always retains a shrub-like form. Both species flower at an early age, but are best when they reach a certain size, for they keep their flowers clean by turning their backs to the weather. To enjoy the upright chalices of the Yulan it is necessary to hover overhead, but for the pendent saucers of *sinensis* and *wilsonii* one must grovel underneath.

The genus was named (by Plumier) after Pierre Magnol (1638–1715), of Montpellier. "He was bred to physic,

but being a Protestant, could not take his degree there, and was obliged to have recourse to some more sensible and Christian university, where such exclusive laws were unknown."[1] He practiced medicine and cultivated botany at Montpellier for many years, but though nominated for a professorship in 1667, "his religion was an insuperable obstacle to his appointment, as that of King Solomon himself would, in the same case, have been,"[2] and he did not attain the professorial chair and the directorship of the Botanic Garden until he became a Catholic in 1694. Meantime, he had published two works, *Botanicum Monspeliensis* (1676) enumerating some 1,366 plants, nearly all collected by himself, and *Prodromus Historia Generalis Plantarum* (1689) incorporating a new scheme of natural arrangement; and the great Tournefort had been amongst his pupils. The only living magnolia he might possibly have seen was *M. virginiana*; little did he know what a superb genus was to perpetuate his name. NOTES: Miss Coats, admittedly, has taken considerable license with the term "shrub" when she included this genus. But the history of the magnolias is so intriguing that it is understandable. There have been so many developments that I must refer you to Neil G. Treseder's fine book, *Magnolias* (1978), for an update on the latest in the classification, history and rapid advancement through breeding (as well as sorting out the confusion).

Among the hybrids offered in the trade, American gardeners may wish to take special note of the "girl series" hybrids developed at the National Arboretum, which, actually, are not all named for females. The group is considered to be among the best because they are less subject to frost than others and because they occasionally produce flowers well into the summer.

Breeding efforts at the Brooklyn Botanic Garden, New York, have given rise to considerable interest in developing magnolias with yellow flowers. One of their selections, 'Evamaria', has buds and flowers that are magenta, suffused with pale orange. Propagation by cuttings, however, has met with limited success, and consequently new hybrid selections are slow to appear in the nursery trade. New selections with good upright habit and potential street-tree use are being introduced by Dr. Frank Santamour at the National Arboretum.

Mahonia

Under this name the botanists have assembled those species of berberis that have pinnate, not entire, leaves; and although Hibberd in 1898

[1] J. E. Smith in Rees' *Cyclopaedia*   [2] Ib.

held that the distinction "served no useful purpose," the Berberis family has now grown to such unmanageable proportions that any method of reducing it is welcome. To the gardener's eye, the Mahonias are very distinct, and fully justify their segregation; they now number over ninety species, of which only about a dozen are in cultivation.

The first to be introduced came from America. When David Douglas visited New York in 1823, the only person he disliked was William Prince, a nurseryman, whom

Mahonia diversifolia
(now *M. aquifolium*)   E. D. Smith (1831)

he found "a man of moderate liberality . . . his nursery covered with weeds, his reputation overrated and prices excessive." Acting on instructions to obtain plants as gifts, or as cheaply as possible, Douglas made most of his purchases elsewhere, with the exception of "one great ornamental novelty, the *Berberis aquifolia*," which had been raised from seeds brought from the far west by Lewis and Clarke's expedition in 1804–6. Actually this was not the true plant, but the closely allied *Berberis* (*Mahonia*) *repens*, the confusion having arisen through a mistake of the botanist Pursh, who thought that the Lewis and Clarke specimens were of the same plant as that collected by Menzies some years earlier. When

Douglas himself explored the Pacific northwest between 1825 and 1827, he found Menzies' plant, the true *M. aquifolia* or Oregon Grape, in the wild, and was able to introduce it on a large scale. By 1838 it was "already to be found in all good collections," and was described as "perhaps the handsomest hardy evergreen we have"[1] At first it was propagated with difficulty by layers, and plants were sold at 10 guineas apiece; when quantities of seed became available, the price dropped to 5s, and in 1914, Bean tells us, good plants could be had at 30s per thousand—or three a penny; those who live in the parts of the country where this shrub has become naturalized could probably get them for nothing. The berries are edible, and are (or were) marketed for culinary purposes under the name of Oregon Grapes. Several garden varieties are grown, including wavy-leaved *undulata*, which originated before the end of the nineteenth century, and in 1960 was awarded the A.G.M.; the colorful *moseri* with its variegated leaves, selected at the nursery of Moser at Versailles, and the tall *magnifica*, raised by Fisher, Son and Sibray of Sheffield.

The twin Asiatic species, *M. japonica* and *M. bealei*, have caused as much trouble by their similarities and differences as if they were Castor and Pollux, and hatched out of Leda's double-yolked egg. When not in bloom they are indeed much alike, except that *bealei* has broader and fewer leaflets than *japonica*, with a larger terminal leaf; but the flowers are quite distinct, those of *bealei* being born in short erect bunches, while those of *japonica* are larger and more widely spaced on drooping racemes, and smell very sweetly of lily-of-the-valley. It all began with a Japanese garden plant found by Thunberg, which he classified, rather strangely, as a holly, and named *Ilex japonica*. When Robert Fortune made his first visit to China (1843–46) he introduced a lax-flowered shrub which was recognized as Thunberg's plant, but removed to its proper genus and renamed *Berberis japonica*. In 1848, during his second trip, Fortune saw in a neglected garden in the district of Hwuy Chow an upright-flowered shrub "surpassing in beauty all the known species of Mahonia;" he succeeded in procuring some plants, which he left in charge of his friend Mr. Beale of Shanghai until he was able to ship them home to the firm of Standish and Noble at Bagshot. In a letter to the *Gardeners' Chronicle* in 1850, he suggested calling his new find *M. bealei*; but two of the principal botanists of the day, Hooker and Lindley, considered it only a form of the previous species, and his letter was printed under the title "An account of the finding of a fine new evergreen shrub called *Berberis japonica*"—thus causing

[1] Loudon

confusion from the start. Other botanists, meantime, recognized *M. bealei* as a distinct species, but thought Thunberg's *M. japonica* to be only a geographical form of the Himalayan *M. napaulensis*. It is hardly surprising that the nurserymen confused the two names, and that for years the upright-flowered shrub was marketed as *japonica*, and the lax-flowered one—a much better plant, Fortune's opinion notwithstanding—as *bealei*. Restoring these species to their original and correct names has not helped, for there is no means of knowing whether a nurseryman has brought his nomenclature up to date, or is still keeping to the old usage, and some of them slide out of the difficulty by listing "*M. bealei japonica*" (or "*japonica bealei*") and leaving it at that. Confusion is compounded by the existence of a kind called *hiemalis*, said to be a variety of *bealei* but probably a hybrid, for its seedlings produce flowers at any angle from the vertical to the horizontal.

It is a relief to turn to some of the unquestioned species: to *M. napaulensis*, introduced before 1850 from Nepal, and called by Hibberd "A beautiful curiosity . . . a horny shrub," and *M. lomariifolia*, seed of which was brought from Tengyueh on the Burma-China frontier by Major Laurence Johnston of Hidcote, in 1931. Both are very handsome shrubs, but unfortunately both are slightly tender, though they sometimes succeed in unexpected places.

The family was named by Thomas Nuttall in honor of the Irish-American horticulturist Bernard MacMahon (1775–1816). MacMahon emigrated from Ireland for political reasons, and in 1796 settled in Philadelphia, in the neighborhood of which he established a nursery where he cultivated over a thousand species of useful or ornamental plants. He had a seed shop and general nursery business in the town, which was looked after by his wife, and which became the meeting place and forum of the most prominent botanists and horticulturists of the day, in whose discussions MacMahon took an active part. He also conducted an extensive correspondence for the exchange of seeds and plants from abroad and from other parts of the U.S.A., and it was to him that President Jefferson entrusted a large proportion of the seeds brought home by Lewis and Clarke. In 1806 he published *The American Gardener's Calendar*, which went into its eleventh edition in 1857.

NOTES: Although a number of cultivars are offered in England, they have yet to appear in the U.S. trade because of quarantine restrictions. Consequently, our nurseries are limited to offering only the two main species, *M. aquifolium* and *M. bealei*.

The Mahonia are dependable shrubs in foundation plantings and are often used in the landscaping of commercial sites because of their robust habit, soil tolerance and ability to withstand abuse. They fit in well with evergreens such as hollies, conifers and other rugged plants. The bright clusters of yellow flowers of early spring are soon followed by blue fruits—added attractions to the crisp, spiny evergreen leaves.

Myricaria, see under *Tamarix*

Myrtus communis
(MYRTLE)   Ferdinand Bauer (1825)

Myrtus

I saw by night, and behold a man riding upon a red horse, and he stood among the myrtle-trees that were in the bottom; and behind him were three red horses, speckled and white . . . And the man that stood among the myrtle-trees answered and said, These are they whom the Lord hath sent to walk to and fro in the earth . . .
Zechariah, i, 8–11

There is an ancient Arab tradition that when Adam was turned out of Paradise, he took with him a grain of wheat, a date stone and a sprig of myrtle, which proves, if not its antiquity, the value set on this fragrant shrub in its native place; for although long naturalized about the Mediterranean, the common myrtle is believed to have originated, like Adam himself, in western Asia. There are numerous other myrtles, however, besides *M. communis*—all evergreen shrubs with fragrant white flowers, many of them natives of the New World.

According to Richard Bradley, the myrtle was introduced to England along with the first orange trees, by Sir Walter Raleigh and Sir Francis Carey when they returned in 1585 after a sojourn in Spain, where they had observed preparations already being made for the launching of the Armada. But we have documentary evidence of at least one earlier importation; in 1562 Lord Burghley wrote to Mr. Windebank in Paris to ask him to procure for him a lemon, a pomegranate and a myrtle tree, with instructions for their culture; and Windebank replied that he had duly dispatched "a lemon tree and two myrtle trees, in two pots, which cost me both a crown, and the lemon tree fifteen crowns . . . it is the best cheap we could get it, and better cheap than other noble men in France have bought of the same man . . ." It seems unlikely that even this was actually the *first* introduction; the Romans, for example, must surely have attempted to establish a shrub so closely associated with their mythology and tradition. Classical literature abounds with stories of people being turned into, and out of, myrtle trees; and the leaf is supposed to be perforated by hundreds of tiny punctures, ever since Phaedra, watching on the hillside while her stepson Hippolytus exercised his horses in the arena, passed the time by pricking the leaves of a nearby myrtle with a hairpin. Above all it was the plant of Venus, who is said to have worn a garland of myrtle when she rose from the sea, and also to have taken refuge behind a myrtle when surprised by satyrs in her bath, "which increased her attachment to this tree."[1] Roman generals who had achieved bloodless victories were crowned with myrtle, Venus being the goddess most averse to war.

The common myrtle is a variable plant, and in 1597 we hear of no less than six varieties being already in cultivation, including a dwarf form and one with white fruit. In 1640 Parkinson added a double-flowered kind, which had recently been "noursed up in the Gardens of the cheife Lovers of rarities," and was probably the one that, according to Evelyn, "was first discovered by the incomparable Fabr. Piereshy, which a mule had cropt from a wild shrub." (Evelyn also had a variety whose leaves were "tip't with white.") In the late seventeenth

[1] Phillips

and early eighteenth centuries the myrtles quickly found a place among the "curious greens" that stood out-of-doors in the summer and were housed in the winter. They were also used for the decoration of empty fireplaces in summer, and sent back to the nursery gardens "to be taken Care of for the Winter at the Usual Price."[1] Myrtles were still universally grown in the early nineteenth century, but the height of their popularity seems to have been reached in the 1770s. Eleven columns of fine print are devoted to them in Abercrombie's *Universal Gardener and Botanist* (1778), where nineteen varieties of the common myrtle are described, and detailed instructions given for their propagation, as practiced by "the setting-gardeners about London, who raise prodigious quantities of myrtle annually for the supply of the markets." At Covent Garden "amazing quantities" were "brought for sale at all times of the year, which sell from sixpence to half-a-crown per pot, according to the size and goodness of the plants."

One of the kinds described by Abercrombie was the broad-leaved Jew's Myrtle, which had three leaves at every joint, instead of the usual pair—"by which particular circumstance this species is in universal estimate among the Jews in their religious ceremonies, particularly in decorating their tabernacles, and for which purpose many gardeners about London cultivate this variety with particular care, to sell to the above people, who are often obliged to purchase it at the rate of sixpence or a shilling for a small branch; for the true sort having the leaves exactly by threes, is very scarce, and is a curiosity."

The hardiest and most floriferous of all the kinds was the broad-leaved or Roman myrtle, and out of a score of varieties formerly grown, this is almost the only one to have survived. It is extraordinary that this once popular plant has now become so rare that a great many gardeners have never even seen one. It is true that it is on the borderline of hardiness, but not more so than many other plants that are much better known. Evelyn thought it flourished better grown outside with some winter protection, than when brought indoors into the "Conserve." Liger in 1706 said that in one place in England there was a myrtle arbor, "which is cover'd every Winter to protect the Plants from the Extremity of the Frosts," and Phillips in 1823 suggested that a specimen plant on a lawn might be protected "by placing a light iron cage over it in winter, that might be covered with moss or panels of some ornamental design."

"No lady would be without her myrtle," wrote Glenny is 1851, "and slips of this will root anywhere and anyhow." Not altogether so; they will not strike if the lady

[1] Fairchild, *The City Gardener*, 1722

is destined to be an old maid, and to ensure success, according to a Somersetshire tradition, it is necessary when planting them to "spread the tail of one's dress, and *look proud*." Formerly myrtle was always included in bridal bouquets—probably owing to its long association with the Goddess of Love—and sprigs from these were frequently planted by the bridesmaids, though it was considered unlucky for the bride herself to do so. Sometimes they were planted close to the door, on arrival at the new home, and many old myrtle trees in the southern counties trace their origin to some such ceremony. But long before the days of bridal bouquets, the fragrant myrtle had been one of the most popular cut flowers; Liger in 1706 spoke of the use of the sprigs "to dress up Basons with variety of Flowers, for the ornament of our Rooms, and for Nosegays."

The distilled water of myrtle flowers was regarded as a beautifier and aphrodisiac, and was the principal ingredient in the eighteenth-century Portuguese "Angel Water." The finest Turkish leather was tanned with the roots, which imparted a delicate scent to the skins; and the berries were used by the Romans to make a sauce to accompany the brawn of the wild boar. Many species of myrtle have edible berries, and one, *M. ugni*, aroused considerable interest on its first introduction as a possible fruit bush. In its native Valparaiso, where it was known to the natives as Ugni and to the Spanish settlers as Myrtilla, it was cultivated for its delicate juicy berries, about the size of large peas, which were said to be "much relished by some, having a peculiarly rich musky flavour," and to make an excellent jelly. The plant was brought to England by William Lobb, when collecting for the firm of Veitch in 1844; trials were made of the fruit and prizes offered by C. Wentworth Dilke, Esq., at the Grand Autumn Fruit Show of the Horticultural Society in 1857. The myrtilla is hardy, and jam has been made of the fruit; in Devon and Cornwall, it produces its small white flowers freely, but its leaves are not fragrant like those of the common myrtle.

"Myrte is a lytell tre so called, the whiche tre bereth a fruyte that is named Myrtylles"[1]—so it appears that, as in the case of the bay, we use for the shrub a name that originally applied only to the berries. According to Lyte, the family name (almost the same in Greek, Latin and all European languages) was derived from a Greek girl called Myrsine, who excelled in beauty and athletics, and was a favorite of the goddess Minerva. It was her function to bestow the crown of honor on winners at the games, "but some of those that were vanquished, were so much displeased with her iudgement, that they slue

[1] *The grete herball*, 1529

her. The which thing as soone as the Goddess Minerva perceived, she caused the sweet myrtle to spring up, and called it myrsine . . . which tree or plant shee loveth as much as ever shee loved the yong Damosell Myrsine." (But there would surely be trouble on jealous Olympus if Minerva took too great an interest in Venus's tree?) Needless to say, in the Language of Flowers it stood for love:

> . . . *myrtle—which means chiefly love; and love*
> *Is something awful, which one dare not touch*
> *So early o' mornings . . .*[1]

NOTES: Because they are so limited in hardiness, the delightful fragrant myrtle is very rare in this country. It is limited to greenhouses and the warmest parts of California.

Nandina

The Heavenly Bamboo, *N. domestica*, has the reputation of being only a semi-hardy shrub, requiring a sheltered position; it thrives in the rock garden at Wisley, where it has attained 6 feet in height, and produces its attractive white flowers; but in this country, though at all times elegant and distinguished, it rarely bears the abundance of red berries that makes it so strikingly decorative a plant in its native China. It has been cultivated for centuries in Chinese and Japanese gardens, and it was in Japan that it was seen by Kaempfer at the end of the seventeenth century. He described it as having a reed-like appearance at a distance, and curious seeds, concave on one side and convex on the other, which he compares to "what are called in the shops Crab's Eyes."[2]

It reached England in 1804, with the first consignment of plants sent home from Canton by William Kerr on the East Indiaman *Henry Addington* (Captain Kirkpatrick), and within a comparatively short time it became "not uncommon in British greenhouses."[3] When E. H. Wilson made his fourth visit to western China in 1910–11, he found the Heavenly Bamboo growing freely in the wild, near Sha-to-tzu in Szechwan, "its elegant foliage and large erect trusses of white flowers with conspicuous yellow anthers making it very attractive. In autumn and winter the masses of scarlet fruit render it extremely beautiful."

In Shanghai this shrub is used as we use holly in

[1] E. B. Browning, *Aurora Leigh* [2] i.e. the scarlet seeds of the tropical climber *Abrus precatorius*, used in the East for stringing into necklaces and rosaries [3] Loudon

England; the branches loaded with red berries, are sold in the streets in winter, and used for the decoration of altars, both in temples and in private houses or on boats. In Japan it is planted close to the door in every garden and courtyard, however small—"but whether for ornament only," says the *Botanical Magazine*, "or that any particular use is made of it, we are not informed." Actually it had a most important function; if any member of the household had a bad dream, it was confided to the "home shrub," which ensured that no harm would follow. The wood is said to have an aromatic flavor and to be considered by the Japanese to be "the most tasty and suitable for toothpicks".[1]

The name is derived from the Japanese *Nan-ten* or *Nan-din*. It may be heavenly, but is not a bamboo; it is a monotypic genus, akin to the Berberis family.

NOTES: No southern garden would be complete without its nandina display of red foliage and bright red berries in winter. Several dwarf cultivars have appeared in large numbers as border plants.

In Japan the array of bizarre types with fern-like foliage, distorted branchlets and white, yellow or crimson fruit is extensive. In fact, now there is even a Nandina Society in Japan. Despite records of its common occurrence in Japanese gardens from the earliest of times, nandina was introduced to Japan from China before the sixteenth century.

The nandinas are best planted in moist garden soil with plenty of humus and then provided with a mulch of pine needles or bark. They tolerate full sun or shade and require no particular maintenance.

Nerium

"Let Woodward, house of Woodward, rejoice with Nerium the Rose-Laurel; God make the Professorship of fossils in Cambridge a useful thing."
Christopher Smart, *Rejoice in the Lamb*, 1756–63

The Oleander, Rose Bay or Rose Laurel is more an indoor than an outdoor plant; even in the mildest localities it is advisable to house it during the winter. Nevertheless it was for generations a favorite with English gardeners, and having been continuously in cultivation here for more than 350 years, deserves at least a brief account. There are only two species, and some botanists consider that one of them, *N. odorum*, is not truly distinct, but is only the Asiatic form of the familiar *N. oleander*, which grows, mostly by coasts and in river

valleys, in the Mediterranean region. The French naturalists Pierre Bélon (known as Petrus Bellonius)[1] observed that it was particularly abundant in Crete, where it grew to so great a size that the wood was occasionally used for building; it was there, too, that the white variety was found.

It is hard to say when the oleander was first introduced; Turner had seen it only in Italy, but Gerard had both white and red varieties in his garden in 1596. Parkinson's plants were grown from seed brought from Spain by "Master Doctor John More"; they must have thriven well, for whereas in 1629 he describes the shrub as having a stem "as bigge at the bottome as a good mans thumbe," in 1640 he says it is "as bigge below as a reasonable man's wrist." The shrub could also be propagated by layers, and Liger (in 1706) warns us that "Most hir'd Gard'ners are apt to lay these Branches of Rose Bays to make a Penny of them: Preferring their own Advantage to their Master's Pleasure, who would delight in seeing a Rose Bay-tree adorned with Branches at the Foot; whereas if 'tis naked there it loses half its Beauty; for which reason Gentlemen should look after their Gard'ners, and see they don't serve their Rose Bay-trees so for their own Interest." By the middle of the eighteenth century several varieties were in cultivation, in different shades of red, single or double, or with striped flowers; besides the white, which was more tender than any of the others. Shirley Hibberd, in *Familiar Garden Flowers* (1898) has a good deal to say about the oleander as a family shrub or household plant, "a sort of patrician laurel." He says it improves with age, if reasonably well treated; and while he knows of some "family oleanders" that are "kept in dark conservatories or lumber-rooms all the winter, and in some obscure corner out-of-doors all the summer," others "are creditable to their owners" winning prizes annually at flower shows. Apparently the shrub can withstand a good deal of rough treatment, so long as it is protected from frost in the winter and given rapid growing conditions—warmth and liquid manure—in spring and early summer. Otherwise the flowers are produced too late to develop properly; "and the question is often asked by cottagers and amateurs—'What does the too late show of blossoms indicate?' To this I reply," says Samuel Wood, "that more stimulant is required early in the season."

Theophrastus, about 330 B.C., described a plant which he called *oenothera*, which present-day authorities believe to have been the oleander, as this shrub answers to his description of a bush with leaves like the almond and a

[1] Bowles, *My Garden in Summer*

[1]"The which hath much haunted and travelled the yland of Crete or Candie"—Lyte

Nandina domestica

(HEAVENLY BAMBOO)   S. Edwards (1808)

Nerium oleander

(OLEANDER)   Ferdinand Bauer (1819)

Olearia gunniana

(DAISY-BUSH)   W. H. Fitch (1852)

Osmanthus delavayi

(DEVIL-WEED)   M. Smith (1912)

red flower like the rose. He says that the root administered in wine makes the temper gentler and more cheerful, but this recipe should be regarded with caution, for all parts of the plant are extremely poisonous to man and beast—even the flowers; and according to some, the very exhalation or perfume is dangerous. Dioscorides said the leaves could be used as a counterpoison against the bites of venomous serpents, but Turner thinks this remedy should be applied only in cases of extreme emergency and when no "triacle[1] or other good herb" is to be had. "I have sene thys tre in diverse places of Italy." he continues, "but I care not if it never com into England, seying it in all poyntes is lyke a Pharesey,[2] that is beuteus without, and within, a ravenus wolf and murderer." William Rhind, in *The Vegetable Kingdom*, tells of a party of French soldiers who went foraging for provisions near Madrid during the Peninsular War. One of them cut some oleander boughs, and stripping off the bark, used the wood as skewers for their meat; with the result that of twelve men who ate of the roast, seven died, and the rest were seriously ill.

The shrub is abundant in Palestine, and it is suggested that the "willows of the brook" mentioned with other "boughs of goodly trees" in *Leviticus* xxiii, 40, may have been the Oleander. In Tuscany it is called St. Joseph's Staff, after the staff that burst joyfully into bloom when an angel announced to Joseph that he was to be Mary's husband.

The Oleander (Latin, *Nerium*) is not to be confused with the Oleaster (Latin, *Elaeagnus*) or with the Olearias or Daisy-bushes. All of them got their confusing names through superficial resemblances to Olea, the Olive, to which none of them is actually related; the Rose-Laurel is a distant cousin of the periwinkle. *Nerium* is derived from *neros*, humid, in reference to the plant's favorite habitat; Pliny says that the Greeks called it Nerion, Rhododendros or Rhododaphne but that it "hath not been so happy yet, as to find so much as a name among the Latines."

> . . . *It is strange that a little mud*
> *Should echo with sounds, syllables and letters*
> *Should rise up and call a mountain Popocatapetl,*
> *And a green-leafed wood Oleander . . .*[3]

NOTES: In the South the Oleander is widely grown in highway plantings and public gardens. There are numerous selections and local varieties. Its poisonous properties are fairly well known, but signs alerting visitors to this fact are still highly visible along the coastal highways in the Southeast.

[1]Treacle: i.e., remedy [2]Pharisee [3]*Talking With Soldiers*, W. J. Turner

Olearia

It is an unfortunate fact that most of the plants of the Antipodes are too tender to grow out-of-doors in Britain; we have very few shrubs in our gardens that come from Australia or New Zealand. On the edge of hardiness are the Olearias or Daisy-bushes, and the shrubby Veronicas; in both cases there are one or two species that are generally reliable, and many that are not.

The hardy member of the Olearia family is *O. haastii*, which was raised by Messrs. Veitch from seed sent to them in 1858 by a New Zealand correspondent, whose name has not been preserved. Five years later it was found growing wild near the glacier of Lake Ohau in South Island by Julius Haast, a German scientist who was appointed geologist to the province of Canterbury in 1861, and who explored the headwaters of all the great rivers in the vicinity. Haast sent a great deal of botanical material to Kew, and when Sir Joseph Hooker published his *Handbook of the New Zealand Flora* in 1867, he named this plant in his honor. Its habitat is in river-gorges on the east side of the Southern Alps (South Island), but it is always rare and local, and there are probably more plants of it in Britain than there are in its native homeland. One of its localities is on the upper ranges of the Rangitata river, described in the opening chapters of Samuel Butler's *Erewhon*. Unfortunately, as so often is the case, the hardiest species in the genus is one of the least attractive. *O. haastii* is very floriferous, but the sheets of white flowers with which it covers itself have the greyish look of inadequate laundering; they outwear their welcome, remaining on the bush long after they are faded and shabby. On the other hand, the shrub has the compensating virtue of a remarkable tolerance of industrial and polluted atmospheres, and the flowers have a pleasing hawthorn fragrance.

Of the other species, *O. gunniana*, we are told, "braves the cold of England" even near London, if grown against a wall. "We wish it had more beauty to recommend it" is the comment of the *Botanical Magazine*, but the R.H.S. *Dictionary* allots it a star of excellence. Seed was sent to England in 1848 by the Tasmanian naturalist R. C. Gunn, "by whom . . . it was first detected."[1] The flowers, like those of most of the olearias, are white; but in the present century a colored form was found on Mt. Seymour by a Tasmanian farmer and gardener, appropriately named Mr. G. J. Weeding. He transferred it to his garden at Eldon near Colebrook, and by the raising and selection of generations of seedlings, succeeded in producing plants with flowers in "all the colours found in

[1] *The Botanical magazine*, 1852

modern Michaelmas Daisies and ranging up to the rich dark purple of *Clematis jackmanii*."[1] Seed of this strain was procured by H. F. Comber when he visited Tasmania in 1929–30, and introduced to England under the name of *O. gunniana* var. *splendens*.

O. macrodonta is a comely seaside shrub, which will endure much wind, but very little frost. In its native New Zealand, from which it was brought in 1886, it grows into a tree 20 feet in height, the wood of which has been used in veneering. Its leaves have a musky scent when bruised. *O. semidentata*, the handsomest of the whole family, is of course one of the most tender—a shrub that everyone praises, but very few can grow. In 1910 Captain A. A. Dorrien-Smith found it growing on Chatham Island, especially in the "Tobacco-country"—so-called because the land was once purchased for a handful of cigars—which contained "the highest ground, worst bogs, and greatest quantity of *Olearia semidentata*."[2] He brought a plant home to his famous garden of Tresco in the Isles of Scilly, where it bloomed in 1913. The type is lavender with a dark purple center, but a rare pink form was also found.

At one time the name Eurybia, "Mother of Stars," was suggested for this genus, and it seems a pity that it should not have been adopted rather than Olearia (because a few kinds have leaves like those of Olea, the Olive) which is confusingly similar to Oleander and Oleaster. Forty or more of the species are endemic to New Zealand; the rest are found in Australia or Tasmania. Several attain tree-size in their own country, where the gardener from the Northern Hemisphere, Mr. A. B. Anderson suggests, "feels that he is seeing something that ought not to be. Daisies are ground-plants, and have no business to perch their flowers on trees."[3] But this is only to be expected in the Antipodes, where everything is naturally upside-down.

NOTES: Due to their nature, the olearias are useful only in southern California.

Osmanthus

A small family, with a relatively short history in the west, though a long one in the East. In China *O. fragrans* has been cultivated for centuries, and was a favorite shrub for planting both in private gardens and in the courtyards of temples, where the sweet-scented flowers were used for offerings to the gods; but it is tender and hardly to be trusted outside a green-house, except in the south. One or two of its tiny flowers, it is said, will fill a whole conservatory with fragrance; in China they are dried, and used to perfume tea. It was introduced in 1771, and was illustrated in the *Natural History of the Tea Tree* (1799) by J. C. Lettsom, who said it was then "not infrequent in gardens near the metropolis."

The next in order was the hardy *O. ilicifolius*, which is chiefly remarkable for looking exactly like a holly tree, but without the red berries. Only a close examination reveals that it bears its leaves in pairs, while the holly's leaves are alternate; its inconspicuous, scented flowers appear in autumn. It was introduced from Japan in 1856, and has several varieties which enjoyed some popularity as evergreens at the turn of the century, but which are rarely seen now.

The important member of the family, so far as British gardens are concerned, is *O. delavayi*—as fragrant as the first species (or nearly) and as hardy as the second. It was discovered by the French Jesuit missionary, whose name it bears, in the mountains near Lan-kong in Yunnan, in 1890; he sent seed to the firm of Maurice de Vilmorin, who distributed it to various correspondents. Only one seed germinated—in the garden of the Paris School of Arboriculture at St. Mandé—and from this all subsequent plants were propagated, with the result that it was a very scarce shrub in cultivation, until further supplies of seed were obtained by George Forrest early in the present century. It received an A.G.M. in 1923, and is now listed by every shrub nurseryman and lauded by every English gardening author. Dr. H. F. Dovaston found it "hardy to zero" in Scotland, and said that a 10-foot specimen "never turned a minute holly leaf" to the hard frosts of 1947; but it is said to produce its scented white flowers more freely in the south and west. It begins to bloom at an early age, and in suitable localities will make a good hedge.

O. delavayi has been crossed with *Phillyrea decora* to produce a bigeneric hybrid which has been named *Osmarea* × *burkwoodii*, after Messrs. Burkwood and Skipwith, in whose nursery at Kingston in Surrey the cross was made. This is rapidly gaining esteem as an evergreen hedging plant.

To the Chinese, the object visible in the moon is not a thorn bush, as we would have it, but an osmanthus bush; Wu Kang, banished to the moon for infidelity, eternally tries to chop it down, but eternally it renews itself. Occasionally on shiny nights osmanthus seeds shower down to earth, but apparently fail to germinate—at least, in England. The genus name, from *osme*, fragrance, and *anthos*, a flower, is self-explanatory.

[1] Comber in the *Gardeners' Chronicle*, 1931 [2] Dorrien-Smith in *Kew Bulletin*, 1915 [3] *The Gardeners' Chronicle*, 1930

NOTES: The dark green, spiny foliage of *O. heterophyllus* (syn. *O. ilicifolius*)—the hardiest member of the genus—often causes this shrub to be confused with American holly. The cultivar 'Gulftide', selected at the U.S. Plant Introduction Station, Glenn Dale, Maryland, is recommended among the improved forms. In the South, a selection with variegated leaves is fairly common. But in warmer regions, *O. fortunei* and *O. fragrans* are far more popular, as both are especially fragrant. A native species, *O. americanus*, has been found in coastal forests from Virginia to Florida but is not cultivated.

These evergreen shrubs are easy to cultivate and can be allowed to grow freely or sheared if used against a building. They can be used as a hedge similarly to privet.

Paeonia

For me, it would be perfect if the litchi had pretty flowers, and the peony bore good fruit.
Yumengying (before 1693), Chang Chao,
trans. Lin Yutang

The Royal and Ancient Order of the Paeonies contains some shrubby species which are among the aristocrats of the garden, both for beauty and rarity; though there is no reason why, in these democratic days, they should not be much more widely grown. Some of them have a reputation, not altogether deserved, for costliness and difficulty; others are comparatively recent, and have not had time to become very well known. These newer introductions include *P. lutea* and *P. delavayi*, discovered in Yunnan in the 1880s by the French Jesuit missionary, Père Jean-Marie Delavay, but not grown in Britain until the present century. Seed of both species was sent by Delavay to the Natural History Museum of Paris, in the garden of which the plants bloomed for the first time in Europe—*lutea* in 1891, and *delavayi* the following year. A plant of the former was sent by Professor Maxime Cornu to Kew, where it flowered in 1900; but a far superior variety of this species was found by Messrs. Ludlow and Sheriff[1] in southeast Tibet, and introduced by them in 1936. The flowers of the normal *P. lutea* hang their heads; in the var. *ludlowii* the much larger blooms are carried erect, their bright color and rounded form reminding us that the peony is first cousin to the buttercup—but these buttercups are nearly 5 inches across. This plant received an Award of Merit in 1954, and is sometimes classed as a distinct species. Seed of *P. delavayi* was collected by Wilson in 1909

[1]And also by Captain Kingdon-Ward

and by Forrest in 1910; in 1934 it too received an award, more on account of its extremely handsome foliage than for its nodding flowers of a deep and somber red, suggestive of Victorian dining room curtains. (A very similar species, *P. potanini* (syn. *P. delavayi* var. *acutiloba*) was introduced by Wilson for Veitch in 1904). *P. lutea* and *P. delavayi* grow readily from seed, and hybrids between them occur both in the wild and in gardens, bearing flowers of unusual brownish-red and coppery shades, which are interesting rather than showy, and poor relations indeed compared to the superb varieties of *P. suffruticosa*.

P. suffruticosa (syn. *P. moutan*) is of course *the* Tree-Peony, so rightly called by the Chinese the King of Flowers. It is known to have been cultivated in China since the seventh century, and is therefore an upstart compared to the herbaceous peony, whose history goes back to the fifth century B.C. Peony festivals and ceremonies were well established by the first half of the eleventh century, and the earliest surviving monograph on the flower, in which more than ninety varieties are mentioned, dates from the same period. In the southern provinces it was cultivated "with the same rage as Tulips have been in Europe,"[1] and grafts of choice kinds were sold for large sums. The tree peony was a favorite subject for decorative design, and became known to the West through Chinese art and the descriptions of missionaries, long before the flower itself was seen in Europe.

A number of attempts to introduce the peony were made by Sir Joseph Banks, who "instructed several persons trading to Canton to enquire for it,"[2] but most of the plants that were dispatched perished on the voyage. A Dr. Duncan of the East India Company brought a living plant to Kew in 1789, but it did not survive for long. In 1794, however, the East Indiaman *Triton* reached port with seven tree peonies on board, two for the King, two for Sir Joseph Banks, and three consigned to Gilbert Slater, Esq. The ship had a rough passage, and was dismasted in the Channel; the double camellias and other plants that she carried arrived in "a very shattered condition,"[3] and two of the seven peonies died, but the rest were successfully established; they all seem to have been double and semi-double pink forms. Ten years later the *Hope* (Captain Prendergass) brought a consignment of plants sent by William Kerr from Canton, including a tree peony for Sir Abraham Hume[4] of Wormleybury, Hertfordshire, which proved quite different from any variety previously grown—so different

[1]*The Botanical Magazine*, 1808   [2] Ib.   [3] *The Gardener's Magazine*, 1827   [4] See under Camellia

Paeonia suffruticosa

(TREE-PEONY)   P. J. Redouté (1813)

indeed that some botanists regarded it as a new species and named it *P. papaveracea*, on account of the poppy-like shape of its seedheads. Its flowers were white or blush-pink, nearly single, with a large purple spot at the base of the petal, "vyeing with the beauty of a Gum Cistus."[1] Hume also procured a pinkish-mauve double kind, about 1817; and for several years these were all that were available. "Collectors who have means of communication with China," wrote Sir J. E. Smith in 1819, "have anxiously endeavoured to obtain some of the other varieties, but in vain."[2] The first attempt to produce new kinds from seed in this country was made in the 1830s by the Earl of Mount Norris at Arley in Worcestershire, who raised some plants from seeds of *P. suffruticosa* var. *papaveracea*, which he thought had accidentally crossed with a herbaceous variety.

When Robert Fortune was sent to China by the Horticultural Society in 1834, he was instructed to look out for tree and herbaceous peonies, including those with blue flowers, "the existence of which is, however, doubtful." He found that each locality had its own particular varieties, but that there was little interchange, so that kinds common in Canton (whence all the previous introductions had been made) were rare in Shanghai, and vice versa. The plants he procured were raised in a series of small nursery gardens about six miles from Shanghai, and propagated by grafting on the roots of a wild peony species; young plants with only one bud were the most valued as they could be easily lifted and marketed, forced to produce a single enormous bloom, and then thrown away. (Fortune was often able to purchase large plants at a cheaper rate than small ones.) The most valuable was the so-called yellow, a white flower with a yellow-tinted center; there was also a mauve, "the color of the wisteria," a black, or very dark maroon, and a very large and double purple which Fortune thought was probably the "blue" which was supposed to have a thousand petals and to exist only in the gardens of the Emperor. Altogether Fortune was able to send home between thirty and forty varieties.

The tree peony had been transported from China to Japan, possibly as early as the eighth century, and was cultivated there with equal enthusiasm; and in 1844, when Fortune's importations were just beginning to reach England, a large collection of Japanese varieties was brought to Europe by Dr. Philippe von Siebold, said to be from the Imperial Gardens of Yedo and Mijako. They were quite distinct from the Chinese kinds, with a larger proportion of single or semi-double flowers.

The stage was now set for European horticulturists to

raise their own strains, and French nurseries in particular were quick to take advantage of their opportunities. (Most of the varieties available today are of either Japanese or French origin.) Early in the present century the firm of Lemoine of Nancy crossed *P. suffruticosa* varieties with the newly introduced *P. lutea*, thus adding to the color range the yellow, peach and flame shades that had hitherto been lacking. (This hybrid strain is sometimes called *P. × lemoinei*). Unfortunately many of the varieties also inherited the weak stems of *P. lutea*, and the enormous flowers are apt to hang their heads unless very carefully staked. The best known of these Lemoine crosses is 'Souvenir de Maxime Cornu', named after the then Director of the Jardin des Plantes. These golden glories are even more expensive than the older types.

Tree peonies, it must be confessed, have never been cheap. One of the Chinese varieties was named "A Hundred Ounces of Gold," the tradition being that it was formerly sold for that sum. A poem written by Po Chui (A.D. 772–846) describes the sale of peonies in the flower-market:

> . . . *The cost of the plant depends*
> *on the number of blossoms.*
> *For the fine flower—a hundred pieces of damask:*
> *For the cheap flower—five bits of silk.*

while a passing peasant muses:

> . . . *A cluster of deep-red flowers*
> *Would pay the taxes of ten poor houses.*[1]

In England, the price at first was 10 guineas, and later 5 or 6; in 1827 the Vauxhall Nursery of Messrs. Chandler and Buckingham had a stock of both 'Moutan' and *papaveracea* for sale, from 5 guineas upwards; "that it should be worth a tradesman's while to do such a thing" comments the *Gardener's Magazine*, "is a gratifying proof of the immense riches and botanical taste of this country." As an American author put it, "the next swift step to financial Avernus is to indulge oneself in Tree Paeonies . . . for just around the corner from Tree Paeonies stands the Poor House."[2] Nevertheless it would be worth selling all one's possessions to buy one, and digging out everything else in the garden to make room for it. Tree peonies are very hardy, but are among the plants liable to suffer from the mildness rather than the severity of British winters, starting into growth unseasonably early in the spring, and Paxton advises planting them under a north wall, "so as to retard their flowering, and lower

---

[1] Rees' *Cyclopaedia*   [2] Ib.

[1] *170 Chinese Poems*, trans. Arthur Waley   [2] *Another Gardeners' Bed Book*, Richardson Wright, 1933

their excitability."[1]

Until the first decade of the present century, the only Moutan peonies known were cultivated plants from Oriental gardens; the wild progenitor had never been found—not even on the hill called Mou-tan Shan, Peony Mountain, because of the tradition that tree-peonies formerly grew there in lavish abundance. In 1910, however, William Purdom found a dark red kind growing among the foothills of the Min Shan, by the upper reaches of the Blackwater River in southern Kansu; and in 1914 Reginald Farrer found a white-flowered form with maroon blotches, growing undoubtedly wild near the little village of Fu-ah-Chieh, on the lower reaches of the same river. Unfortunately Farrer was not able to collect seed of his find; but in 1926 Dr. J. F. Rock sent to the Arnold Arboretum at Harvard seeds of a peony answering to Farrer's description—and incidentally, almost exactly to that of Sir Abraham Hume's variety *papavera-cea*—which he found growing in the courtyard of the Cho-ni Lamasery in southern Kansu, the plants having been brought by the lamas from the Min Shan. Two years after Dr. Rock's visit the lamasery was destroyed, peonies and all, by Mohammedans, and all the lamas killed; but within ten years it was rebuilt, and it is pleasant to record that Dr. Rock was able to send back to the restored institution seed of the peony that had been originally brought from there. Rock's first consignment of seed had been widely distributed by the Arnold Arboretum, and plants raised from it flowered in several countries of Europe in 1938. This variety comes true from seed, and is now regarded as the original wild type of the species, Purdom's red-flowered plant (not in cultivation) being classed as var. *spontanea*.

The family is named after "that good old man, Paeon, a very ancient Physition who first taught the knowledge of this Hearbe"[2] and who is said to have used the roots of the common herbaceous peony to cure a wound given to Pluto by Hercules. "We presume its virtues are altogether reserved for such august occasions, they having never been made manifest on any other, so far as we can learn."[3] *Mou-tan*, one of several Chinese names for the flower, means Male Vermilion.

NOTES: Herbaceous peonies are popular because they are easy to cultivate and are relatively inexpensive. The Tree-Peony, *P. suffruticosa*, despite its reputation as being difficult to grow, is a magnificent plant. There are problems with propagation, precise cultural practices are required and they are slow to bloom; yet the rewards are great for the persistent gardener.

[1] *The Flower Garden*, 1882–84 [2] Lyte [3] Smith, in Rees' *Cyclopaedia*, 1819

The Tree-Peony requires a soil rich in humus and a well-drained location with some shade. While they are tolerant of cold (they originated in a rigorous climate), a mulch is desirable to keep the soil temperature cool in winter and to help prevent moisture loss in summer—both essential for good root growth. They are grafted plants, so deep planting—to encourage rooting along the stem—is recommended. If you follow these and other suggestions from reliable peony growers or The American Peony Society, your Tree-Peony will last for decades.

Paliurus

A small family related to Rhamnus and Zizyphus, one member of which, *P. spina-christi* (Christ's Thorn), was formerly much planted by the pious and is still occasionally grown, because of the ancient tradition that it was from this plant that the Crown of Thorns was made. It grows in central Europe and the Near East, including Palestine and Judea. Turner had only seen it in Italy, but Gerard had a specimen that he had raised from seed, and quotes the opinion of Petrus Bellonius (Pierre Bélon) that it was the plant used in the Crucifixion, because "in Judea there was not any thorne so common, so pliant or so fit to make a crown or garlande of, nor any so full of cruel sharpe prickes." The thorns are arranged in pairs, one of each pair being straight and the other hooked; Evelyn called them "terrible and irresistible spines, able almost to pierce a coat of mail." Later botanists put forward the theory that the Crown of Thorns was more likely to have been made from *Zizyphus spina-christi*, which is less tall and easier to pick, and which grows abundantly at Jerusalem and Golgotha. The fruit of the latter is a black, edible berry, whereas the paliurus has an extraordinary seed-vessel which forms another of its claims to interest: it has been variously described as "flat and broad, very like unto small bucklers, as hard as wood" (Gerard) or as resembling a "low-crowned, wide-brimmed hat." These seeds used to be sold in the herb shops of Constantinople under the name of *Xallé*, and were prescribed by the hakims for many complaints; but the uses for the shrub mentioned by Dioscorides were magical rather than medical: "And it is said that the branches thereof, being laid in gates or windowes, doe drive away the enchantments of witches . . . If any take up Rhamnus, the moone decreasing, and beare it, it is profitable against poyson & against naughty men, and it is good for beasts to beare it about them, and to put about shipps, & it is good

Paliurus australis
(now *P. spina-christi*)
(CHRIST'S THORN)   Ferdinand Bauer (1819)

Pernettya angustifolia
(now *P. mucronata*)
(PRICKLY HEATH)   W. Fitch (1841)

Parthenocissus quinquefolia
(VIRGINIA CREEPER)   J. Abbott (1797)

Passiflora spp.
(PASSION-FLOWER)   E. Twining (1849)

against ye paine of ye head, & against devills and their assalts." Turner adds that the leaves are good for "wild fires, and great hot inflāmationes."

Bean considered the shrub worth growing for its abundant, though small, greenish-yellow flowers, its curious fruits and its perfect hardiness; but even in Parkinson's day it was cultivated rather for its interest than for its beauty. The name of Paliurus, which goes back to the time of Theophrastus, is said to have been taken from that of a town on the coast of Tunisia, now called Nabil. NOTES: While frost hardy from Pennsylvania, Christ's Thorn is rarely grown as an ornamental in this country.

Parthenocissus

. . . The only sound to be heard was that of the vain Virginia Creeper at her toilet, drying her scarlet leaves, and now and then letting fall a drop of water, which struck the flags with a pleasing "plop."
Jean Santeuil, Marcel Proust, trans. Hopkins, 1955

Here is another of the plants that cannot be discussed until the muddle of its nomenclature has been explained. The original species was discovered in North America before 1629, and was thereafter simply and logically known as the Virginia Creeper. Parkinson called it "the Virginia Vine, or rather Ivie," and as an Ivy it was officially classified for the next 160 years, until de Jussieu suggested that it should be removed to the Vine family, and *Hedera quinquefolia* became *Vitis quinquefolia*. In 1803 Michaux placed it in the new genus of Ampelopsis, but many subsequent botanists did not agree with this arrangement and continued to class the plant under Vitis, until Planchon in his monograph on the vines (1887) not only recognized the new family, but divided it into two, Ampelopsis and Parthenocissus, owing to a minute difference in the structure of the flowers. The Virginia Creeper can thus tell over a name on each of her leaves' five fingers—*Hedera quinquefolia, Vitis quinquefolia, Ampelopsis hederacea, Ampelopsis quinquefolia* and now *Parthenocissus quinquefolia*. No wonder she is vain.

The most obvious difference between the Virginia Creeper and the vines lies in the adhesive discs the former produces at the end of its tendrils. Some, though not all, of the Parthenocissus species have these suckers, and they also occur in one or two other plant families, but the remainder of the vines have only spiral claspers. As Parkinson noted, the tendrils of *P. quinquefolia* divide into "foure, five or six or more short and somewhat

broad claws, which will fasten like a hand with fingers so close thereunto, that it will bring part of the wall, morter or board away with it, if it be pulled from it." The disc develops when the tendril comes into contact with a hard object, and is said to vary in size according to the roughness or smoothness of the surface; once attached, the tendrils contract spirally and draw the branch close.

This mechanism enables the plant to climb quickly to a great height "it will shoot up so high, that whatever Height the Ridge of the House may be of, 'twill soon surpass it";[1] and as it will grow in almost any soil or situation and endure with apparent imperturbability the smokiest atmosphere, it soon became appreciated as a cloak for urban ugliness. By 1722, according to Fairchild, there was hardly a street, court or alley in London without it; and although Miller (1759) carps because it is not an evergreen and therefore makes "but an indifferent Appearance in Winter," Abercrombie (1778) praises it for covering unsightly buildings "with its beautiful five-leaved foliage in summer," and Hibberd (1870) describes a house in Piccadilly "where the Virginia Creeper is trained in festoons across an extensive flat frontage . . . in its way worth a note of admiration, as one of the lesser lions of London." (None of these authors, by the way, makes any mention of the autumn color, though some of the later writers speak of it.) Phillips tells of a house in London which was covered by this climber, until the owner cut it down "to prevent having his house indicted as a nuisance for harboring sparrows, whose twittering commenced too early in the morning for those whose evening parties begin at midnight."

The original Virginia Creeper has to a great extent been superseded, not altogether with advantage, by later introductions, especially *P. tricuspidata*, more familiar perhaps as *Ampelopsis veitchii*, the name under which it was originally distributed. It was collected by John Gould Veitch in Japan in 1860–2, and distributed by the Exeter firm of the same name in 1868; Richard Oldham sent a plant to Kew about the same time, and it is now more or less naturalized by the side of the Thames. This too climbs by means of suckers, but has variable leaves, sometimes entire, sometimes, on the same plant, three-foliate, on which account it was at one time called *Vitis inconstans*. A less familiar, but more distinguished member of the family, *P. henryana*, was found by Dr. Augustine Henry in China about 1885, and introduced by E. H. Wilson for the firm of Veitch in 1900. Each of its dark leaflets is attractively marked down the center with pink and greyish-white, but it is slightly tender, or

[1] Liger

at least, less bone-hardy than the other two, though it can ramp well enough when well-situated. A fourth species, *P. vitacea*, is not self-clinging like the first three; Bean thought it handsomer than *quinquefolia*, and suggested that it should be trained along a rafter or other horizontal support, to "send down a thick curtain of branches." It was brought from eastern North America in 1824.

For once the Greek name of the family is a translation from the English—*parthenos*, a virgin, *kissos*, ivy—a reversal of the usual procedure that derives the English name from the Greek or Latin. " 'Tis called the Virgin Vine," says Liger, pleasingly though erroneously, "because, if one may say so, 'tis a Maid, and has hitherto brought forth nothing."

NOTES: Both the native Virginia Creeper, *P. quinquefolia*, and its Japanese counterpart, *P. tricuspidata*, are excellent climbing plants. *P. tricuspidata*, our common Boston Ivy, is especially vigorous on brick walls, wooden structures or trees. In autumn, the glossy leaves turn scarlet.

There are no special cultural requirements other than regular garden conditions. Perhaps its greatest failing is its ability to cling tenaciously to structures and invade eaves and other structural openings.

Passiflora

When the first of these flowers to be seen by a European was discovered in Central or South America, some pious and imaginative Spanish missionary discerned in its strangeness the emblems of Our Lord's Passion. The three stigmas represented the three nails (two for the hands and one for the feet); the five stamens the five wounds; the corona, the Crown of Thorns, or alternatively the Halo of Glory; the five sepals and five petals of the corolla, the Apostles, Judas and Peter omitted; and the five-lobed leaves and curling tendrils, the hands and whips of His persecutors.[1] An account of this wonderful *Flos Passionis* or Passion-flower, compiled from drawings and descriptions brought to Rome from Mexico, appeared in a treatise on the Cross of Calvary published by Jacomo Bosio in 1610; the plant itself quickly followed, for it bloomed in Europe before 1622 and in England before 1629. The Protestant Parkinson expressed much righteous indignation at what he considered a superstition of the Jesuits, who had "caused

[1] Other religious interpretations of the various parts have also been made. Incidentally the leaves of *P. coerulea* have as often seven fingers as five; those of *P. edulis* and *P. incarnata* are three-lobed, and those of *P. quadrangularis* entire

figures to be drawne, and printed, with all the parts proportioned out, as thornes, nailes, speare, whippe, pillar etc., in it, and all as true as the Sea burnes, which you may well perceive by the true figure, taken to the life of the plant, compared with the figures set forth by the Iesuites, which I have placed here likewise for everyone to see"; but the drawing to which he took such exception, though admittedly crude, was probably intended to be diagrammatic rather than representational.

The plant grown and illustrated by Parkinson was not the hardy blue passion-flower which is the most familiar today, but the tender *P. incarnata*, from the southeastern United States, described as "a pretty thing of semi-herbaceous habit," which requires a greenhouse to bring its flowers and fruit to perfection. Apparently live plants, and not just seeds, were imported, for Parkinson says, "I have seene roots that have beene brought over, that were as long as any rootes of *Sarsa parilla*, and a great deale bigger, which to be handsomely laid into the grounde were faine to be coyled like a cable."

We first hear of the blue passion-flower, *P. coerulea*, as being grown by the Duchess of Beaufort in 1699. It is the only member of the family to approach outdoor hardiness; although a native of central and western South America, it can successfully be grown on sheltered walls. In the north, it cannot be relied upon to produce its striking but fugitive flowers. Johnson suggested that in cold districts the shoots "might easily be wrapped together and defended in winter by a mat." Not so very easily; for they will grow 12 to 15 feet in a season, and "if not well looked-to and nailed up, will get into great confusion, and become rather ugly than otherwise."[1] This passion-flower was very popular in the London suburbs in the 1890s, and the white variety named 'Constance Elliot' was grown before the turn of the century.

In the London area and the south and west, *P. coerulea* not only flowers, but fruits. The orange-yellow berries are large and ornamental, and as long ago as 1875 were recommended for indoor decoration, as they last well when cut. They are edible, but not very palatable, even with wine and sugar; they do not compare for flavor with the fruit of some of the other species.

These are only a few examples from a very large genus, containing over three hundred species, a number of which were cultivated in nineteenth century greenhouses. "The possessor of a spacious stove-house may find much to interest him in the culture of passifloras," wrote Hibberd; but today such fortunate specialists could probably be counted on the toes of a two-toed sloth, and those who wish to see a variety of passion-flowers must

[1] Cobbett

visit Brazil, where, it is said, they "climb to a height of sixty feet, forming festoons from tree to tree, which are spangled with these brilliant stars in the most superb manner."[1]

In the Language of Flowers, the passion-flower stands for Religious Superstition. Parkinson gives it the alternative name of Maracoc.

NOTES: Mostly tropical and sub-tropical vines, the passion-flowers are limited to Florida and similar climates. The flowers are so striking, however, that some northern gardeners will take the time to grow them under glass.

Pavia, see under *Aesculus*

Pernettya

Several members of this family are grown by rock-garden enthusiasts—including the Chilean *P. furens*, whose berries if eaten cause wild excitement, acute mania and sometimes death—but the only species suitable for this book is the 3-foot *P. mucronata*, the type and original of the family, and one of the first shrubs to be developed and cultivated entirely for the sake of its ornamental fruits. It was first recorded by the gentleman whose name it bears, Dom Antoine Joseph Pernetty (1716–1801), who took part in de Bougainville's expedition to the Straits of Magellan and the Falkland Islands in 1763. Pernetty himself called the plant *Bruyère à feuilles pointues*, or Prickly Heath, and this is preserved in the Latin specific name, *mucro* meaning a sharp point. "Cranberries both white and red" (i.e., Pernettyas) were also found in Tierra del Fuego by Sir Joseph Banks and Dr. Solander, on their exploratory voyage with Captain Cook in 1770; but this plant of the circumnavigators was not introduced until 1828, when seed was sent to Edinburgh Botanic Garden by James Anderson, who sailed on H.M. Survey Ship *Adventure* (1826–30) as botanical collector. (The *Adventure* and her sistership the *Beagle*, were then engaged on a survey of the coast of South America; on her next trip, in 1831, the *Beagle* carried a young naturalist named Charles Darwin.)

Plants raised from Anderson's seed flowered in 1830, but the new shrub was little noticed until taken up by an Irish nurseryman, L. T. Davis of Hillsborough, County Down. About 1850 he started to raise seedlings from the variety *angustifolia* (at one time classed as a distinct species), the hardiest and most prolific sort then in cultivation. These seedlings won several awards when he began

[1] Phillips

to exhibit them, from 1878–82, and are still among the best varieties, the berries ranging in color from white and rose to carmine, mauve and deep purple.

Loudon in 1838 praised the plant for the "neat appearance and dark colour of its foliage," but does not mention the fruit; it is possible that he never saw it, for the pernettya is unisexual, and has caused a great deal of trouble to unsuspecting gardeners by requiring a male plant to fertilize each harem of females. The Davis varieties, however, are self-fertile, and so is Bell's Seedling, a handsome kind with large fruits of a mahogany-red. About 1928 a pernettya growing in the Wisley Garden crossed with a nearby *Gaultheria shallon* to produce an accidental bigeneric hybrid, whose correct name is *Gaulthettya* × *wisleyensis*, though it is more often seen listed as *Gaulnettya* 'Wisley Pearl'. Other hybrids between the two families are known to occur in the wild.

The object of de Bougainville's expedition was to found a new French colony in the Falklands, to compensate for the recent loss of the French territories in Canada. Pernetty, an Abbé in the Benedictine order, accompanied the expedition in the dual capacity of natural historian and officiating priest, and recorded his impressions in his *History of a Voyage to the Malouine (or Falkland) Islands, etc.*, translated into English in 1771—a work, said Lindley, "remarkable for its interest, as well as for its candour and exactness."

The plant was not given its present name until 1825, many years after Pernetty's death. Botanist's honors are usually posthumous, like wreaths on coffins; doubtless they would have preferred to have had both flowers and favors when they were still alive.

NOTES: Pernettyas, natives of Chile and Argentina, can only be cultivated in warmer regions of the country. They are not common nursery material.

Philadelphus

P. *coronarius* was introduced to Europe along with the lilac, by Ogier Ghiselin de Busbecq, Ambassador from the Emperor Ferdinand to Soleiman the Magnificent, on his return from Turkey to Vienna in 1562. The two shrubs long remained inseparable companions, and this led to a confusion in nomenclature that has lasted to this day. At first they were classified together, and shared the name of Syringa, derived from *syrinx*, a pan-pipe, because the wood of both shrubs, hollow and pithy like that of the Elder, was used by the Turks to make pipes; and we find them together in Gerard's *Herball*, under the names of Blew Pipe and

Philadelphus inodoratus
(MOCK ORANGE)   M. Catesby (1843)

White Pipe tree. Herein is demonstrated the extraordinary strength of oral tradition; for though the name of Philadelphus was given to the White Pipe by Bauhin as early as 1623, and confirmed for the genus by Linnaeus in 1735, the shrub is still familiarly called "Syringa."

The lack of improvement under cultivation was of less importance as the paramount property of the Mock Orange was its scent, which some find intoxicating and others insupportable. Gerard was among those who disliked it. "I once gathered the flowers," he writes, "and laid them in my chamber window, which smelled more strongly after they had lien togither a few howers, with such a ponticke and unacquainted savor, that they awaked me from sleepe, so that I could not take any rest till I had cast them out of my chamber." Out-of-doors, they hold all the concentrated sweetness of early summer, when "the air at some distance will be replete with the odoriferous particles of these fragrant flowers," but indoors, Marshall continues, they are "very improper for chimneys, water-glasses etc. in rooms, for in those places their scent will be too strong, and for the ladies in particular, too powerful." The late Mr. E. A. Bowles found that they gave him hay-fever, and removed all bushes of *P. coronarius* from his garden, except for the variegated

forms, from which he had the flower buds removed before they opened. Mock Orange flowers have been used in perfumery, and to flavor tea; the leaves will impart a flavor of cucumber to summer drinks.

For years—decades—centuries, the original *P. coronarius* was the only species generally grown in British gardens, though a number of others were introduced, first from America and then from the Orient. (There are about forty members of the family, but they hybridize so freely, that in gardens it is difficult to keep the species true.) By no means all are scented, and among the fragrant kinds there is much variety of perfume; the second species to be introduced was one with no scent at all, and it was therefore named *P. inodorus*. A single specimen was found in South Carolina about 1726, by Mark Catesby; Miller had a plant before 1734, and also received several consignments of seed from Dr. Thomas Dale of Charleston,[1] but the species proved difficult to propagate, and has always been rare, both in cultivation and, apparently, in the wild. A long pause was broken by the arrival of two handsome tall-growing American kinds, *P. pubescens* in 1800 and *P. grandiflorus* in 1811; after this, new philadelphuses came thick and fast, including several from China and one from Japan. In this case, however, the oriental species have not played so large a part in the horticultural development of the genus as the American ones.

Among the most important and distinct of the latter are *P. coulteri* and *P. microphyllus*. The former was named after the Irish botanist Dr. Thomas Coulter, who explored central Mexico in 1831–3; it is thought by some to be only a variety of *P. mexicanus*, which was one of the earliest known members of the genus, having been described by Hernandez in his *Thesaurus* (1651) under its Mexican name of *Acuilotl*. Both *mexicanus* and *coulteri* were introduced in 1840, and are about the only members of the family that are not absolutely hardy; but *coulteri* is distinguished by having a purple spot at the base of each petal, which it has passed on to its numerous descendants. *P. microphyllus* is the smallest member of the genus, and was introduced to England and to Europe from its home in Colorado and Arizona by Professor Sargent of the Arnold Arboretum, about 1883.

These two species, *P. microphyllus*, and *P. coulteri*, are of great importance as parents of the modern garden hybrids. The first of these was raised by M. Lemoine of Nancy. As soon as he received *P. microphyllus* from Professor Sargent he crossed it with the old *P. coronarius*, and produced *P. × lemoinei*, which Bean called "one of the greatest successes ever achieved by the hybridiser's

[1] Nephew to the botanist Samuel Dale

Phillyrea latifolia

(EVERGREEN PRIVET)   Ferdinand Bauer (1806)

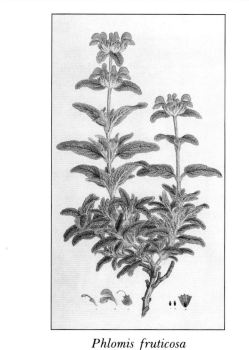

Phlomis fruticosa

(JERUSALEM SAGE)   Ferdinand Bauer (1827)

art," but which has in turn been surpassed by some of the later kinds. When the new hybrid flowered, the same grower crossed it again with *P. coulteri*, and the result (before 1891) was *P.* × *purpureo-maculatus*, from which all the modern varieties marked with pink or purple on the petals are ultimately descended. Another of Lemoine's successes was the semi-double 'Virginal', which received an F.C.C. in 1911, and is still one of the most popular of shrubs. Since then a great many other varieties have been raised, in France and elsewhere; and for those that regard the besmirched, blowsed or scentless kinds with distaste, there are plenty of others that have retained the original purity of form and color, with larger and more liberal flowers, smaller leaves and better size and habit than the original *P. coronarius*, which is now quite superseded.

"Philadelphus" was a name used by the ancient Greeks for an unidentified shrub, and may have been derived from Ptolemy Philadelphus, King of Egypt— "but on what account, we are left to surmise."[1] Literally translated, it means "brotherly love"—hence the sect of the Philadelphians and the name of the capital of Penn-

[1]Phillips

sylvania, neither of which, unfortunately, has any connection with the shrub. The family belongs to the Saxifrage Order—far removed from the Lilacs, which are members of the Oleaceae—and its nearest relations are the Deutzias, Hydrangeas, Ribes and Escallonias, families not, on the whole, remarkable for perfume, with which the Philadelphus is so liberally endowed. In the Language of Flowers, the Mock Orange is made to signify Memory, because "when we inhale this penetrating odor, it seems to follow us everywhere for a considerable time."[1]

NOTES: With more than fifty species and hybrids to choose from, the Mock Orange has been a classic garden shrub for over fifty years, mainly due to the remarkable efforts of the Lemoine Nursery in France between 1894 and 1927. Since the Second World War, many of the fine selections have now been lost, even from most botanical collections. Nurseries offer only three or four cultivars at a time. Among these, 'Minnesota Snowflake', a hardy hybrid developed in Minnesota more than fifty years ago, is a perennial favorite. There are several species native to the United States, one group in the southeast

[1] Phillips

and another in the southwest. The various Lemoine hybrids involve both North American and Asiatic species.

The Mock Orange has simple cultural requirements: a good garden soil, ample moisture that mulching will provide and a sunny location. There is a Mock Orange for most U.S. conditions and, while it is not an exciting plant when past flowering, the stark white flowers in early summer are interesting.

Phillyrea

Let Phillyrea on your Walls be plac'd,
Either with Wire, or Slender Twigs made fast.
Its brighter leaf with proudest Arras *vies*
And lends a pleasing Object to our Eyes.
Then let it freely on your Walls ascend
And there its native Tapestry extend.
Of Gardens (1712) R. Rapin

Up to the middle of the eighteenth century the number of evergreen shrubs available to the British gardener was extremely small. Miller in 1724 listed no more than twelve that were commonly stocked by nurserymen—even at a time when "greens" were in great demand. The Phillyreas were proportionately valued; their dense growth and tolerance of clipping rendered them particularly suitable for formal styles of gardening, whether of the Elizabethan or the William and Mary period. There are two main species, *P. angustifolia* and *P. latifolia; P. media,* formerly regarded as a third, is now classed as a variety only. Both are natives of southern Europe, and all three, including var. *media,* were brought to England before 1597. They were used "for the raising of *Espalier* Hedges, and covering of *Arbors,* being always of incomparable Verdure";[1] and Hanmer (in 1659) said that the phillyrea had replaced the pyracantha as a hedge plant. Silver- and gold-variegated sorts were grown by the mid-seventeenth century, and by 1728 five varieties were to be had—"the True, the Plain, the Bloach'd, the Dutch silver-leaved, and the Dutch guilded"—the first of which would make "a most beautiful Tree, and especially when bred up in the form of a Cone."[2] By 1778 the numbers had increased to three species (counting *media*) and seven varieties, "all delightful evergreen shrubs of a free easy growth . . . no good shrubbery . . . should be without a collection of these fine ever-greens, disposed in the most conspicuous points of view."[3]

But a decline was at hand. English tastes in garden design were changing; new and more interesting shrubs were pouring in; and by 1838, according to Loudon, many of the varieties were only to be found in botanic gardens—the nurserymen had ceased to grow them, owing to lack of demand. From then onwards, references to the phillyreas practically disappear, and today they are seldom listed, though they may be more often grown than seen, for they are the sort of shrubs the eye is apt to pass over without notice. A comparatively new species, *P. decora,* which was introduced from Lazistan on the shores of the Black Sea in 1868, is said to be the most handsome of the genus, but even this kind is chiefly recommended as a foil or background for other shrubs. It has the distinction, however, of being one of the parents of that useful hybrid, *Osmarea* × *burkwoodii.*

The phillyreas are members of the Olive Order, and their small flowers are fragrant. *P. angustifolia* has sometimes been called Evergreen Privet or Mock Privet, but the genus never really acquired an English name, though the spelling of the Latin one is a trap for the unwary, and many variants occur. (London and Wise spelled it "Phyllyaroea"; Celia Fiennes plumped frankly for "filleroys.") This name, a Latin version of the Greek one used by Dioscorides, is derived by some from the Greek *phyllon,* a leaf (i.e., a leafy shrub, the flowers being inconspicuous), while others say it commemorates Phillyra, the mother of Chiron the Centaur, who was changed into a tree—a fate which frequently befell the heroines of Greek myth, who seem to have taken to metamorphosis much as heroines of a later date took the veil.

NOTES: While they can be grown for their foliage (the flowers are inconspicuous) in mild or warm regions as they are in Europe, the evergreen privets have not received much attention here.

Phlomis

Some very handsome labiates belong to this family, but only one is hardy, a shrub, and in general cultivation—*P. fruticosa,* or Jerusalem Sage, described by Dioscorides in the first century A.D. as "bearing high rods and treeish, the leaves like to Sage . . . a yellowish flower, like gold."[1] In our gardens the "rods" (shoots) are not very "treeish," rarely exceeding 3 or 4 feet, but the shrub still produces its terminal "roundels, or crownets, of yellow gaping flowers, like those of dead Nettle, but much greater."[2] Gerard grew it (and also a herbaceous mauve-flowered species with the pleasing name of *P. herba-venti,* the Plant of the Winds), but the narrow-

[1] Worlidge, *Systema Agriculturae,* 1675   [2] Langley   [3] Abercrombie

[1] Goodyer's translation   [2] Gerard

leaved plant which Parkinson in 1640 described as a variety of *P. fruticosa* is now given specific rank as *P. lychnitis*. "Lychnitis" was a name used by Apuleius for the Mullein, and like Lychnis (from *lychnos*, a lamp) and Phlomis (from *phlogmos*, flame) was applied to the plant concerned because it had woolly leaves "fit to make candlewicks." Some of the early botanists actually classed *P. fruticosa* as a Mullein, much to Parkinson's scorn. "So why . . . they should call it . . . Verbascum Mullein, I see no cause more than that the leaves in both are wooly like Mullein, and may serve as a weeke for Lampes . . . but that is not a sufficient cause in my judgement, to make them of the tribe of Mulleins, other things not concurring as the flowers whereof I have spoken before. Let others of knowledge bee judges herein."

Parkinson was no better satisfied with the English name of French Sage, which Gerard had used: "whereas it is as great a stranger in France as in England; Yet they doe with this as with many other things, calling them French, which come from beyond the seas . . ." The more recent name of Jerusalem Sage is no improvement, for the shrub does not grow in Palestine or Judea, though fifteen other phlomis species are found there. It is a native of the Mediterranean region, and needs a dry sunny spot in this country to show it at its best.

Besides its handsome flowers—"a finer yellow can hardly be conceived than the colour of which they are possessed"[1]—the shrub has semi-evergreen, or rather ever-grey leaves, of what Gerard calls a "strong ponticke savour," which "make a pleasing show in winter," and decorative seedheads. The leaves, says Miller, are "greatly recommended by some Persons to be used as Tea for Sore Throats."

NOTES: While hardy in southern states, these shrubs are not cultivated to any extent elsewhere.

Phyllostachys, see under *Arundinaria*

Pieris

One would expect to find nine species in this family, named after the Pierides, the Muses, but there are only four of garden importance—all evergreen shrubs with white flowers—and one or two others not in cultivation. All four are worth growing, though, as usual their beauty is in inverse ratio to their hardiness, the tenderest kind being the handsomest. They also follow closely the pattern established by so

[1] Marshall

many of the families of similar American and Asian distribution; the American species being the hardiest and the first to be introduced, but surpassed in beauty by the oriental sorts.

Three different dates are given by respectable authorities for the introduction of *P. floribunda* from the southeastern United States—1800, 1806 and 1812. It is said to have been one of the importations of John Lyon, after whom the nearly related family of Lyonia is named. Little is known about him, except that he was a Scot who emigrated as a gardener before 1796 and became an enthusiastic collector of American trees and shrubs, returning to England in 1806 and again in 1811, and that before 1816 he "fell a victim to a dangerous epidemic amidst those savage and romantic mountains, which had so often been the scene of his labours."[1] His Pieris is very hardy, and holds its panicles of small white flowers sturdily upright, with what Dr. Stoker called an air of chilly self-righteousness, whereas all the oriental kinds are more or less drooping.

The first of these, the Japanese *P. japonica*, although described by Thunberg in 1784 under the name of *Andromeda japonica*, was not apparently a very early introduction; details are lacking, but it is known to have been in cultivation by 1870. It is hardy, but grows larger in mild districts; blooming as it does in March and April, the flowers are sometimes damaged by bad weather. Seed of a much more recent discovery, the Formosan *P. taiwanensis*, was sent to the Arnold Arboretum by E. H. Wilson in 1918, and distributed by Professor Sargent with his customary generosity. Plants raised from this seed, sown in 1919, gained an F.C.C. when they were exhibited at the Royal Horticultural Society in 1922; for this Pieris has the great merit of beginning to bloom at a very early age. It has proved hardy in the south and west, and may be so elsewhere.

But the belle of the family is undoubtedly *P. formosa*, a Himalayan species introduced before 1858. In the southwest of England this Pieris will grow to a height of 20 feet, but elsewhere in Britain it is rather tender and requires a well-sheltered situation. (Although less precocious than *P. taiwanensis*, it begins to bloom when still relatively small.) One of its charms is the reddish tint of its young shoots, and this is considerably more pronounced in the large-flowered variety *forrestii* (at first thought to be a distinct species), seed of which was sent home by Forrest from Yunnan about 1910, when he was collecting for A. K. Bulley, the founder of the seed and nursery-firm of Bees Ltd. In sheltered gardens the fiery-red new foliage of this variety comes out at the same

[1] Nuttall, quoted by Bean

Pieris forestii
(CHINESE PIERIS)   M. Smith (1909)

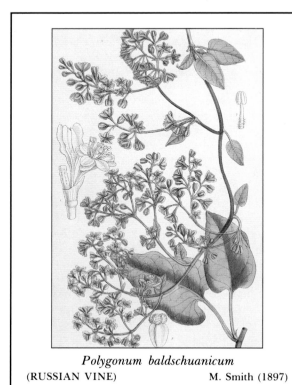

Polygonum baldschuanicum
(RUSSIAN VINE)   M. Smith (1897)

Poncirus trifoliata
(HARDY ORANGE)   A. Barnard (1885)

time as the relatively large white flowers; in colder districts, the flower-buds, formed in autumn, are liable to drop before spring, but the red young growth still renders the shrub ornamental. Possibly we are only at the beginning of the development of this family; a hybrid has already been raised—it appeared as a chance seedling at the Sunningdale nursery in Surrey—which is said to combine the red foliage of *P. formosa* var. *forrestii* with the hardiness of its other parent, *P. japonica*; it has been given the appropriate name of 'Forest Flame'.

The earlier discoveries were classed as Andromedas, until a new genus was founded for them by David Don in 1834. Much rearranging has been done by later botanists, who still seem only half-satisfied with the results. The Nine Muses were called the Pierides, either because they were born in a part of Thessaly called Piera, or from Pierus, their sacred mountain, also in Thessaly; but there is also a story that the name originally belonged to the nine daughters of Pierius (a rich Thessalian), who challenged the Muses to a musical contest, and failing, were turned into magpies. Don gave no reason for his choice of "Pieris," which is also the family name of the Cabbage-White butterfly.

NOTES: The andromedas are among our finest broad-leaved evergreen shrubs wherever they are hardy. Early on, we grew only the unselected Japanese species, *P. japonica*. But now there are at least ten improved forms, ranging from those with brilliant red flushes of new growth and those with more perfect flower panicles to variegated and dwarf forms. As Miss Coats points out, we continue to look for varieties that have brilliant spring foliage. More recent hybrids with other oriental species are now being offered by the nursery trade. Our one native species, *P. floribunda*, which is known as the Fetterbush, comes from the Appalachian mountains. Although difficult to propagate, it is beginning to appear in catalogues.

P. floribunda is usually propagated from seed because cuttings are usually difficult to root—nurserymen are willing to use the seed method in order to make it more widely available.

Polygonum

Although the Polygonum family is a large one, there is only one species that needs concern us here, and that is the rampant climber *P. baldschuanicum*, which in late summer covers whatever ugliness may be underneath with a smiling innocent show of not quite white flowers, and is never content with its own

territory, but must encroach on that of its neighbors. It is called Russian Vine, not on account of these characteristics, but because it comes from Bokhara. Albert von Regel, the son of Dr. E. A. von Regel (a German botanist and for thirty-seven years director of the St. Petersburg Botanic Garden), found it growing on the banks of the River Wachsch or Vaksh, at the foot of Mount Sevistan, in 1883. Seed was sent to Kew from the St. Petersburg

Potentilla fruticosa
(CINQUEFOIL)   J. Sowerby (1793)

garden a few years later, and the plant bloomed for the first time in England in 1896. Mr. Jason Hill suggests that it can never do enough to show its gratitude to the gardener who transferred it to more genial conditions, and that is why, having submerged his bungalow and garage, it tries to "follow its owner affectionately on his way to the station in the mornings."[1]

The name of the genus is derived from *poly*, many, and *gonu*, knee joints, because of the knotted stems of several of the herbaceous kinds. The specific name of the Russian Vine is derived from Baldschuan, or Baljuan, a town and district of Bokhara near which the plant was found.

NOTES: *P. aubertii*, the Fleece Vine, a handsome climber,

[1] *The Curious Gardener*, 1932

is the only species regularly cultivated in the United States. Despite its rampant habit, it is useful; it grows rapidly and produces clusters of small white flowers in August.

Poncirus

For those who would like to grow orange blossoms in the garden, *P. trifoliata* (syn. *Aegle sepiaria*) the Japanese Bitter Orange, though still not suitable for bridal wreaths, is a step nearer the objective than the Mexican Orange-blossom, *Choisya ternata*. Both are members of the same Natural Order, the Rutaceae, and the Mexican plant shows its affinity both to orange and rue in the rather rank smell of its bruised leaves; but the Poncirus is very closely allied to the true oranges, and was originally classed as a Citrus (*C. trifoliata*) by Linnaeus. In 1824 it was transferred by de Candolle to the family of the Bengal Quince (*Aegle marmelos*, "a delicious Indian fruit possessing high medicinal qualities.")[1] but it differs from the Aegle in its fruit, so it was put in a corner by itself as a monotypic genus, by the French botanist C. S. Rafinesque in 1838. Nevertheless, the relationship with the orange is close, and the shrub has been crossed with the China Orange, *Citrus sinensis*, to produce a hardy, or near hardy, hybrid called the 'Citrange', raised in America about the beginning of the present century, whose fruit can be used for marmalade.

The Hardy Orange was first discovered in Japan by Engelbert Kaempfer in 1690–2, and described in his *Amoenitates Exoticae*, 1712; but although cultivated in Japanese gardens, it is actually a native of China. It was not brought here until 1850, when it was introduced by Robert Fortune. It was slow to attain popularity, for the *Botanical Magazine* reported in 1880 that it was still practically unknown in English gardens, although perfectly hardy, free-flowering and sweet-scented. Perhaps gardeners were dismayed by its formidable thorns, which make it such a valuable and "boy-proof" hedge-plant— hence de Candolle's specific name of *sepiaria*, from *saepes*, a hedge. E. A. Bowles observed that the first flowers to open usually had stamens, but no stigma; then a few perfect (hermaphrodite) flowers appeared, and the last buds had stigmas but no stamens. If for any reason the female flowers were delayed until the pollen ones had fallen, no fruit was born that season. The small gold oranges are ornamental but quite inedible; they are apt to be eclipsed, at least for a time, by the golden glory of the autumn leaves. Kaempfer says that from the dried

[1] Johnson

rinds of the fruit, mixed with other spices, "is prepared the celebrated medicine *Kĭ Kŏkve*, by which name the common people also call the fruit itself"—but he does not tell us to what use the medicine was put.

"Poncirus," the name chosen by Rafinesque, is derived from the French *poncire*, a quince. "Aegle," though harder to pronounce, would have been much more poetic and appropriate, for it was the name of one of the nymphs, daughters of Hesperus, appointed to guard the golden apples given by Juno to Jupiter on their wedding day, and known thereafter as the Golden Apples of the Hesperides.

NOTES: The Hardy Orange, *P. trifoliata*, is a very effective hedge plant. While fully hardy, it is too little seen in this country. Given sun and well-drained soil, it will produce handsome foliage, showy flowers and orange-like fruits.

Potentilla

No introduction was needed for *P. fruticosa*, the Shrubby Potentilla or Cinquefoil, for it is a member of British flora, though a rare one; with so many of our garden shrubs laboriously transported by human agency from foreign lands, it is pleasant to record that this one came to us, as it were, under its own steam. It is very widespread in the northern hemisphere, and is known to have grown in Britain, in localities far south of its present range, in the late Glacial age. Shrubby potentillas are plants of the north and the cool uplands, and on the slopes of the Himalayas they grow like heath or ling, from 4 inches to 4 feet in height. Farrer found Tibetan hillsides on which their flowers ranged in color from white through cream, butter, saffron and canary yellow to almost orange, with the white forms more common on the lower slopes and the yellow ones at the higher levels.

Unfortunately it is hard to determine just what kind these Asiatic potentillas were, for the shrubby species in the family do not differ very greatly, are extremely variable, and hybridize with great freedom, so that botanists have found it more than usually difficult to decide where one species stops and another begins. Farrer thought the plants he found might be crosses between a yellow *P. fruticosa* and a white *P. davurica*, which had given rise to hybrids "primary, secondary and tertiary, to the confusion of gardeners and the multiplication of unstable species." The latest authority on the genus, Mr. H. L. J. Rhodes, considers *P. davurica* (syn. *P. glabra*) a species very variable in habit, hardiness and color, which has crossed with the closely allied *P. fruticosa* (whose range it overlaps) to produce *P.* × *friedrichsenii*, and with the more distinct *P. parviflora* to produce *P.* × *rehderiana*.

Our own *P. fruticosa* was first discovered growing on the banks of the Tees in Yorkshire by one Johnson, who supplied the description published in Ray's *Historia Plantarum*, 1688; it was "commonly cultivated in the Nursery Gardens as a flowering Shrub"[1] before 1759. The first of the *davurica* forms was sent from St. Petersburg to the firm of Loddiges in 1822. It came originally from Siberia, and the sender was Joseph Busch, the son of John Busch, a gardener of German descent, who owned at one time a nursery at Hackney, but who sold it to Loddiges in 1771, and set sail for Russia to become gardener to Catherine the Great and lay out the gardens of the Tsarskoe Selo. Most of the garden varieties grown today are derived more or less directly from *P. davurica*, said to be more floriferous and of better habit than *P. fruticosa*, but one or two, such as 'Gold Drops' and var. *Farreri*, are more closely related to *P. parviflora*.

The potentilla family is a large one, with more than 350 members, mostly herbaceous perennials. Some of them are British wildflowers, very ornamental in their place but too invasive for the garden, such as tormentil, cinquefoil and goosegrass or silverweed. The latter has roots like small parsnips, formerly eaten by children in the north; Ray in 1677 said that the boys in Yorkshire called them "Moors," and Anne Pratt, that in time of dearth they were used instead of bread for weeks together by the islanders of Tiree and Coll. They are regarded as delicacies in the Himalayas and Farrer was regaled with them more than once by his Tibetan hosts.

The name is derived from *potens*, power; some thought that coupled with *illa* it meant "Plant of little power," but the diminutive seems more likely to apply to the plant—"Small herb of great potency"—for the common cinquefoil, *P. reptans*, was used in very ancient times to drive out the demon of fever, and was also believed to be good against witches. Its five-fingered leaf was a popular emblem in heraldry, for it symbolized the five senses, "and he that can conquer his affections, and master his senses," says Guillim, ". . . may worthily and with honour beare the *Cinquefoile*, as the signe of his *fivefold victorie* over a stronger Enemy than that *three-headed monster* Cerberus."[2] Nineteen or twenty examples of cinquefoil leaves occur among the carvings in the Chapter-house of Southwell near Newark, which was built about 1330. Rhodes rightly considers the name of Shrubby Cinquefoil cumbrous, and prefers the German *fingerkraut* or Finger-bush; but it would be a pity, all the same, to jettison the Norman-French "cinquefoil," so old

[1] Miller   [2] *A Display of Heraldrie*, 1632

and rich in tradition—even though the traditions are associated with a different member of the family. The specific name of *fruticosa* means shrubby, and has not—unfortunately, in this first cousin of the strawberry—anything to do with fruit.

NOTES: Only *P. fruticosa* is used as a shrub in our gardens to any extent. The genus is circumpolar and is most useful in northern gardens because they are so hardy and adapt well to continental climates. The flowers appear in mid-summer and range from white, bright yellow, to orange. Several cultivars are offered.

The genus includes many species from the Arctic. I have collected the one most commonly cultivated, *P. fruticosa* within 100 miles of the Arctic Circle near Yakutsk, Siberia. But I would find similar plants in Alaska, Canada, and down into the northern tier of states.

The potentillas can be grown in poor gravelly soils and once established they require no particular care.

Prunus

A Trie than I sie than
Of CHERRIES on the Braes;
Belaw to I saw to
Ane Buss of bitter SLAES.
The Cherry and the Slae, Alexander Montgomery, 1597

Lurking among the Plum, Cherry, Almond, Apricot and Peach trees that comprise this large and valuable family are a number of dwarf and shrubby species that, except for our own wild blackthorn or sloe, are not so familiar as they might be. The blackthorn (*P. spinosa*) is not only wild, but fierce—Dr. Withering thought its thorns were poisonous, especially in autumn—and the type is not usually admitted into gardens; but there is a double form, valuable for its profuse and early blossom, which Loudon in 1838 said was found "a few years ago" at Tarascon, and a purple-leaved one that is sometimes used to make a handsome and unusual hedge—both to be found today in nurserymen's lists. "An eldern stake and a blackthorn ether Make a hedge to last for ever," goes the country saying; but the blackthorn has many uses besides that of a hedge plant. The juice of the fruit will dye linen a reddish color which eventually washes out to a durable light blue, and Godwin suggests that the sloe-stones found so abundantly on some Neolithic sites—almost a wheelbarrow-load from one mound at Glastonbury—represent fruit used for dyeing rather than food; though one would like to think that our Stone Age ancestors solaced themselves with some early form of sloe gin. Doubtless primitive man made his cudgels of blackthorn, as the Irishman his shillelagh. All parts of the plant were later used in medicine, and the dried leaves were quite extensively employed to augment or replace China tea. At a time when more so-called port was drunk in England alone than was manufactured in the whole of Portugal, sloe-juice was one of the largely-used adulterants, and, says Richard Brook, "many a fop when taking his 'poortwind' and tapping

Prunus japonica
(JAPANESE BUSH CHERRY)   S. Edwards (1815)

his boots with his beautiful blackthorn stick, is little aware that both 'wind' and stick have the *same origin*."

The "flourish" is an old Scots word for tree blossom, especially that of fruit trees ("The borial blastis . . . had chassit the fragrant flureise of evyrie frute tree far athourte the fieldis . . ."[1]), and the flourish on the blackthorn usually coincides with a spell of particularly bitter weather, which has become known as the Blackthorn Winter—a phrase that was recorded by Gilbert White as being in use among the country folk of Selborne in 1775. The shrub was named blackthorn on account of its very

[1] *The Complaynte of Scotland*, 1548

dark-colored bark, and to distinguish it from the lighter white-thorn or May. In France it is called *Mère des Bois*, on account of its rapid spread by suckers, and the shelter these afford to young tree-seedlings.

This shrub has a number of distant cousins, including the American Sand-cherry (*P. pumila*), the Russian Dwarf Almond (*P. tenella*), the Chinese Flowering Plum (*P. triloba*) and the Japanese Fuji Cherry (*P. incisa*). There is also a central European species, the Ground Cherry (*P. fruticosa*), the earliest of all to be introduced (1587), which Loudon suggested might be grafted to make a standard tree "at once curious and ornamental," but which is little grown now. The same treatment was suggested by the graft-happy Loudon for *P. pumila*, which was introduced to France from Canada in the mid-eighteenth century, and to England in or before 1756; but the two most important species for general garden use— both perfectly hardy—are *P. tenella* and *P. incisa*, with 230 years between them.

According to Aiton, *P. tenella* (syn. *Amygdalis nana*, the dwarf Almond) was introduced by James Sutherland, Intendant of the Edinburgh Physic Garden, and listed in his catalog of 1683; by 1759 the shrub was common in the London nurseries. It is a native of Russia; "the vast plains of the Volga being annually set on fire," we are told, "it never rises to any height, but is low and shrubby, creeps very much at the root, and impedes the plough";[1] but in English gardens it may become 5 feet tall. Marshall in 1785 mentions a double variety, which he says is matchless, while both are "in the first esteem as flowering shrubs"; and though Cobbet (1833) thinks that "their not bearing flowers and leaves at the same time, is a remarkable illustration of how much flowers borrow effect from foliage," few today would agree that this is to their detriment. A particularly fine form of this plant, with large carmine flowers, collected in Roumania by Lady Martineau, received an Award of Merit in 1929; it is now regarded as indistinguishable from the earlier var. *gessleriana* (1864) and the later 'Fire Hill' (A.M. 1959), which has been called one of the most gaudy plants of the spring garden.

The beauty of the Fuji cherry, *P. incisa*, is delicate rather than gaudy; on the east and south slopes of Mount Fujiyama it is said to be so abundant and so lovely when in flower that it is worth a special journey to see. Although described by Thunberg as early as 1784, it was not brought to this country until the present century; Kew received its first specimen from the Arnold Arboretum in 1916, but the plant seems to have been grown at Edinburgh a year or two previously. It will stand hard

[1] Martyn

pruning, and is the species used by the Japanese for training as dwarf trees in pots. It also hybridizes fairly easily, and Mr. Collingwood Ingram, who praises it for its "kindly disposition and convenient size," has used it as one of the parents of a race of hybrids called × *incam*, the best of which, 'Okamé', received the coveted A.G.M. in 1952. *P. incisa* itself has received its awards—A.M. in 1927, A.G.M. in 1930—and an early flowering form, var. *praecox*, raised by Messrs. Hillier of Winchester, gained an A.M. in 1957.

There are also two oriental species (in the peach or apricot group) which are not happy in the open, but which can be grown against a wall or in a cool greenhouse—*P. japonica* and *P. triloba*. Both are used for early forcing. *P. japonica*, in spite of its name, is actually a native of China, and was grown about 1808 by Charles Greville, Esq., "in his botanic garden at Paddington."[1] *P. triloba* is better described by its Chinese name Yu Ye-mei, the Elm-leaved Peach, than its Latin one, as its leaves are only occasionally three-lobed; it was introduced from Pekin by Robert Fortune in 1855. It is this plant that produces the long sprays of double pink or white flowers, looking as though made of muslin, which we see in the windows of expensive florists in the spring.

Most of us assume that "Cherries and Plums are never found But on the Plum and Cherry tree";[2] but the botanist assures us that they are also found on laurels, for both the Common and Portugal laurels are members of the Prunus family. The early history of the common or Cherry-laurel, *P. lauro-cerasus*, is unusually well documented. A native of eastern Europe and Asia Minor, it made its first appearance in cultivation in the garden of Prince Oria at Genoa, where it was seen by Pierre Bélon, in the course of his "tedious travell" between the years 1546–50. In 1576, the French botanist Clusius (de l'Ecluse) then in the service of the Emperor Maximilian at Vienna, received a consignment of rare trees and shrubs from Dr. David von Ugnad, the Emperor's ambassador at Constantinople; all were dead on arrival except the horse-chestnut and the *Trabison curmasi*—the date or plum of Trebizond, our "laurel"—which was with difficulty nursed back to life and in due course propagated and distributed. The plant was not known to Gerard in 1597, but Parkinson in 1629 described a flowering specimen in the Highgate garden of the late Master James Cole, "which hee defended from the bitternesse of the weather in winter by casting a blanket over the toppe thereof every yeare, thereby the better to preserve it." It has been suggested that Cole may have got his plant from Clusius, with whom he corresponded;

if so, it must have been received before 1606, the year of Clusius' death. Evelyn, on the other hand, repeats the tradition, which he had from "a noble hand," that the laurel was introduced from Civita Vecchia by Alethea, Countess of Arundel, on her return from a visit to Italy in 1614. By whatever means, before 1633 the plant had "got into many of our choice English gardens, where it is well respected for the beauty of the leaves, and their lasting, or continual, greennesse."[1]

At first, the laurel was regarded as a tender plant, to be grown in pots or tubs and housed in the winter; Par-

Punica granata
(POMEGRANATE)   Ferdinand Bauer (1825)

kinson said it would make a tree if "pruined from the lower branches," and Evelyn, that such standards resembled "the most beautiful headed Orange for shape and verdure." In his day—1662—it was sufficiently common to be used for hedges; a purpose for which Evelyn considered it unsuited, "the lower branches growing sticky[2] and dry, by reason of their frequent and unseasonable cuttings." But fashion was against him, and the plant continued to be tortured into sheared globes and pyramids until the end of the century, when a change of fashion permitted a freer use of it in woods and wilder-

[1] *The Botanic Register*, 1815   [2] *The Waterman*, Charles Dibdin, 1774

[1] Gerard's *Herball*, Johnson's edition (1633)   [2] i.e., twiggy

nesses. Miller, in 1759, pointed out that if planted "pretty close together in large Thickets and permitted to grow rude, they will defend each other from the Frost and they will grow to a considerable Height." Between 1743 and 1746 John, the fourth Duke of Bedford, planted what must surely have been an excessively dreary wood, of a very large extent, entirely of laurels. It was much admired by his contemporaries, and was reported as being in a flourishing condition in 1776, but little trace of it now remains. "Nous n'irons plus aux bois, les lauriers sont coupés . . ."[1]

The popularity of the laurel in the eighteenth century does not seem to have been seriously affected by the discovery of its poisonous properties, due to the presence in its leaves of hydrocyanic acid. As early as 1662 it was recorded that the leaves imparted an almond flavor and were used for that purpose in cookery. A water distilled from them seems to have been particularly popular in Ireland, and in 1731 a Dublin physician, Dr. Madden, published a paper in the *Philosophical Transactions* describing the accidental deaths of two women through the use of laurel-water as a cordial. The decoction had been "for many Years in frequent Use among our Housewives and Cooks, to give that agreeable Flavour[2] to their Creams and Puddings. It has also been much in Use among our Drinkers of Drams; and the Proportion they generally use it in, has been one part of *Laurel-Water* to four of *Brandy*. Nor has this practice (however frequent) ever been attended with any apparent ill consequences"—until September 1728, when the unfortunate women appear to have partaken of a brew containing a larger proportion than usual of prussic acid. Some experiments subsequently carried out by Dr. Madden on dogs confirmed its lethal properties; and there was a famous case in 1780, when the extract was used to poison Sir Theodosius Boughton by his brother-in-law, Captain Donaldson, who was executed for the crime. (Nevertheless laurel-water was extensively prescribed by the celebrated Dr. William Baylies (1724–87) who found that it had a "remarkable power of diluting the blood."[3]) By 1838 the "odour of bitter almonds" so familiar to the amateurs of detective fiction, had become unpopular with the would-be murderer, as being too easily recognizable, both by victim and police; but a jam-jar containing a few bruised laurel-leaves is still a valuable aid to the young entomologist.

It was a change of fashion in gardens rather than poisons, that brought about a decline in the esteem in which the laurel was formerly held. It was displaced by the introduction of many other evergreen shrubs, especially rhododendrons, and by the end of the nineteenth century was classed by Robinson with privet and elder, as "hungry rubbish of the shrubbery." Today there is a faint revival of interest in its good qualities, particularly as shown in some of its many varieties. At least fifteen of these have been named, differing in leaf and habit, and ranging from *latifolia* with the broadest leaves to *zabeliana* with the smallest.

There is much less to be said about the Portugal Laurel, *P. lusitanica*, which, hardier than the common laurel and a much nobler plant, has only a meager literature. Perhaps it should be classed as a tree rather than a shrub, for "by Experience we find, that when it is in a proper Soil, it will grow to a large Size"[1] with "a not inconsiderable, though very short, trunk."[2] Collinson says it was brought from Portugal in 1719 by Thomas Fairchild, "a famous gardener for rarities, at Hoxton, and was for some years kept in a greenhouse; it was exposed by degrees, and has since been found to endure all weathers"; but actually a specimen was grown in the Oxford Botanic Garden in 1648, which survived until it was cut down in 1826. Birds, especially pheasants, are fond of the berries, and Phillips suggested that it might be more frequently planted in woods, as "it is well-known that birds which feed on berries have a much finer flavour than those that are bred in corn-lands."

The flowers of the common laurel are fertilized by flies and beetles, which seem, like the Irish, to relish a smack of prussic acid about their drams; the berries do not share the harmful properties of the leaves, and can be eaten, as Miller affirms, "in great Quantities without Prejudice." They can be used for jam, and Evelyn said that a wine was made from them, "to some, not unpleasant." It was the resemblance of this fruit, "as large as any *Flanders*, and of a very blacke shining colour and very sweete," that led Parkinson to classify this laurel, correctly, among the cherries. Linnaeus amalgamated cherry, bird-cherry, apricot and plum into one genus, which Miller thought "exceeded the Boundaries of Nature," on the grounds that cherry and plum will not graft on one another, as they should do if truly members of the same family. The cherry-laurel has long ago supplanted the true bay-laurel as the emblem of victory, and its leaves are much in demand for formal wreaths. Adorned with branches of this plant, the mail coaches carried through the land the news of Trafalgar and Waterloo; "And the coachman would be wearing laurel, and the guard would be wearing laurel; and then they would know, then they would know!"[3]

[1] Théodore de Banville, 1890   [2] Of almonds or peach-kernels
[3] Phillips

[1] Miller   [2] Cobbett   [3] Kenneth Grahame, *Dream Days*, 1898

NOTES: Call them what you will, trees or shrubs, the flowering cherries and allied species are among the finest of our spring flowering plants.

Little has occurred since the beautiful 'Okame', developed by Collingwood Ingram, was introduced into the United States. In the 1980s, however, the National Arboretum sent collector Roland Jefferson to the Orient to collect seeds and cuttings throughout the entire range of the flowering cherries. This vast genetic resource collection will be the basis for future improvements in the flowering cherries. Of these, the lovely *P. incisa*, native to the slopes of Mt. Fuji, will no doubt play a central role in this effort to enhance the flowering cherries, as it did in Ingram's pioneer breeding work.

P. laurocerasus will grow in sun or light shade and in any good garden soil.

Punica

It may be something of a surprise to learn that the exotic-sounding Pomegranate (*P. granata*) has been grown in England since at least the sixteenth century. It is a native of Persia and Afghanistan, a hawthorn-like shrub or small tree of about the same degree of hardiness as the peach; and having been cultivated since prehistoric times, is now widely naturalized in all the countries bordering the Mediterranean, where it has acquired a wealth of folklore and legend. The story of Persephone, doomed to stay in Hades for three months of every year because while there she nibbled seven pomegranate seeds, is one of the most familiar.

The date of the entry of the pomegranate to England is uncertain; Turner reported in 1548 that it grew in the garden of the Duke of Somerset at Syon House, but it has been suggested that it may have been cultivated still earlier by some of the monastic communities. (Gerard had a number of young plants which he had raised from seed, and in 1597 was "attending God's leisure for flowers and fruite.") By the early seventeenth century the shrub was being grown as a plant of ornament and several varieties were already in cultivation. A very fine double sort, with blossoms "as large as a double Province Rose . . . of an excellent bright crimson colour, tending to a silken carnation,"[1] was introduced from the Continent by John Tradescant before 1618, and grown in Lord Wotton's garden at Canterbury. Parkinson considered it necessary to plant it against a wall "and defend it conveniently from the sharpness of our winters, to give his Master some pleasure in seeing it beare flowers," but

[1] Parkinson, 1629.

Evelyn found the pomegranates "easily enough educated under any warm shelter, even to the raising of hedges of them. . . . They supported a very severe winter in my garden, 1663, without any trouble or artifice." In 1698 Celia Fiennes saw at Ingestre in Staffordshire (apparently out-of-doors) "a fine pomegranate tree as tall as myself. The leafe is a long slender leafe of a yellowish green edged with red and feeles pretty thicke, the blossom is white and very double"—yet the white and double varieties are reputed to be less hardy than the type. By 1759 the number of kinds had risen to six, including one with a striped red and white flower, and the dwarf form, *nana* (grown by Miller before 1731). The double red, Miller said, would flower "for near three months successively, which renders it one of the most valuable flowering trees yet known."

The double kinds, of course, bear no fruit; the singles would, Evelyn said, produce in a good situation "a pretty small pome." Actually there are numerous records of pomegranates fruiting in Britain, but they rarely ripen sufficiently well to be eatable. Miller reported that his own specimens bore a great quantity of fruits "which have arrived to their full Magnitude; but I cannot say they were well-flavoured, however, they made a very handsome Appearance upon the Trees." In 1765, a Mrs. Gaskry of Parson's Green near Fulham had a good crop of pomegranates, "near two dozen on each tree, of a remarkable size and fine ruddy complexion, of the size of middling oranges. One that was split showed the redness and ripeness within."[1] Other good fruiting years were 1874, when a tree on the front of a house in Bath was loaded with fruit, and 1911. The *Botanical Magazine* for 1816 makes the point that the pomegranate might be more generally grown, if it were not for the fact that it flowers and fruits best on a warm wall, and such situations are usually required for more profitable fruits.

Perhaps it is fortunate that pomegranates are not often perfected here; we are told that they "should be eaten cautiously, lest they throw the blood into a state of putrefaction"[2]—advice that should have been given to Persephone. It would be wiser to reserve them for other uses: for their astringent properties in physic, for tanning and dyeing leather, and "to make the best sort of writing Inke, which is durable to the world's end."[3]

The Romans received the pomegranate from Carthage, and named it on that account *Malus punica*, from the old name of the district. Miller, however, derived the name from the "Punicean or red Appearance" of both flowers and fruit. *Granata* is probably derived from *gra-*

[1] Collinson, quoted in *Transactions of the Linnaean Society*, vol. X
[2] Bryant, *Flora Diaetetica*, 1783   [3] Parkinson, 1629

num, a grain; and Gerard says that Granada was so-named "of the great multitude of Pomegranats which be commonly called Granata," and which are supposed to have been planted by the Moors. The fruit has been made the emblem of democracy, because of the numer-

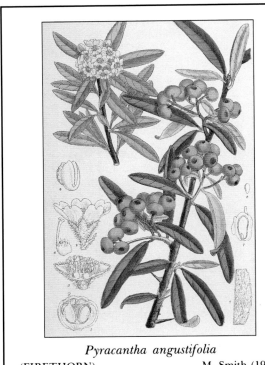

Pyracantha angustifolia

(FIRETHORN)   M. Smith (1910)

ous seeds, which are the valuable part, and the worthless "crown," the remains of the calyx, which is retained on the top of the fruit. Yet this same crown is supposed to have furnished the pattern for that of King Solomon.
NOTES: We usually think of the pomegranate for its edible fruit. But in warm parts of the South and Southwest, it is frequently used as an ornamental shrub for its bright, decorative flowers. These range in color from white to bright scarlet and may be single or double. There is a dwarf form, 'Nana', being offered in many southern nurseries.

Pyracantha

The family of the Pyracanthas or Firethorns, like that of the Berberis and the Cotoneaster, includes one European species which has been cultivated for

centuries, and a number of Asiatic ones which are relatively new to gardens in the West.

The old-established one is *P. coccinea*, which was introduced from south Europe at some unknown date—before 1629, but presumably after 1597, since Gerard does not mention it. Parkinson, who called it the "ever greene Hawthorne or prickly Corall tree," considered it a fine ornament, either "noursed up into a small tree by itself, by pruning and taking away the suckers and under branches" or grown as a hedge. Evelyn in his *Sylva* (1662) suggests it might be used among the "preciouser sorts of Thorn and robust Evergreens adorned with Caralin Berries" for agricultural hedges—"and then, how beautiful and sweet would the environs of our fields be!" The "Paricanthas" that Celia Fiennes noticed in the gardens of Hampton Court about 1702, seem to have been grown as shrubs or small trained trees, and it is not until late in the eighteenth century that we hear of them being grown on walls, as they so often are today. Abercrombie (1778) said that the shrub was "most commonly trained against walls or fronts of houses . . . that it may exhibit its berries more ornamentally," and Marshall (1785) that it was chiefly used to hide unsightly buildings, "as by its evergreen leaves, closely set, it will not only keep from sight whatever cannot regale that sense, but will be to the highest degree entertaining by the profusion of berries it will produce, and which will be in full glow all winter." He adds that there are "few towns which have not a house or two whose front is ornamented with them being trained up to a great height."

About 1874 M. Lalande, a nurseryman of Angers in northwest France, raised from seed a much improved variety of pyracantha, which in 1882 was named *P. coccinea* var. *lalandei*. Grown as a bush or tree in the open, it is said to produce its orange-scarlet berries more freely than the type, which fruits best against a wall; it is now the most widely grown variety, and has produced in turn the yellow-berried form, var. *aurea*. Robinson speaks of a dwarf variety called *pauciflora*, but this seems to have disappeared.

The most important of the Asiatic species are *P. angustifolia*, *P. atalantoides* and *P. rogersiana*, all from China. The first, *P. angustifolia*, was raised for the first time in Europe by the French firm of Vilmorin, from seed sent by the Jesuit missionary Père J. A. Soulié in 1895. It was introduced to Kew (presumably from China) by a Lieutenant Jones in 1899; some of Vilmorin's plants were also imported, a year or two later. It is a rather tender species, best against a wall; but plants raised from a later consignment of seed sent home by George Forrest proved hardier than those of the original importation. (Bushes grown in the open fruit more freely than those confined on a wall, but are particularly vulnerable to frost when heavily loaded with berries.) *P. atalantoides*, the largest in height, foliage, fruit and flowers, was introduced by E. H. Wilson in 1907, and called at first *P. gibbsii* after the Hon. Vicary Gibbs, who grew a fine 20-foot specimen in his garden at Aldenham, but it was later found to be identical with a species already discovered and named in 1877. *P. rogersiana*, the smallest and youngest of the three, was raised by Mr. C. Coltman Rogers from seed sent home by Forrest in 1911, and distributed by J. C. Williams of Caerhays. It was exhibited by Mr. Rogers at the R.H.S. in 1913, when it gained the Award of Garden Merit, and the R.H.S. *Dictionary*, a work not in general given to encomiums, says that it is "of quite dainty beauty" when young, and the best kind for small gardens. There are forms with orange or yellow berries, instead of the usual red.

Pyracanthas are grown for their berries alone, so it is worth in this connection considering the tastes of the birds. Blackbirds, unfortunately, seem especially fond of the fruits of *P. coccinea* and all its varieties, and their berries are among the first in the garden to be eaten; wall-shrubs, especially those in built-up areas, are allowed to keep their splendor longer than bushes planted in the open. The fruits of the Chinese kinds seem much less attractive, to European birds at least; even in its native land the berries of *P. angustifolia* remain untouched until taken by migrants in April.

At the start of its career the pyracantha was called by the name we use today, but "M. Tournefort and Father Plumier" says Liger in 1702 "make it another kind of Tree, and rank it with the Medlar." This classification was confirmed by Linnaeus, who named the shrub *Mespilus pyracantha* in his *Species Plantarum* in 1753. Later it masqueraded in turn as a Crataegus (Hawthorn) and as a Cotoneaster, to both of which families it is allied. Liger seems to think the name is derived from Pyrus, as he translates it "Thorny Pear"; but actually it is from the Greek *pyr*, meaning fire, and *acanthos*, a thorn—all members of the family are spiny, thereby differing from the Cotoneasters—so that the English name of Firethorn is a literal translation from the Greek.

NOTES: As shrubs, wall plants, and for fruiting displays, the pyracanthas are used more widely in the United States than anywhere else. Credit for this recent interest is due to the late D. Egolf of the National Arboretum who devoted his career to this genus and several other woody plants (crab-apple, crape myrtle, viburnum). The most familiar of his pyracanthas is 'Mohave', a profusely flowering selection with masses of orange berries. It is

both hardy and resistant to fire-blight and scab diseases. Another outstanding American breeder, Dr. E. R. Orton, Jr., of Rutgers University, released a hardy pyracantha, 'Fiery Cascade', with a much more spreading habit than the upright 'Mohave'.

Rhamnus infectoria
(AVIGNON BERRY)   *Flora Dalmatica* (1842)

Pyrus, see under *Chaenomeles*

Rhamnus

Hicum, peridicum, all clothed in green
The King could not tell it, no more could the Queen.
So they sent to consult wise men from the East
Who said it had horns, though it was not a beast.
Traditional riddle (Buck's horn or Buckthorn)

W. J. Bean remarked of the Buckthorn family that there were few groups containing so many species (about a hundred) that possessed so little of garden value. The genus is included here only on account of the Alaternus, and that only because of the reputation it formerly enjoyed, for like the phillyrea it was once a popular evergreen, and is now almost forgotten. It is not unlike the phillyrea in appearance, and as the old authors are never tired of telling us, received its name of *R. alaternus* because its leaves are alternate, while those of phillyrea are in pairs. It is a native of Europe, and was a great favorite with John Evelyn, who says that although lately received from the hottest parts of Languedoc, it would thrive "as if it were an indigene and natural," unless a very severe winter was followed by "a tedious eastern wind in the spring, of all the enemies of our climate the most hostile and cruel . . . I have had the honour," he goes on, "to be the first who brought it into use and reputation in this kingdom, for the most beautiful and useful of hedges in the world . . . and propagated it from Cornwall to Cumberland. The Seed grows ripe with us in August, and the honey-breathing blossoms afford an early and marvellous relief to the bees." In spite of his claims, however, Evelyn was not the first to introduce the alaternus, though he may have speeded up its distribution; for it was known to Parkinson in 1629, when Evelyn was only nine years old. It was then fairly new and rare, finding place "in their Gardens onely, that are curious conservers of all natures beauties"; and Gerard, thirty years earlier, knew it only by repute.

The alaternus was a useful and popular shrub when formal gardens were in fashion; Liger says it was used in the borders of parterres, "where it grows sometimes in the form of a Bush, sometimes in the form of a Bowl or Ball, and sometimes in another Figure"; four varieties were cultivated, including one with variegated leaves. By the start of the nineteenth century, however, it had fallen out of favor, though it was still used, especially in towns, as an evergreen covering for walls, or to afford variety in ornamental plantations. In spite of Evelyn's praises, Martyn thought it unsuitable for hedges, for it grew too vigorously, requiring clipping three times in a season, "which is both expensive, and occasions a great litter in a garden." It never really regained its early popularity, though it is still available today.

English Buckthorn, *R. cathartica*, is said to be a good hedge plant, but as far as ornament goes, the highest praise it receives is that it is a well-looking tree and sometimes striking when in berry. The fruit, as the specific name implies, is cathartic—so powerfully so, that the very flesh of birds that have fed on the berries is reputed to be purgative; the bark of a related American species is the source of *cascara segrada*. Several members of the family are used in dyeing; Clusius reported that the fishermen of Portugal dyed their nets with the bark of the alaternus, and Duhamel said that its wood was used for fine cabinet making.

The family name is derived from the Celtic *ram*, meaning a tuft of branches.

NOTES: In the United States, only the species *R. frangula* is commonly cultivated. The celebrated tall-hedge buckthorn, 'Columnaris', developed by the Cole Nursery of Painesville, Ohio, in 1935 is an excellent narrow, upright hedge plant of exceptional hardiness.

Rhododendron

For lime-free gardens the rhododendron is the King of Shrubs, as the Rose the Queen—unbeatable for beauty, variety and adaptability; the vast genus looms like a Himalayan peak among the lesser families, and fills with awe the nonspecialist who has the temerity to approach it. Estimates of its numbers vary greatly, as many rhododendrons originally thought to be species proved on closer acquaintance to be only varieties; even so, the *Rhododendron Handbook* for 1952 listed 750 species, and Captain Kingdon-Ward thought a further hundred might yet be found in unexplored valleys of China and Tibet. They include prostrate shrublets a few inches high, and trees, such as *R. giganteum*, that may attain 90 feet; and range from *R. nivale*, which is reputed to be the highest-altitude shrub in the world, and is buried in snow for eight months of the year, to the epiphytic species that grow like orchids in the tropical forests. The genus erupts in a great whirlpool of species in the western Himalayas, and from thence outliers extend over large tracts of the northern hemisphere (including North America, but not, since prehistoric times, Britain) and down into tropical Asia.

The inaccessibility of the family's headquarters accounts for the comparatively late arrival of a large number of the species. Fortune's travels in China did not approach the main center of distribution, though he collected a number of cultivated kinds; Hooker made a remarkably clean sweep of the southern fringes of the region; the French Jesuit missionaries of the late nineteenth century sent home specimens and a little seed; but the fountain head of the family remained untouched until the period 1904–54, when first Wilson, and then Farrer, Forrest, Rock, Ludlow and Sheriff and Kingdon-Ward made their collections, and new species came pouring in, until those who financed the expeditions must have felt like the sorcerer's apprentice, unable to reverse the spell and put a stop to the deluge that threatened to overwhelm them. Forrest alone made 5,375 gatherings of rhododendron seed, representing 260 different species—and all to be crammed somehow into British gardens.

But this is the end of the story, and we must go back to the beginning, which, for us, starts in Europe, where fossil evidence shows rhododendrons to have existed in the Miocene age. The first species known to science were the Alpine ones, *R. ferrugineum* and *R. hirsutum*, which were described under various names—Ledum, Cistus, Chamaecistus—by Clusius and other sixteenth-century Continental botanists. The name "Rhododendros" (meaning Red- or Rose-Tree) was originally used by the Greeks for the Oleander, and seems to have been first associated with the present family by the Italian botanist Andreas Caesalpinus in 1583, who described *R. ferrugineum* as having "handsome flowers so like the Rhododendron (i.e., Oleander) that some people call it Alpine Rhododendron."[1] The genus was officially established and christened by Linnaeus in 1753, who created at the same time the separate genus of Azalea, taking as the type a native of our own country, since reclassified as *Loiseleura procumbens*.[2] Early in the nineteenth century it was realized that there was no essential difference between these two families—indeed, one of the first hybrids (*R. × odorata*, about 1820) was an azalea-rhododendron cross, of the sort later called an azaleodendron; and all are now classed together as rhododendrons. For the purpose of this article, however, it is convenient to retain the old familiar division into evergreen "rhododendron" and deciduous "azalea."

The first of the evergreen kinds to be cultivated was the Alpine Rose, *R. hirsutum*; it was described by Gerard only at second-hand, but was grown in the Tradescant's garden before 1656. It is curious that the first three American species should have been introduced before the second and commoner Alpine kind, *R. ferrugineum*, was brought to England. Martyn says that this rhododendron was cultivated by Miller in 1739; he also mentions a rare white form.

It is a pity that the first evergreen American species was named *R. maximum*, for though undoubtedly the biggest of the few rhododendrons then known, it has long been surpassed both in size of plant and flower. (It rarely exceeds 10 feet in cultivation, though larger in the wild.) Plants of the "Rock Rose of Pennsylvania," the "Rose Bay of the Carolinas," were sent by John Bartram to Collinson and others in 1736, but were slow to establish themselves; Catesby, in 1747, reported that they had not

yet bloomed, and that it seemed to be "one of those *American* Plants that do not affect our soil and climate." (Probably it was not then realized that an acid soil was essential.) Collinson's specimen flowered eventually in 1756; and in 1760 he proudly reported that he had plants in bloom from seed sown only seven years before.

The shrub that we think of as *the* rhododendron, the ubiquitous *R. ponticum*, was a comparative latecomer—counting azaleas, sixth on the list; it did not reach England until 1763. It was first discovered by the French botanist Tournefort during his travels in the Near East in 1700–02, and received its name because the country in which it grew, on the southern shores of the Black Sea, corresponded with the ancient kingdom of Pontus. About the middle of the century it was found by Claes Alstroemer, a pupil and correspondent of Linnaeus, growing "near a Carmelite Convent, named Cuervo, between Cadiz and Gibraltar . . . on the sides of rivulets, in consort with the Nerium or Oleander";[1] and it was from Gibraltar that plants were eventually imported. One hardly thinks of this shrub as a cherished pot plant. But the *Botanical Magazine* of 1803 reports that it had been found to force remarkably well, and that large numbers were brought yearly to the London markets "to ornament our houses in the spring." It was also honored by being made the subject of one of the plates in Thornton's *Temple of Flora* (1801). Later, as it became more common, it was planted by landowners literally by the mile; and responded to this encouragement with such good will, that in some places it has become a troublesome weed, costly and difficult to eradicate.

Two species, both of great importance in the breeding of hardy hybrids, came from opposite directions in the first decade of the nineteenth century—*R. caucasicum* and *R. catawbiense*. The former was discovered growing on "the most elevated parts of Mount Caucasus, on the verge of the region of perpetual snow,"[2] and a plant of it was sent by the Russian traveler, Count Apollos Apollosovitch Mussin-Puschkin, to Sir Joseph Banks in 1803; seeds or plants were also procured from St. Petersburg by the firm of Loddiges, and flowered for the first time in 1808. Early in its career *R. caucasicum* was crossed with the closely allied *R. chrysanthum* from Siberia (1796)—a difficult and unrewarding species in cultivation, but the only one then known which had yellow flowers. Interest was then concentrated on their straw- or sulphur-colored offspring, and other improved *caucasicum* varieties, and the true type was allowed to die out, and has never been reintroduced. Among its hybrids was the early-flowering 'Christmas Cheer', raised in Bel-

[1] *De Plantis*, Lib. XV Cap. XVII [2] "Azalea" comes from the Greek word meaning "dry"; curious, when one considers that the only species known to Linnaeus when he chose the name (including the Loiseleura) were swamp-lovers

[1] Collinson (Dillwyn) [2] *The Botanical Magazine*, 1808

gium by a nurseryman called Vuylesteyke, and 'nobleanum', grown by Michael Waterer in 1835 and named in compliment to the head of the rival firm of Standish and Noble.

In 1799 John Fraser and his son were in America, collecting plants for the Emperor Paul of Russia, and "it was their good fortune to discover and collect living specimens of the new and splendid *R. catawbiense*" on the summit of "the Great Roa or Bald Mountain ... on a spot which commands a view of five states"[1]—a discovery surely not very difficult to make, for a hundred years later this rhododendron still covered thousands of acres in the area with thickets so dense that they could be penetrated only by following cattle or bear tracks. These two collectors brought plants of the new rhododendron to Britain in 1809, from the source of the Catawba River in the same neighborhood. It bloomed in 1813 in the nursery of Lee and Kennedy, and was at first considered disappointing; the flowers, which had been reported to be scarlet, turned out to be "hardly more shewy than those of *Rhododendron maximum*."[2] Soon, however, it was found to be "a first-class rhododendron to breed from," and Bean went so far as to call it "perhaps the most valuable evergreen shrub ever introduced, as a parent of hardy hybrids."

Breeding began with these four kinds—*maximum, ponticum, caucasicum* and *catawbiense*—all of them members of the "ponticum" series, and all but *ponticum* now extinct in gardens, superseded by their numerous progeny. All were comparatively pallid in their pinks and mauves; but color came soon afterwards from the Orient, with the arrival of the first Himalayan species, the blood-red form of *R. arboreum*. It was found by a Captain Hardwicke in 1796, and it was hoped "that the seeds, which that gentleman has liberally distributed in England, will enrich our collections with this noble tree."[3] It does not seem to be recorded whether any plants were successfully raised from this first importation; it was another twenty-nine years before the species flowered in England, and a second introduction was "almost certainly" made by Dr. Francis Buchanan Hamilton, about 1815. In sheltered places in the south and west this rhododendron will attain a height of 40 feet, but it is tender, and in most counties is best in a greenhouse. Many of its hybrids, however, are surprisingly hardy, and it was eagerly seized upon by breeders soon after its first flowering, which took place at the Grange, Northington, near Alresford, in 1825, and, probably in the same season, at Highclere in Hampshire, the seat of the Earl of Carnaer-

von. Here, about 1826, was raised the celebrated hybrid 'altaclarense',[1] from *R. arboreum* crossed with one of the old *catawbiense* × *ponticum* hybrids—the first real forerunner of our garden rhododendrons of today. "The history of this superb plant deserves to be particularly described," wrote Lindley in the *Botanical Register* for 1831, "as it not only shows how great the power of man is over nature, but holds out to us a prospect of the most gratifying kind in regard to the future gayness of our gardens."

The raising of hybrid rhododendrons had therefore made considerable progress before Sir Joseph Dalton Hooker made his celebrated Himalayan expedition in 1847–51, in the course of which he collected forty-three species, many of them new, and sent home seeds, and also drawings and descriptions, from which his father, William Jackson Hooker, compiled the sumptuous *Rhododendrons of the Sikkim Himalayas* (1849), the first of many monographs on the genus. Hooker's introductions included some of the best of the rhododendron species: dainty *ciliatum*, yellow *campylocarpum*, long-leaved *falconerii*, tubular *cinnabarinum*, spectacular *thompsonii*, and best of all, *griffithianum*. An inferior form of the latter had previously been found by Dr. William Griffith, a botanist who had made many arduous and perilous expeditions in the service of the East India Company; but Hooker's plant was so much finer that he took it at first for a new species, and named it *R. aucklandii*, after George Eden, Lord Auckland, who was Governor-General of India from 1835 to 1841; Sir Joseph had an amiable habit of dedicating rhododendrons to friends and helpers, much as a matador dedicates his bulls. Plants raised from the seed he sent home in 1849 flowered in 1858, at Kew and in the nursery of Messrs. Gaines at Wandsworth. This species has been called the loveliest of all wild rhododendrons; its white flowers are 5, sometimes even 7 inches across, and very sweetly scented. Alas! "Birds of Paradise and *Rhododendron griffithianum*," as the late Mr. Charles Eley put it, "can never become inhabitants of ordinary English gardens"—unless in the very warmest and most sheltered localities. It has, however, been extensively used as a parent, and while first-cross hybrids are still rather tender, its grandchildren of the second generation are hardy and beautiful. Six years later came Robert Fortune's one wild Chinese rhododendron, *R. fortunei*, allied to *griffithianum*, and more hardy, though less exquisite. Fortune collected the seed in 1855 in the high mountains west of Ning Po, and sent it to Messrs. Glendinning of Chiswick. It too became of importance as a parent, but again the original species is now rare.

[1] W. J. Hooker, *Companion to the Botanical Magazine*, 1836 [2] *The Botanical Magazine*, 1816 [3] Smith, *Exotic Botany*, 1804

[1] *Alta*, high, and *clarense*, clear, = Highclere

From then on, hybrids and hybridists arose in bewildering profusion; their history will be found in the many books devoted to the subject. As a single example, the firm of Waterer of Knap Hill may be quoted, which specialized in rhododendrons and azaleas for four successive generations. The John Waterer of Bagshot who raised the famous variety 'Pink Pearl' (before 1897) was a collateral of the same family; the rhododendron's parentage was from a *griffithianum* hybrid crossed with an *arboreum* hybrid, so it has the blood of two of the most splendid species at a sufficiently far remove to have escaped their tenderness. Another hybrid that should perhaps be mentioned is the popular *R. × praecox*, raised about 1860 by Isaac Davis and Sons of Ormskirk, from Hooker's *R. ciliatum* crossed with the Siberian *R. dauricum*, the earliest rhododendron to bloom. (According to Aiton, this species was introduced about 1780 by Anthony Chamier, who was a government official of French extraction, and a friend of Dr. Johnson.) The parentage of many rhododendron varieties is unknown, either through negligence or secrecy on the part of their raisers; some even arose by accident, while others were bred as carefully as a racehorse. In 1901 Sir Edmund Loder of Leonardslee in Sussex crossed an especially fine and sweet-scented form of *R. fortunei* with a particularly good *R. griffithianum*, growing in a neighbor's greenhouse; the result was *R. × loderii*, which has never been surpassed, though the same cross has been made on other occasions by the same and other breeders. It was not, of course, the first *griffithianum* hybrid—one was raised by an Edinburgh firm in 1869—but it is thought by many to be the finest.

Of the flood of Chinese species, large and small, introduced in the present century, the most important so far has been *R. griersonianum*. It was found by Forrest in 1917 and named by him after R. C. Grierson, a customs officer stationed at Tengyuan (Yunnan) and Forrest's host during his stay in the region. The geranium-red flowers of this species so excited the breeders that it was "crossed with everything in sight," and by 1952 had been made the parent of a hundred and twenty-two hybrids— a larger progeny than that of any other rhododendron.

For the history of the Azalea, we must go back again to the seventeenth century, for the Swamp Honeysuckle, *R. viscosum*, was among the Virginian plants sent by the Rev. John Banister to Bishop Crompton, and was described by Plukenet in 1691. It seems to have died out soon afterwards, and was reintroduced by Collinson in 1734, along with with *R. nudiflorum*, called by the Dutch settlers in Pennsylvania the Pinxter-bloom[1] or Whitsun

[1] German, *pfingsten* = Pentecost

flower. *R. speciosum*, at first thought to be only a variety of the preceding species, was introduced about 1789 "either by Mrs. Norman of Bromley, Kent, or Mr. Bewick, of Clapham, Surrey (both celebrated for their collections of American plants),"[1] but was little known until after the sale of Bewick's collection, when a specimen of this azalea was purchased for the sum of twenty guineas. North America, though comparatively poor in

Rhododendron ciliatum
(RHODODENDRON)   J. D. Hooker (1849-54)

evergreen rhododendrons is rich in deciduous azaleas, one of the most spectacular being *R. calendulaceum*, a very variable species which some botanists think to be a natural hybrid. William Bartram called it the Fiery Azalea, "as being expressive of the appearance of its flowers, which are in general of the color of the finest red lead, orange and bright gold, as well as yellow and cream color . . . the clusters of blossoms cover the shrubs in such incredible profusion on the hillsides, that suddenly opening to view from the dark shades, we are alarmed at the apprehension of the hill being set on fire."[2] Bartram sent dried specimens to Sir Joseph Banks in 1774, but the plant itself does not seem to have reached Europe until

[1] *The Botanical Magazine*, 1792   [2] Bartram, *Travels through North and South Carolina etc.*, 1792

introduced to France by Michaux in 1806; it was cultivated in at least two English nurseries by 1812.

Meantime the Pontic Azalea (*A. pontica* or *R. luteum*) had at last arrived—the plant that we regard as the basic or typical azalea, just as R. *ponticum* from the same region is the typical rhododendron, so that it seems improper that in each case several other species should actually have got here first. It was discovered by Tournefort at the same time as its companion rhododendron, but did not arrive in Britain until considerably later, when three importations were made within a short time of each other. In 1793, Peter Simon Pallas[1] sent seeds from the Crimea to the firms of Lee and Kennedy and of Bell at Brentford; early in 1798 Mr. Anthony Hove[2] sent a plant to Mr. Watson, a nurseryman of Islington, which flowered the same year—in a heated greenhouse—and was illustrated in the *Botanical Magazine*; and in 1803 a specimen was sent by Mussin-Puschkin to Banks, along with that of *R. caucasicum*, already mentioned. It is therefore surprising to read in Martyn's 1807 edition of Miller's *Dictionary*, that this rhododendron was not yet cultivated in Europe. Now it is naturalized in places, though not so abundantly as *R. ponticum*; both species have been extensively used for grafting, and have often elbowed themselves in as usurpers, when the varieties grafted on them have died.

The breeding of azalea varieties now proceeded briskly, both here and abroad. One of the most important Continental pioneers (about 1815) was P. Mortier, a baker of Ghent, whose secret was to force the late species (using, perhaps, the heat from his ovens) and retard the early ones, in order to make crosses not previously attempted. His work was followed up by other local nurserymen, until the raising of azaleas became a considerable industry in the district. Many of these "Ghent" azaleas, 500 of which were in commerce by 1850, were imported by English nurserymen and used to enrich their own strains. The situation was then further complicated by the addition of three more species—two from the Orient, and a belated American. The first, the Japanese *R. molle*, was grown by Loddiges as early as 1824, and reintroduced by Fortune in 1845, but unfortunately proved too tender to be satisfactory out-of-doors, though lovely in a greenhouse; a Chinese form of the same plant was named at first *R. sinense*. In 1830 Siebold brought from Japan a fine and hardy azalea which he thought was a variety of *R. molle*, but which was later classified as a new species, and named *R. japonicum*; and it is this plant rather than the true (but tender) *R. molle*,

that was the predominant parent of the "Molis" azaleas, first bred by the firm of Koster at Boskoop in Holland.

In 1851 William Lobb sent to Messrs. Veitch seeds of the white, late-blooming *R. occidentale* of California, the only American azalea to be found west of the Rockies; it flowered in 1857, and though at first thought of little value, was used with success by Anthony Waterer to impart fragrance, size of truss and autumn color. Some of the fine varieties raised by the Waterers, father and son, later became the basis of Mr. Lionel de Rothschild's famous Exbury strain, which caused such a sensation when exhibited at Chelsea in 1937. From these and many other breeders we have inherited azaleas in all the delicious fruity colors—lemon, orange, apricot, peach, tangerine, cherry and strawberry with butter, honey and cream to match.

A group that is becoming of increasing garden importance is that of the dwarf evergreen or semi-evergreen azaleas, similar in type to the "Indian" azaleas of the greenhouse—a race of mixed parentage, descended principally not from *R. indicum*, but from the more tender *R. simsii* (neither of which is a native of India). *R. indicum* itself is very nearly hardy, and some of its allied species, such as *R. obtusum*, are quite suitable for outdoor use. Most gardeners are now familiar with the dwarf "Kurume" azaleas, largely descended from *R. obtusum* var. *kiusianum*. In Japan they are garden plants of considerable antiquity, and a few of them reached England by way of China in 1844; but the main introduction was the work of E. H. Wilson. He saw and admired these azaleas in the nurseries near Tokio, in 1914; on that occasion he was only able to bring home herbarium specimens, but he ordered a collection to be sent to John S. Ames of Massachussets, which arrived in 1917. This collection aroused so much interest in America that in 1918 Wilson made a special visit to Kurume (the center of production of this type of azalea, the Ghent, as it were, of Japan) where he made a careful choice of the fifty best sorts from a collection of 250 in the nursery of Mr. Kijiro Akashi, which were sent to the Arnold Arboretum, near Boston. Complete collections of the "Wilson Fifty" are rare in this country, but there is one at Wisley, raised from cuttings planted in 1946; the varieties differ greatly in size and degree of hardiness, but all demand full sun. Breeding is still going on, using such species as *RR. kaempferi, malvatica* and *oldhamii*, to produce strains, like the American "Glen Dales" (over 450 of them) that will be more satisfactory in colder districts, where the Kurumes are shy to flower. These tend to be taller plants, and may attain 10 feet or more in height.

[1] German-born naturalist, employed in Russia by Catherine the Great   [2] Polish-born collector for Kew

The catalogue of a nurseryman who specializes in rhododendrons may now offer us a choice of over 1,100 species and varieties—a far cry from Miss Walker of Drumsheugh near Edinburgh, who in 1860 had a collection of fifty sorts—tender and hardy—and who yet managed to have a rhododendron in bloom in every month of the year. The trend in this day of small gardens is naturally towards the dwarf varieties rather than the tree-like species that were so magnificently planted in the gardens of the south and west under Hooker's persuasion and encouragement; but there are many gardens, large and small, in which these shrubs have been so effectively used that England may well become, as Kingdon-Ward prophesied, as famous for its rhododendrons as Holland for its bulbs.

When Phillips made *R. ponticum* the emblem of "the dangers that lurk about the imperial purple," he was thinking of the old story of the poisonous properties attributed to rhododendron honey. The remnants of the vanquished army of Cyrus, led by Xenophon, beat a fighting retreat from the plains of Babylon to the shores of the Black Sea, and arrived in 40 B.C. in the neighborhood of Trebizond. After a victorious encounter with the Colchians they halted at a deserted village, where the honey they robbed from the local beehives (had the bees, too, deserted?) produced alarming symptoms of vomiting, purging, delirium and coma. Fortunately the Colchians did not counterattack, and in a day or two all the victims recovered. Pliny thought that the source of this honey was the poisonous Oleander, and Tournefort, guided by local opinion in the region, that it was delivered from *R. ponticum*; but later travelers attributed it to *R. luteum (Azalea pontica)* also common in the neighborhood, whose honey has been proved to have deleterious effects if taken in quantity. Anthony Hove, who traveled in south Russia and Asiatic Turkey in 1796, said that by the River Dnieper this azalea was known as "the stupefying shrub," and on the Dniester was "regarded by the common people as intoxicating, and used in the cure of various diseases."[1] He tells of a Tartar farmer who lived entirely on the profits of azalea-honey, sold in Constantinople and Trebizond for medical use. Cases of poisoning are said to occur regularly in the Caucasus, at the season when the new comb honey is taken, and are sometimes fatal to children; a few mild cases have occurred in England. The poisonous property in the nectar can be dissipated by heating, and usually vanishes as the honey ages and ripens; the trouble arises from the use of fresh comb-honey containing open uncapped cells. Some rhododendrons are very rich in nectar, which they secrete

[1] Quoted in *the Botanical Magazine*, 1799

by the teaspoonful; even the common *ponticum*, when grown indoors in a pot, was observed to produce a large drop of nectar in each flower, which crystallized, as the flower faded, to a substance resembling "the purest sugar candy."[1] In Tibet, candied rhododendron-blooms are among the choicest and costliest delicacies; Lobsang Rampa described how his mother prepared them for a party, sending out servants on horseback weeks before, to gather the right kind of blooms, and how his father grumbled, "We could have bought ten yak with calves for what you have spent on these pretty flowers."[2]

In Europe, where yak-saddles made of rhododendron-wood are little in demand, this conspicuously ornamental family is not conspicuously useful. One species, however, was used for a time in medicine—*R. chrysanthum*, which brought from its native Siberia a reputation for the cure of rheumatism and pains in the joints. Before the end of the eighteenth century it was "very generally employed in chronic rheumatisms in various parts of Europe,"[3] and had been successfully introduced into the Edinburgh pharmacopoeia. The leaves in large doses are powerfully sedative and narcotic, and the decoction of them, as employed by the Siberians, was "said to occasion heat, thirst, delirium, and a peculiar sensation in the parts affected."[4] "It is not probable," remarks Martyn, "that it will ever be a favourite remedy in this country;" but as late as 1838 it was still frequently employed instead of colchicine in the treatment of rheumatism and gout. The leaves of *R. ferrugineum* in Italy and of *R. maximum* in America have been used for the same purpose; but the family is suspect, and several kinds are known to be harmful to grazing animals. *R. ciliatum* is particularly deadly; two shoots of it, nibbled by a baker's horse in a gentleman's drive, had an almost fatal effect. The strongly aromatic scent of some of the alpine kinds is said to occasion headaches.

There are remarkably few literary allusions to the azalea or rhododendron; the noble sonorous word is appropriately used only by George Borrow, in a casual reference to a Duke of Rhododendron, who made a speech at an agricultural dinner. "This is the Latin," says Cobbett, "and the only English name of one of the handsomest shrubs that we have any knowledge of;" and it is perhaps a measure of the respect in which we hold the family that we have not demeaned it with a trivial or degrading diminutive. The enormous genus was grouped into forty-three series and subseries by Sir Isaac Bailey-Balfour of Edinburgh and his helpers, and the results published in *The Species of Rhododendron* in 1930;

[1] *The Botanical Magazine*, 1803   [2] *The Third Eye*, 1956   [3] Woodville, *Medical Botany*, 1793   [4] Ib.

but the classification of the family is continually being revised, and some of Sir Isaac's nomenclature has already been superseded.

NOTES: Rhododendrons and azaleas are two distinct groups of plants to gardeners, but all are rhododendrons to the botanist. There is a bewildering array of species; new varieties appear almost daily. Spectacular advances in color, hardiness, and other garden aspects of this great genus continue to be made. Each group of plants has its societies of fierce supporters. In southeastern United States, we are fortunate to have the greatest wealth of deciduous azalea species in the world as well as one of the most important species of rhododendrons, *R. catawbiense.*

But it is the evergreen azalea group, generally native to Japan and subject to over 300 years of development, that is most popular as a garden plant in the milder parts of this country. They are becoming ever more complex and their histories hard to follow, what with the continuous naming of new cultivars that bring only minute variations, which do not necessarily mean advancements.

The cultural history of evergreen azaleas was documented much earlier in Japan than were the true rhododendrons or deciduous azaleas in their introduction to Europe. After all, the Japanese were dealing with native plants of great religious significance. A monograph on azaleas by Ito Ihei that covered every major azalea species of Japan—plus those introduced from China and Korea—was published in 1692, some 150 years before similar events occurred elsewhere.

Azaleas native to America were collected and sent to Europe since Colonial times; they became the basis of the many deciduous hybrids. The Japanese azaleas became known only as greenhouse plants or were cultivated only in our most southern plantation gardens. It was not until the 1930s that evergreen azaleas reached any significant use in our gardens. In hardy evergreen azalea development, *R. kaempferi,* the most widely distributed azalea in Japan, became the basis of the breeding work by B. Y. Morrison (Glenn Dale Hybrids) and Joseph Gable. Surprisingly, this important species did not reach the United States until 1892 when it was collected by C. S. Sargent of the Arnold Arboretum, and from there made its way into England in 1895.

There are many books on the culture of rhododendrons and azaleas. Perhaps the hardest task for the gardener is choosing which species or hybrids are best suited to the individual climate. Reliable nurserymen or various local authorities are often the best sources for information. There is no question but that these shrubs require a well-drained, acid soil. Generally azaleas, particularly the evergreen ones, fit better in the foundation planting because of their more dense habit. The true rhododendrons are generally large plants; they tend to become "leggy" and difficult to prune. But when grown, these handsome evergreens offer great variation in leaf size, and a wide range of colors and flower sizes, assuring their place among our most desirable ornamental plants.

Rhus

Members of the Sumac family are remarkable for the brilliance of their autumn color, for their commercial usefulness, and in some cases for their extreme toxicity. Several species which pass under noncommittal names are even more virulent than those labelled *toxicodendron* ("Poison-tree") and *venenata,* and gardeners would be wise to regard every Rhus with caution until it is proved to be harmless. Nevertheless, some of the species have been used for centuries for dyeing and tanning, and others for the production of lacquer, varnish or wax.

Turner in 1548 said that of the three kinds of Rhus mentioned by Pliny, he could identify one with certainty, "which is called of the Potecaries Sumache" (i.e., *R. cotinus*), and possibly another, the *Rhus coriaria,* which he had seen in Italy. This was called the Currier's shrub (*coriarius*—a tanner or currier) because leaves and bark were used for tanning, especially in the production of Turkish and Moroccan leathers. Other parts of the plant were used in various ways, but the sap is poisonous, and the shrub too tender to be grown out-of-doors in this country. The other European species, *R. cotinus,* the Venice or Silken Sumac (known to Pliny by the loopy name of *Coggygria*) was important as a dye-plant, its twigs yielding the yellow color known as "young fustic." It has been grown in our gardens since before 1656, for the sake of its "hair-like bunches of purplish flowers"[1] which gave this species its name of Wig Tree or Smoke Tree, and which persist through much of the winter—"on account of which singular oddness," says Marshall "this shrub is valued by some persons." Parkinson noted that it had leaves "of a Rossenlike scent, not unpleasant . . . growing to be of an excellent Rose colour, in the end of the summer," but perhaps the best effect is made by the purple-leaved varieties, of which there are several, some particularly deep and rich in color. A still more striking autumn display is produced by the American *R. cotinoides* or Chittam Wood, very similar to *R. cotinus* but rather taller, and exceptionally brilliant in the turning leaf. It

[1]Marshall

Rhus coriaria

(TANNER'S SUMAC)   Ferdinand Bauer (1819)

Ribes sanguineum

(FLOWERING CURRANT)   M. Hart (1830)

was found by Nuttall in 1819 and sent to Kew by Professor Sargent of the Arnold Arboretum in 1882. It is said to have been nearly exterminated in the wild through being extensively cut, especially during the American Civil War, for the sake of the dye that was obtained from the wood.

R. cotinoides, however, was a comparative latecomer. The first American species to be introduced, *R. typhina*,[1] was familiar to Parkinson in 1629; he said it had received its name of "Buckes horne tree of Virginia" because "the young branches that are of the last yeares growth are somewhat reddish or browne, very soft and smooth in handling, and so like unto the Velvet head of a Deere, that if one were cut off from the tree, and shewed by its selfe, it might soone deceive a right good Woodman." (We now call it the Stag's-horn Sumac.) It is a dioecious species, with male and female flowers on different plants; Marshall must have been referring to the female form when he said that the flowers made no show, but the large tufts of scarlet seeds had "an uncommon appearance" in the winter. (The male plant, once classed as a different species, has a larger flower plume.) The

pulp and skin of the berries is said to have "a very grateful acidity, of which Mr. John Banister, a very Curious Naturalist, and one who long resided in Virginia, says, they make Vinegar of it there, and use it to season their Meat;"[1] the roots were used as a febrifuge and the gum as a remedy for toothache. Unfortunately the root of this handsome shrub "shooteth forth young suckers farre away, and round about"[2] which will make light of a concrete path, if one should happen to impede their progress.

R. glabra, allied to *R. typhina* but more compact and less invasive, was introduced from America in or about 1726, when it was grown by the nurseryman Christopher Grey at Fulham. It was handsomely depicted in Catesby's *Natural History of Carolina*, where it is noted that it needs a warm summer to produce its full effect. It has a cut-leaved garden variety called *laciniata* (F.C.C. 1867) which with high feeding and hard pruning will produce leaves a yard long; Robinson says they combine "the beauty of the finest Grevillea with that of a Fern-frond . . . and in the autumn the leaves glow off into a bright colour, after the fashion of American shrubs."

---

[1] *Typhina*—similar to *typha*, the bulrush

[1] Petiver, *Philosophical Transactions*, 1712   [2] Parkinson, 1629

These four well known species—*cotinus, cotinoides, typhina* and *glabra* are supposed to be innocuous, but even they should be regarded with suspicion, in view of the extreme toxicity of the next two—*R. toxicondendron* and *R. radicans*, the American Poison Oak and Poison Ivy, so closely allied that some botanists think the latter only a climbing form of the former. (*R. radicans* climbs like the Virginia Creepers, by means of aerial rootlets; *R. toxicodendron* grows about 2 feet high and spreads by underground stolons.) Strange to say, some people are immune to the poison, while to others it is deadly. In 1668 a Mr. Richard Stafford wrote from Bermuda "You shall receive of Captain *Thomas Morly*, the Commander of our *Magazeen-Ship*, such things as I could at present procure. Among which you shall find of the Leaves and Berries of that Weed you enquire after, which we call *Poyson-weed*, growing like Ivy. I have seen a Man, who was so Poyson'd with it, that the skin peel'd off his Face, and yet the Man never touch'd it, onely look'd on it as he pass'd by: but I have chawed it in my mouth, and it did me no harm. It is not hurtful to all."[1] Kalm tells of two sisters, the one immune, and the other so sensitive that she suffered even when she stood in the wind that blew directly from the shrub. (The Indians attempted to placate it by talking to it kindly and calling it "My Friend.") Contact with poison ivy, or at second-hand with logs, boots, tools or clothes that have been brushed by the plant, can cause itching, painful blisters, fever, loss of sleep and sometimes permanent disfigurement; Bean tells of a gardener who had been taking cuttings for propagation, having to spend several months in the hospital from the "almost corrosive" effects of the sap. The poison, *toxicodendrol*, is insoluble in water, so cannot be washed off, and many of the remedies applied only spread the infection; the only antidote appears to be immediate treatment with alcoholic solution of lead acetate.

In spite of its well-known virulence, no pains were spared to introduce the plant. Parkinson apparently had it before 1640; Mr. Stafford's attempt to send seed from Bermuda has already been quoted; Collinson received it from Bartram in 1760 and later made a note that it had "shot 7 feet last year" Loudon in 1838 reported that both kinds were frequent in collections, and at the end of the century *R. radicans* was in commerce, with its dreadful identity concealed under the name of *Ampelopsis Hoggii*. Many of those who write about it give no warning of its dangers; Parkinson, for example, only says of the "Trefoil Ivy of Virginia" that it yields a white milk without any taste, ". . . which, after it hath abidden a while, will

[1] *Philosophical Transactions*

change to bee as black as Inke, and is therefore held fit to colour the haire or any other thing." (It makes an indelible marking-ink for linen.) Marshall says that *R. toxicodendron* may become troublesome by spreading "in a manner not becoming a well-ordered shrubbery or wilderness," but says nothing about its poison; and even Johnson in the *Cottage Gardener's Dictionary* does not mention it.

More deadly, if possible, than the Poison Ivy is the Poison Ash or Varnish Tree, *R. vernix* (syn. *R. venenata*) also an American, kindly introduced in 1713, and considered by Bean one of the most dangerous of hardy trees. Cases of poisoning have also been traced to objects lacquered with varnish from an oriental species, *R. verniciflua*, and marketed while the lacquer was still "green." But these are trees, not shrubs, and seldom grown here.

It is not always easy to recognize a Rhus at sight, they are so diverse in appearance; some, like *R. cotinus*, have entire leaves; others, like *R. radicans*, trifoliate leaves similar to those of the kidney bean; while the majority have pinnate leaves after the manner of *R. typhina*. Some botanists make a separate family of the first section, and name the Venice Sumac or Smoke-tree *Cotinus coggygria*. "Sumac" is the Persian name, and Rous or Rhus (meaning "red") the Greek one.

NOTES: Our native sumacs, *Rhus copallina, R. glabra* and *R. typhina*, native to eastern states, are among the most colorful wild shrubs in the autumn. They present an outstanding foliage display. The brilliant leaves—from orange to scarlet—brighten roadside plantings across the country. Perhaps because they are so common in the native landscape, they are infrequently planted in gardens. But there are several selected forms with finely divided leaflets worthy of consideration.

The sumacs often grown large and are not suited to the average small garden. Furthermore, the branches are brittle and break easily. Probably they are best enjoyed in nature.

We recognize *R. cotinus* by the name *Cotinus coggygria*, the Smoketree, because of its billowy, pink fruiting panicles in the summer. It is a large shrub, exceeding 15 feet, and should be planted in an open space to be effective. It is hardy throughout the eastern states where it has been grown since Colonial times.

Ribes

The ornamental currants are particularly associated with that hardy and intrepid collector, David Douglas, who introduced no less than fourteen species and varieties—three from California and the rest from the vicinity of the Columbia River. Yet of the three species which alone of the large genus merit garden space as flowering plants, all had been discovered, and two had already been introduced, before Douglas collected them. It was he, however, who sent or brought home the large consignments of seeds that established the shrubs securely in British gardens.

R. sanguineum, the Flowering Currant, was found in 1793 by Archibald Menzies, a surgeon-naturalist in the Navy, who sailed to the Pacific coast of North America with Captain Colnett's expedition (1786–90) and again with Captain Vancouver (1790–95). He had not the facilities for bringing home many seeds or plants, and was obliged to confine himself chiefly to herbarium specimens, so the actual introduction of this universal favorite fell to Douglas, who found it growing near Fort Vancouver within a month of his arrival in the spring of 1825. Plants raised from seeds he sent home flowered in 1828, when they were barely two years old; and the Horticultural Society, for whom Douglas was collecting, considered this acquisition sufficient in itself to justify the entire expense—some £400—of the three year expedition. Yet the original plant was poor and pallid compared to the improved garden varieties we now enjoy. These include the form '*splendens*', distributed by the firm of Smith of Newry in the 1890s, and described as the nearest to a genuine blood color in this rather misnamed species; and a fine variety called Pulborough Scarlet, which occurred as a chance seedling in a bed of the type, some years before the Second World War.

R. aureum, the Golden, Buffalo or Missouri Currant, was discovered by Lewis and Clarke in 1804–6, and introduced to England by Thomas Nuttall in 1812. It is grown for the sake of its yellow flowers with their spicy fragrance, which some describe as resembling cloves, and others, cinnamon; but the name might also apply to the leaves, which in autumn turn the color of the gold that Douglas was the first to find in California, and never thought it worthwhile to mention. Mr. C. R. Falwasser says[1] that in the Yorkshire village of Mexborough this shrub was known as Cinnamon-bush, and the tradition was that it must always be grown in a bottle. Cuttings taken in spring were inserted in a bottle of water on a window sill, and when they had developed roots the bottle was smashed and the slips planted out.

Neither of these ornamental currants is of any value as a fruiting shrub. The berries of *R. sanguineum* are said to have an unpleasant musky flavor, unattractive even to birds; they are rarely produced here. The fruit of *R.*

[1]In the *Gardeners' Chronicle*, September 4, 1958

aureum is normally black, and its taste is described as mawkish, but there is a form with yellow berries (var. *chrysococcum*) which Douglas seemed to find palatable, especially when grown in dry sandy places or on limestone. He was "much refreshed" by this fruit, which he found "in great quantity and of excellent flavour;" a welcome relief in wanderings which often brought him to the very verge of starvation. The two species (*aureum* and *sanguineum*) have been crossed to produce a hybrid of doubtful value, *R. × gordonianum*, which was raised at Shrublands Park near Ipswich, about 1837.

On his second visit to the Pacific coast Douglas visited California, arriving at Monterey on December 22, 1830. "Early as was my arrival on this coast, spring had already commenced," he wrote. "The first plant I took in my hand was *Ribes speciosum* . . . remarkable for the length and crimson splendour of its stamens, a flower not surpassed in beauty by the finest fuchsia; and for the original discovery of which we are indebted to the good Mr. Archibald Menzies in 1779."[1] It had already been introduced to England by Mr. Collie, surgeon-naturalist of the appropriately named *H.M.S. Blossom*, which visited California briefly in 1826; and a further consignment of seed was sent by Douglas himself in 1834. This species belongs to the gooseberry, not the currant section of the family, and has formidable thorns, which will draw blood if they can. It dislikes the cat-and-mouse nature of English winters, the leaves and flowers appearing so early that they require some protection against spring frost; and if the shrub can be placed above eye level (as Loudon suggested) the small but numerous pendent flowers can be better appreciated.

Not all the members of the family come from America; several, including the parents of the dessert currants and gooseberries, are natives of Europe, and some belong to Asia—for example, *R. laurifolia*, with its large, oval, evergreen leaves and drooping racemes of greenish flowers in February; a plant that only a very imaginative botanist would place as a member of the currant family. It is rare even in the wild and, was found by E. H. Wilson on the sacred mountain Wa Shan in Szechwan, in 1908.

"Ribes" is derived by some from an ancient Arabic name, properly belonging to a species of rhubarb; but de Candolle traces the name of this predominantly northern family to the Danish *ribs* or Swedish *risp*. Although not so rank as the Black Currant, the Flowering Currant has an odor which leads some to call it Incense Bush, and others—Ginger Muck.

[1]Quoted by Hooker, *Companion to the Botanical Magazine*, 1836. The date seems likely to be an error for 1789

Robinia hispida

(ROSE ACACIA)   S. Edwards (1795)

Romneya coulteri

(CALIFORNIA BUSH-POPPY)   M. Smith (1905)

NOTES: Members of the currant family, while very popular in England, are cultivated as ornamentals here. The Flowering Currant offers beautiful red flowers that will enhance any southern garden.

Robinia

R. *hispida*, Rose Acacia. "I never saw any of these trees but at one place near the *Apalatchian* mountains, where Buffalos had left their dung, and some of the trees had their branches pulled down, from which I conjecture they had been browsing on the leaves . . ." So wrote Catesby in his *Natural History of Carolina*, and depicted accordingly, along with a spray of the blossom, what he asserted was "a perfect likeness of this awful creature"—one fierce eye rolling, the other invisible—drawn from a living bison calf that had been brought over to England. The plant, however, had to be drawn from a dried specimen or from earlier sketches, for when Catesby had returned to the spot at the right time to get some seeds, he found, to his great disappointment, that the woods had been burned for miles around by "the ravaging Indians," and the acacias totally destroyed. In 1741, the shrub was introduced by Sir John Colliton of Exmouth, from his plantation in Carolina; but in 1748, when Catesby was making his drawing, there was no specimen available within easy reach of Loudon, although both Collinson and Miller had them a year or two later.

A picture of the plant (but without the buffalo) appeared in the *Botanical Magazine* for 1796, where it is described as being "of ready growth, disposed to blow even when young, and not nice as to soil or situation . . . the flowers . . . are large and beautiful, but without scent." We are advised to keep it humble—i.e., pruned low—and to plant it in a sheltered place, as its branches are easily broken by high winds. During the following century it was frequently grown as a standard, grafted on stocks of the common False Acacia (R. *pseudo-acacia*) in spite of the fact that such standards were vulnerable to frost as well as to wind; but the improved garden varieties were considered to deserve "a place on a conservative wall"—at least, in the colder districts. Three such varieties were grown before 1863, including the handsome var. *macrophylla*, said to be the finest member of the genus.

Among the many virtues of the Rose Acacia is its propensity to flower intermittently all summer, or to produce a second crop of flowers in the autumn. "It is well known," says Phillips, "that most plants will continue to give out blossoms, if their flowers are cut off before seed is formed; which seems like the instinct of fowls that continue to lay eggs in the nest that is plundered." But like many other plants that increase by suckers, the Rose Acacia rarely produces seed, even in the wild.

The family name commemorates Jean Robin (1550?–1629), gardener to Henry IV of France. He had charge of the royal gardens of the Louvre, laid out by Henry about 1590, and also had a small plot of his own at the west end of the Ile de la Cité. By 1600 he had more than a thousand different kinds of plants in cultivation, and supplied the flowers which the King's Embroiderer, Pierre Valet, illustrated in his *Jardin du Roi* (1608)—a work primarily intended as a pattern book for the flower loving Queen Marie de Medicis and her ladies. Robin also exchanged many plants with Gerard and Parkinson, and in 1606 was the first to grow a representative of this exclusively American family—the tree named by Linneaus *Robinia pseudo-acacia*. When it was first discovered this tree was classed by some botanists as a mimosa, and named accordingly *Acacia americana Robinii*; others, more doubtful, called it *Pseudo-acacia*, and False Acacia it has remained ever since. Although their flowers appear so different, robinias and mimosas belong to the same Natural Order—the Leguminosae.

NOTES: The robinias are interesting legume trees native to the eastern United States but with sufficient insect problems that they are not specially valuable garden trees. The black locust, *R. pseudoacacia* grows to 75 feet and can become a weed tree. While popular in the English garden, it has not created the same interest here.

Romneya

This family has the unique distinction of having been named after an astronomer—if indeed the great moon-white flowers require any reflected glory other than their own. *R. coulteri*, the Matilija Poppy or Californian Bush Poppy, was discovered by an Irish botanist, Dr. Thomas Coulter, a pupil of de Candolle, who went to Mexico in 1824 as physician to a mining company; after various exploits and explorations in Mexico he visited California and Arizona in 1831, and returned to Ireland in 1834. (In California he unexpectedly encountered David Douglas, to their mutual delight; it must have been rather like the meeting between Livingstone and Stanley.) Coulter brought back a vast number of herbarium specimens, to add to an already large collection of European plants, and had not completed their arrangement and classification at the time of his death in 1843. It was his successor at Trinity College, Dublin, Professor W. H. Harvey, who found among Coulter's specimens this "fine papaveraceous plant, which I soon ascertained to be distinct from any hitherto recorded from that country; and closer examination . . . proved it to belong to a new and curious genus."[1] There was already a family named "Coulteria" by de Candolle, so the new plant could not be called after its discoverer; instead, Harvey suggested that it should be given the name of Coulter's great friend and compatriot, Dr. T. Romney Robinson, who after a brilliant academic career had been appointed astronomer-in-charge at Armagh Observatory, and whose great work *The Places of 5,345 Stars (Principally Bradley's Stars) Observed at Armagh from 1828 to 1854* was published in 1859. Again, there was already a genus "Robinsonia"—a small family of composites, so named because they were mostly indigenous to Robinson Crusoe's island of Juan Fernandez—so the new flower fortunately had to be called "Romneya." Fortunately, because the famous astronomer's father had been a portrait painter, and had named his son after his master, George Romney; so the shrub actually gets its name, though indirectly, from an artist who would surely have delighted in its great glistering flowers of silky white. The plant itself does not seem to have reached England until about 1875, when it was cultivated by the nursery firm of E. G. Henderson and Son.

There is a second species of Romneya, *R. trichocalyx*, but so like the first, that the two were grown together for years before anybody noticed the difference. Attention was called to it in 1898 by Miss A. Eastwood, Curator of the Herbarium of the Academy of California; she sent seeds to Kew in the autumn of 1902, and in the same year the plant was also observed in the garden of Mr. H. C. Baker at Almondsbury near Bristol. It is rather more upright in growth than *R. coulteri*, less tall, slightly hardier, the leaves grayer, the flowers less sweetly scented and the buds rounder and more bristly—none of which are differences to be told at a glance, or without direct comparison between the two plants. Just to add to the confusion, a surely unnecessary hybrid has been raised between the two, by Mr. W. B. Fletcher of Aldwick, Sussex, and distributed as *R. × hybrida*.

NOTES: This California native thrives in gardens of its homeland, and in England, but is too tender for the rest of America.

[1]*London Journal of Botany*, 1845

Rosa

Somehow my heart always opens and shuts at roses . . .
Mrs. Caudle's Curtain Lectures, Douglas Jerrold, 1852

The Rose is not a family whose history has been neglected. Extensive researches have been made on the subject by many learned specialists, though it is so ancient a flower that, even so, much of its story remains obscure. It is not a particularly large genus, but a very complex one, owing to the variability of the species, the ease with which they hybridize, and the prolonged interference of man: for roses have been cultivated for more than three thousand years. There are now thought to be about 120 species (all natives of the northern hemisphere), comparatively few of which have taken part in the evolution of the garden strains. At first these increased but slowly; Parkinson in 1629 thought "the great varietie of Roses . . . much to be admired" when he had in his garden "thirty sorts at the least," not counting the wild kinds; more than a hundred years later the number had only risen to forty-six, and by the end of the eighteenth century, when the first English monograph on the flower was printed,[1] had still not attained three figures. But in the first quarter of the nineteenth, the numbers began to rocket skyward, and Loddiges' catalog for 1826 listed 1,393 species and varieties. Since then the raising of new kinds has gone on with varying rapidity, and in the peak period between 1925 and 1935 some two hundred new roses were produced every year.

Every European and oriental country has its own large body of rose literature and legend, but lest there should be room in this book for nothing else, I will limit myself as far as possible to the story of the rose in Britain. "Sweet is the rose, but grows upon a brere," and we have, to begin with, five English wild briars—*arvensis, pimpinellifolia, rubiginosa, villosa* (doubtful, but accepted as a native by the latest authorities) and the Dog-rose, *canina.* The latter has now been split up into a number of closely related species, bringing the count up to thirteen; there are also four naturalized kinds, and a great many wild hybrids, distinguishable only by those botanists who, like B'rer Rabbit, were "bred en bawn in a brier patch."

These native roses have only played a very minor part in the development of the garden forms, but their flowers serve to demonstrate the basic rose pattern of sepals and petals. Those edges of the sepals of the dog-rose that are covered when the flower is in bud are smooth, while the free edges are fringed; this gave rise to a popular medieval riddle of which there are several versions,

Of us five brothers at the same time born
Two from our birthday ever beards have worn;
On other two none ever have appeared
While our fifth brother wears but half a beard.[1]

This pattern is explained by the fact that the leaves of the rose like those of many other alternate leaved plants, are arranged on the shoot in such a way that a thread tied to the base of each leaf will make two complete spiral turns around the stem before reaching the sixth leaf, which is exactly above the first. If a strip of paper with short broad leaves projected at regular intervals is wound twice around a pencil until the sixth leaf is over the first and if this spiral is gently telescoped until all the leaves are bunched together, and the sixth then removed, it will be found that the remaining five overlap in the same way as the sepals of the dog-rose—two completely free, two covered, and one half free and half covered. This is called a quincuncial phyllotaxy, and almost reconciles one to the science of botany.

The first record we have of a cultivated English rose dates from the reign of William Rufus (1087–1100). In an attempt to see the young Matilda—descendant of the vanquished Saxon kings, and subsequently married to his younger brother, Henry I—William entered the cloister of the nunnery at Romsey where her Aunt Christina was keeping her concealed, with the excuse that "he only wanted to admire the roses and other flowering plants."[2] The white and red roses, afterward so famous, "assumed by our precedent Kings of all others, to bee cognisances of their dignitie,"[3] both came to us originally from France. The white rose was the badge of Eleanor of Provence, who married Henry III in 1236, and descended from her to her son, Edward I—though he seems to have transformed it into a golden one. Her second son, Edmund, first Earl of Lancaster, became by marriage Count of Champagne, and resided much in Provins in the first years after his wedding in 1275; he brought back from there a red rose which he subsequently adopted for his emblem. The badges were therefore in the family generations before the famous brawl in the Temple gardens that sparked off the Wars of the Roses. (There is no mention of this incident in Shakespeare's authority, Holinshed, and it is possibly no more authentic than that of the gardeners who painted the

[1] Lawrance, *A Collection of Roses*, 1799. Ninety sorts are described

[1] "The five brethern of the rose seems a wonderful disposure in nature . . . and most disagree in the enquirement, but the disposure of the calicular leaves gives some resolution." Sir Thomas Browne [2] Quoted in *The Leaves of Southwell*, Pevsner, 1945 [3] Parkinson, 1629

Duchess's white roses to make them red, in *Alice in Wonderland*.) The wars ended with the marriage in 1485 of Henry VII and Elizabeth of York, when the two badges were combined to form the conventional Tudor rose. The red and white Damask rose known as "York and Lancaster" is said to have been discovered at that time in the garden of a monastery in Wiltshire, and was regarded as a portent—an event so convenient that it arouses a certain scepticism, but this rose was certainly in cultivation, under the same name, before 1551.

Gerard described some seventeen roses in his *Herbal* of 1597, the most important of which were *RR. gallica, moschata, damascena, alba* and *centifolia*—the Five Ancestors of the rose's family tree.

R. gallica is the only red rose native to Europe—none of the others achieve more than a deep pink—so early references to it are comparatively easy to identify. It is believed to have been the sacred rose of the fire worshipping Medes and Persians, and to have been grown by them for use in their religious ceremonies, in the twelfth century B.C.; which must surely make it the oldest of cultivated roses. It was probably also the Rose of Miletus of the Romans, and grown wherever their empire extended. There is no record of the date of its introduction to England; it may have been a rose of this kind that William Rufus admired, brought over by Benedictine monks in the seventh century, or even by the Romans themselves.

Thibaut IV, "le Chansonnier" (1201–1253), Count of Champagne and Brie and King of Navarre, is said to have brought a rose or roses back to Provins on his return from the Crusade he led in 1239–40. The *Roman de la Rose*, written about 1260, mentions "roses from the lands of the Saracens," and it is possible that several sorts were included in various crusaders' various baggage; but it seems likely that the one Thibaut introduced was the famous Provins or Apothecary's Rose, *R. gallica* var. *officinalis*. This was a red, semi-double flower, whose petals (unlike those of the Damask roses) had the property of retaining, and even increasing, their color and scent when dried; moreover, their properties were tonic and astringent, while those of other roses were laxative. Provins was famous for the culture of these medicinal roses from the thirteenth century to the nineteenth; and it was possibly this rose that Edmund of Lancaster—whose wife, Blanche, was the widow of Thibaut's younger son, Henry the Fat—brought home with him about 1279. All that can be said with certainty is that "the red rose is the badge of England, and hath growne in this country for as long as ye minde of man goeth."[1]

[1] A monk of St. Mary's Abbey, York, 1368; quoted by Shepherd

Another rose of this group that is supposed to be of great antiquity is the striped Rosa Mundi, *R. gallica* var. *versicolor*, which is traditionally associated with Fair Rosamund, mistress of Henry II, who is believed to have been murdered by the jealous Queen Eleanor about 1176. Actually there is no written record of such a rose before 1583, when a similar flower was grown on the Continent under another name; and we hear nothing of it in England until the mid-seventeenth century. Hanmer in 1659 called it a new rose, which had appeared in Norfolk as a sport on a branch of the common *gallica*, a few years before; and in 1662 Thomas Fuller, speaking of Norwich, says that "the rose of roses (*rosa mundi*) had its first being in this city." It seems probable that this rose "of the world" (*mundi*) became linked with Rosamund only on account of the name—just as a later gardener transformed it into Rose of Monday.

R. moschata, the Musk Rose, is also easily recognized, as it is the only climbing species among the Five Ancestors—hence its choice by Shakespeare for Titania's bower. It is fundamentally an oriental rose, and its home is in the Himalayas, but it has been cultivated for untold centuries in Mediterranean and Near Eastern countries, and is one of the sources of attar of roses. Hackluyt says it was procured from Italy and names in the same sentence three kinds of plum brought "by the Lord Cromwell after his travell"—whether he also brought the rose is not clear, but if so, that would date it at about 1513. (It was known to Turner in 1551, and Gerard had both single and double forms.) It was much appreciated by the Elizabethans for its scent—a fragrance more perceptible by night than by day, and residing, as Parkinson observed, rather in the stamens than in the petals of the flowers. Unlike that of the red and damask roses, the fragrance of the musk rose is airborne, and perceptible from afar—when weather conditions permit; for the truth is that this rose is slightly tender, and is not at its best in Britain. In a warmer climate it is not only more fragrant, but will bloom two or three times a year—a quality that was to prove of great value to its progeny.

A famous, but for long unsuspected descendant of the Musk Rose was the Autumn Damask. Doubtfully accepted by earlier botanists as a species, or at least a group, the researches of the late Dr. Hurst revealed that the name of *R. damascena* covers two hybrids of slightly different parentage; the Summer Damask, from *R. gallica* × *R. phoenicea* (a Syrian wild rose, otherwise of no garden importance) and the Autumn Damask, sometimes called *R. bifera*, the Monthly or Four Seasons Rose, from *R. gallica* × *R. moschata*, which inherited from the latter parent the property of producing a few blooms at

Rosa Pumila. *Rosier d'Amour.*

Rosa gallica pumila
(FRENCH ROSE)   P. J. Redouté (n.d.)

odd seasons. (This second crop was not very profuse, but was highly valued at a time when no other rose bloomed more than once a year.) The crosses must have occurred at a very remote period, most likely in the wild, for roses of this description which bloomed twice a year were known to the ancient Greeks. Damask roses have long been regarded as the best kinds to use for the manufacture of rose water and attar of roses, and were already cultivated on a large scale for this purpose in Syria in the tenth century; but they seem to have been comparative latecomers to western Europe. The Spanish Dr. Monardes, who wrote a treatise on the roses of Persia and Alexandria in 1551, said the Damasks had only been known in the west for about thirty years, and were called *Damascenae* because they came from Damascus, always famous for its roses. (This seems to dispose of the tradition that they were brought from that city by a crusader.) Hackluyt said the Damask Rose was introduced to England from Italy "by Doctour Linaker, King Henry the seventh and King Henry the eights Physician." Linacre died in 1524, which fits well enough with Monardes' estimate of the date of introduction of this rose to Europe, and makes the alleged appearance of its 'York and Lancaster' sport in 1485 seem even more questionable.[1] An interesting variety of the Damask was raised from a hip which was plucked at the tomb of Omar Khayyám by a traveling artist for the *Illustrated London News* in 1884, and sent to his editor, Bernard Quaritch. It proved to be a new and very distinct form, and a cutting was soon afterward planted on the grave of Khayyám's translator, Edward Fitzgerald, in Boulger churchyard near Woodbridge, where after some vicissitudes, its descendant still flourishes.

The remaining ancestral roses are also hybrids. *R. alba* is now believed to be descended from *R. damascena*, its parent on the female side being a white form of the Dog-rose, *R. canina*—the one side entrance by which this species has slipped into the mainstream of our garden roses. The White Rose is a very ancient flower, known and cultivated by the Romans, and often represented in the paintings of the Italian Renaissance. Although occasionally found wild in Britain, it is not a native; the date of its introduction is not known, but it is generally assumed that the white rose of York was an *alba*. Later it became the badge of the Jacobites, partly because of its age old association with the Royal House of France, and partly because it was the emblem of secrecy. This goes back to a very early date—when Cupid bribed Harpocrates with a golden rose to keep silence about the

affairs of his mother, Venus; or perhaps when the Romans, relaxing after dinner in their rosy garlands, indulged in gossip that was not to be repeated in the morning. It became customary in northern countries to suspend a rose over the table at any private and momentous conference; and later it was carved on the ceilings of banqueting halls and over the doors of confessionals. Why a *white* rose should become particularly associated with this symbolism is not explained, but in various works on the Language of Flowers it is always a white rose with two buds that stands for secrecy. (A dried white rose, on the other hand, means "Death preferable to loss of innocence.") Long after the last echoes of the 'Fifteen and the 'Forty-Five had died away, devoted adherents of the lost Jacobite cause still wore a white rose on the birthday of the Old Pretender, the 10th of June. Not all the *alba* roses are white, thought most are light in color; the best known variety is probably the 'Maiden's Blush', known on the Continent as *Cuisse de Nymphe* (Nymph's Thigh) and when particularly deep in color, as *Cuisse de Nymphe Émue*. Turner in 1551 knew this variety as the Incarnacion Rose—"carnation" in those days meaning flesh-colored.

The last of the Five Ancestor roses is an imposter—an upstart no more than four centuries of age, compared to the twenty at least of its companions. For years it was believed that our *R. centifolia* was the Hundred-leaved Rose mentioned by Theophrastus and Pliny, though there was a significant gap in its records between the Romans and the Renaissance; but recent research has shown it to be a complicated garden hybrid derived from four different species (*gallica, phoenicia, moschata* and *canina*)—probably by way of a cross between an Alba and an Autumn Damask; and no unmistakable reference can be traced earlier than about 1580. It probably came from the east by way of Austria, but its origin was claimed both by Holland and France; in old books it is usually called either the Dutch Hundred-leaved or the Provence Rose, which led to a great deal of confusion with the much older Provins Rose, especially as both were frequently spelled "Province;" it is sometimes difficult to tell which flower is intended. Later a prosaic generation dubbed it the Cabbage Rose—but no cabbage of that shape would pass muster today, when we expect these vegetables to resemble the delicate pointed bud of the Hybrid Tea. The scent is rich and sweet, and in the nineteenth century this type of flower was considered the very epitome of the rose. "All others are varieties of roses," wrote a lady gardener in 1839, "but this grand flower is the rose itself."[1]

---

[1] Bacon, writing in 1627, said that damask roses had "not been known in England above a hundred years, and now are common"

[1] Louisa Johnson, *Every Lady Her Own Flower Gardener*

As might be expected in so very double a flower, *R. centifolia* is sterile; but as if to compensate for its lack of seed, it has produced a number of bud-sports, the most famous of which was the moss rose.[1] Again the early history is obscure—there is an unconfirmed report that it was known in Carcassonne in the 1690s; but we are on firm ground in 1720, when it appeared on Boerhaave's list of plants in the physic garden at Leyden, and in 1724 it figured in the catalog of Robert Furber, nurseryman, at Kensington. It proved difficult of propagation, and did not become really widespread until the end of the century. In 1785 Marshall said that the moss rose had been "sought after of late more than any of the others," but that if the moss had been common, it would have been regarded as an imperfection, as "this mossy substance has a strong disagreeable scent, and is possessed of a clammy matter." But to the Victorians, three-quarters of a century later, the moss added the one beauty that the rose had hitherto lacked. "When it was found," as Bunyard puts it, "that the Rose could look cosy as well as beautiful, no wonder that hearts were stormed. Cosiness lay at the very centre of Victorian taste." It was principally in England that the varieties of the moss rose were developed; most of the early ones originated as mutants from the Cabbage Roses, to which they occasionally revert, but a few derive from the Damasks, and a mossy form of the sweet briar has been recorded.

These Five Ancestor roses and their varieties, together with a few species of lesser importance to which I shall return presently, were the only kinds grown in gardens up to the end of the eighteenth century. Only one of them, the Autumn Damask, flowered more than once in a year, and many curious recipes are given in old books to obtain roses out of season. The sudden and spectacular advance in rose growing in the early nineteenth century was due to the arrival of certain Chinese garden roses, which brought with them the precious "remontant" or perpetual blooming property—a characteristic which seems peculiar to the roses of central Asia, where it occurs in several species, including, as we have seen, the Himalayan musk rose. The roses of China have probably been cultivated at least as long as those of the Near East, and the double China Rose (misleadingly named by Linnaeus *R. indica*) appears in paintings of the tenth century, looking very much as it does today. The original wild species, rather surprisingly a vigorous climber, was only discovered in 1885, and was named *R. chinensis*.

A pink China Rose appears to have been grown by Philip Miller as early as 1752, but no more is heard of the species until 1789, when two new varieties arrived—Slater's Crimson and Parson's Pink. The first was brought from a Calcutta garden by a captain in the employment of the East India Company, to Gilbert Slater Esq., a Director of the Company, in whose greenhouse it flowered about 1791; and "as he readily imparted his most valuable acquisitions to those likely to increase them, the plant soon became conspicuous in the collections of the principal Nurserymen near town."[1] It was a dwarf bush rose with a dark red, semi-double flower, and not being identical with Linnaeus' specimen, was given the name of *semperflorens*. Two accounts are given of the introduction of the pink variety, not necessarily incompatible; one, from Aiton, that it was procured by Sir Joseph Banks in 1789, and the other, from Andrews, that it bloomed first in the garden of Mr. Parsons at Rickmansworth, in 1793. It proved hardier than the red form, and Parson's Pink China soon became the Common Blush, or simply the China Rose.[2] By 1823 it was in every cottage garden—"As the smallest cutting of this rose will grow," said Phillips, "we are not without the hope of seeing it creep into our hedgerows."

Another important Chinese rose quickly followed—the first Tea rose, bought in the Fa-te nurseries near Canton by an agent of the East India Company and sent to Sir Abraham Hume of Wortleybury in 1808. Hume's Blush Tea-scented China was given the name of *R. odorata*, but is now known to have been a hybrid between *R. chinensis* and *R. gigantea* (a magnificent but tender climbing species from the Himalayas) with the latter predominating. Its English name has puzzled generations of gardeners, unable to detect any scent of tea in its fragrance; but it seems it was the odor of the bruised *fresh* tea leaf it was supposed to resemble. This rose is now extinct; and though we know from pictures what it looked like, its scent is gone forever. Nearly twenty years later, in 1824, a yellow tea rose of similar parentage (also now extinct) was bought from the same nursery by John Damper Parks, then collecting for the Horticultural Society, to add a further range of delicate shades to the breeding that was already going on.

From these Four Chinas, crossed with the flowers already developed from the Five Ancestors, all the main groups of nineteenth century garden roses were evolved. There is no room here for the details; in brief, Parson's Pink crossed with the Musk Rose yielded the Noisettes,

---

[1] *R. centifolia muscosa*. It seems perverse that the Musk Rose should be called *moschata* and the Moss Rose *muscosa*

[1] *The Botanical Magazine*, 1794   [2] Said to be the original of Moore's *Last Rose of Summer* (1813)—which must, of course, have been a remontant variety

and with the Autumn Damask, the Bourbon Roses; these two strains were crossed with the two Chinese tea roses, to produce the Teas, while meantime the Hybrid Perpetuals had been built up from a very complex ancestry, excluding the tea roses but including almost everything else. (By no means all of them were perpetual blooming; Robinson grumbled at the name being given to "Roses that flower the shortest time.") The Teas were exquisite, but tender—a greenhouse race; so to gain hardiness they were crossed with the Hybrid Perpetuals to produce the Hybrid Teas—chief origin of the roses of today. Each strain began by being quite distinct, but through repeated crossings in the attempt at "improvement," their individualities became merged into a general rosiness—"A rose is a rose is a rose is a rose," as Gertrude Stein so truly said. Such of the roses in the older groups as still survive have been rescued from oblivion like so many vintage cars, and are highly prized by a certain section of the "excitable and exacting fraternity"[1] of rosomaniacs.

The crimson China rose which had so large an influence on the development of the race was a dwarf, almost a miniature, variety—Curtis said that it would "grow in so small a compass of earth, that it may be reared almost in a coffee cup"—and the same group has produced some real midgets in the diminutive Fairy Roses. Again accounts differ as to their origin, and probably refer to two different plants; one raised in 1805 at Colville's nursery from Parson's Pink China and named *R. indica pumila*, and the other introduced from the island of Mauritius when it was taken from the French in 1810; the illustration in the *Botanical Magazine* was drawn from a specimen "communicated by Mr. Hudson at the War Office." This tiny single rose was named *R. indica minima* or *R. lawranceana*, in honor of Miss Mary Lawrance (author of *A Collection of Roses*, 1799), who was then at the height of her fame as a painter of flowers. There were sixteen varieties of these miniature roses by 1836; one was later described as having "fine full-blown and very double flowers, and the half of a common hen's eggshell would have covered the whole bush without touching it."[2] Of this race was the rose discovered by Major Roulet during the First World War, growing in window boxes in the Swiss village of Mauborget, and distributed about 1922 as *R. roulettii*. The Fairy Roses are now very popular in the United States, where some 160 varieties are cultivated.

The Tea and China roses were not the only kinds to reach us from the Orient as the eighteenth century gave way to the nineteenth; there were several others that were also of importance. *R. rugosa*, the Japanese Rama-nas Rose, was a very old garden flower both in China and Japan; two forms of it are said to have been distributed by Lee and Kennedy in 1796, but it does not seem to have been widely known until reintroduced from Japan by Siebold, about 1845. Its first successful hybrid did not appear until 1887, but it is now becoming important as a parent (or rather, grandparent) through the work done by Kordes on its hybrid, 'Max Graf'. The other oriental newcomers were all climbers: *RR. banksiae, bracteata, multiflora* and, later, *wichuriana*.

Of the four forms of the Lady Banks' rose, the first to arrive was the double white; it was procured by William Kerr from a garden near Canton and sent to Kew in 1807, where it was named by the botanist Robert Brown, in honor of the wife of the Director, Sir Joseph Banks. The double yellow form was procured by Parks for the Horticultural Society, from Calcutta Botanic Garden, and sent home along with the yellow tea rose in the East Indiaman *Lowther Castle* in 1824. The single yellow was recorded, but not introduced, by Dr. Clarke Abel at Pekin in 1816; it eventually reached us by way of the Florence botanic garden and the Riviera garden of Sir Thomas Hanbury, in 1871. The single white was unknown here until 1905, when a very old rose which seldom flowered, growing on Megginch Castle, Strathay, was found to be of this kind; it was believed to have been brought back from China by Robert Drummond of Megginch in 1796. If so, the survival of this specimen for over a hundred years is remarkable, for the Banksia roses are tender here—the whites more so than the yellows. Unfortunately it is the white variety that is fragrant, the yellow having little or no scent; "it is a little white rose," says Cobbett, "and bears its flowers in bunches, and yields to nothing in point of odour, except the Magnolia Glauca . . ." All travelers in China are enthusiastic about this rose, the scent of which drove Farrer "perfectly wild with drunken rapture," and which will grow to a great size; but it does better on the Riviera than here.

R. bracteata, Macartney's Rose, was brought home by Sir George Staunton, secretary to Lord Macartney, on his return from the latter's embassy to China in 1793. It is a strong growing rose with formidable thorns, and has made itself a naturalized nuisance in the American south, where it is called the Chickasaw Rose. It does not hybridize readily, but is the parent of one very famous variety—the rose 'Mermaid', raised in 1917 and still first favorite among climbing roses. The Bramble Rose (*R. multiflora*), on the other hand, has a numerous progeny. Cultivated forms of this climber had long been grown in Chinese gardens; a variety with double pink flowers was

[1]Hibberd   [2]Quoted by Shepherd

introduced by Thomas Evans of the East India Company, in 1804, and was followed by *R. multiflora* var. *platyphylla*, the Seven Sisters Rose, sent home by Charles Greville in 1815. ('Seven Sisters' was the Chinese name, said to have been given because seven colors could be seen in the flower cluster at the same time; it was also called 'Ten Sisters, Older and Younger'.) The wild type—'Snow on a Visit' to the Chinese—was sent from Japan to France in 1862; with its profuse bunches of small, blackberry-like flowers it is not one of the most attractive species, but it proved of great value both as a stock on which to graft other varieties, and as a parent. It often happens that when climbing roses are self-fertilized, dwarf forms will eventually appear among their seedlings; and in this way *R. multiflora* became the ancestor of the first of the Polyantha roses, as well as of numerous climbing kinds. Among the latter was a Japanese hybrid (*multiflora* × *chinensis*) sent in 1898 by Robert Smith, then Professor of Engineering at Tokio University, to Thomas Jenner, who called it The Engineer; but it was eventually marketed by Turner of Slough under the name of 'Crimson Rambler'. Apart from this famous variety, the climbing roses raised from *R. multiflora* were superseded soon afterward by the introduction and use in breeding of *R. wichuriana*. This species was discovered in 1861 by a German botanist, Max Ernst Wichura, growing on rocks beside a Japanese river. Plants he brought back to Europe failed to survive, and Wichura himself died in 1866; but the rose was again procured by the botanic gardens of Munich and Brussels in 1886, and named in his memory. It eventually reached Britain—via America—in 1890. Its great hardiness, long flowering period, and shiny, disease resisting foliage made it a valuable parent; a variety raised by the firm of Jackson and Perkins of New York was distributed in 1901.

All this time, from the sixteenth century to the nineteenth, a number of roses had been existing on the borders of the mainstream, in single species or small groups, that contributed little or nothing to the development of garden roses as a whole, though some of them are being pressed into belated service by the rosarians of today. Such, for example, were *RR. cinnamomea, hemispherica* and *virginiana*, and the Ayrshire, Scottish and Sweetbriar roses. The Cinnamon Rose was grown by Gerard in single and double forms, and is remarkable for the early season of its bloom. It flowers about Whitsuntide, and is sometimes called *R. majalis*—the 'Rose of May' to which Laertes compared Ophelia. It kept its place in gardens at least until the nineteenth century, for Phillips wrote in 1823, "It is a favourite with our fair, as it may be worn in the bosom longer than any other rose, without fading, while its diminutive size and red colour together with a pleasant perfume, adapt it well to fill the place of a jeweller's broach." Not even Gerard, however, could detect any resemblance to cinnamon in the scent of the flower; the reason for the name was a mystery even in the sixteenth century, though Bunyard suggests it may have been given on account of the cinnamon brown color of the young shoots. Its only hybrid offspring was the Frankfurt Rose, a very distinct flower, known to Parkinson and still available, but without further descendants. Also without progeny was the double yellow *R. hemispherica*, a beautiful but erratic bloomer, whose vagaries exercised generations of gardeners from the time when Clusius saw its replica in a paper model of a Turkish garden exhibited in Vienna before 1583, and procured a plant through a correspondent in Constantinople. It was twice introduced to Britain—by Nicholas Lete in 1586, and by John de Franqueville in 1595; but as the buds of this rose fail to open if so much as touched by a drop of water, it is not really suited to our climate. Another uncommonly chaste flower was the neat little *R. virginiana* (St. Mark's Rose, or *Rose d'Amour*) the first species to be introduced from North America. The single form was known to Parkinson in 1640, and the double, now believed to be a hybrid, to Miller in 1768.

Albion, according to Pliny, received its name either from "its white cliffs washed by the sea, or the white roses with which it abounds." This "English, unofficial, rose" was probably *R. arvensis*, more common in the south than in the north, which by a curiously roundabout route became the ancestor of the climbing Ayrshire Roses. In 1767 John, Earl of Loudon, received a parcel of seeds from Canada or Nova Scotia, including some rose hips. These, sowed at Loudon Castle, produced a climbing rose of great vigor, capable of growing 30 feet in a season, which was later distributed by nurserymen of Kilmarnock and Ayr. It is believed to have been a hybrid between *R. arvensis* and *R. sempervirens*,[1] and of garden origin, since neither species is a native of America. A number of garden varieties were raised, but appear now to be extinct. The Ayrshire Rose should not be confused with the Scotch rose, *R. pimpinellifolia* (syn. *R. spinosissima*) in some of its forms the smallest of all wild roses, and even more prickly than the Scotch thistle. This native plant was not at first thought worthy of cultivation, but four varieties were grown in the latter part of the eighteenth century; Marshall, in 1785, remarked that "in winter, they will be full of heps that have the

[1] A Mediterranean species, which has been identified as the "dog"-rose that scratched Locris, in fulfilment of an oracle that he would be bitten by a wooden dog

appearance of black-berries; and if the weather be mild, the young buds will swell early, and appear like so many little red eyes over the shrub, which is a promise of the reviving season." In 1793 Robert Brown, a nurseryman of Perth, collected a number of wild variants from the Hill of Kinnoul, and from them raised eight good doubles by 1802; from these seedlings another Scots nurseryman, Richard Austin of Glasgow, produced over a hundred sorts by 1822. Other firms, both English and Scottish, followed suit, and at one time there were about three hundred of these Scotch Roses, probably not very distinct; a few of them still survive, the best known being 'Stanwell Perpetual'—thought to be a hybrid, with an Autumn Damask parent—found as a chance seedling at Stanwell in Middlesex before 1799, and distributed by Lee and Kennedy in 1838. Although *R. pimpinellifolia* is a very widespread and variable species, ranging from Iceland to Mongolia, it was not used by Continental rosarians until about 1931, when the German grower, Wilhelm Kordes, crossed some of its forms with various Hybrid Teas to produce the famous "Frühlings" series (Frühlingsgold, Frühlingsmorgen, etc.) which seem likely to play an important part in the gardens of the future.

For another line of breeding Kordes used as a parent another rather neglected rose, a descendant of the sweetbriar (*R. rubiginosa*, syn. *eglanteria*.) The Eglantine of Chaucer, Shakespeare, Spenser and innumerable lesser poets derived its name, via the Old French, from the Latin *aculeatus*, prickly, because it is "armed with the cruellest sharpe and strong thornes, and thicker set, then is any other Rose, either wilde or tame"[1]; but in spite of this "its fragrant odour highly obliges us to plant it in all parts of our Gardens and Wildernesses."[2] A few varieties were cultivated from an early date; Parkinson had a double form in 1629, and in 1728 Batty Langley spoke of a red flowered kind which made "very beautiful headed plants,[3] which are very proper to be raised in Pots, to adorn the Ladies' Chimneys and perfume the Air of their Chambers with its pleasant and most delightful Odour." The type was liberally planted for hedges on account of its fragrance, and also to supply sprays of foliage for nosegays. Late in the nineteenth century the pretty variety called Janet's Pride was found growing, far from other roses, in a Cheshire hedge; and it is said to have been the sight of this flower that suggested to Lord Penzance the possibility of raising some good hybrid sweetbriars. He started to work on the strain about 1884, and the Penzance Briars made their debut in 1894–5. Unfortunately the delicious leaf scent is not transmitted unless the sweetbriar is the seedbearer,

not the pollen parent, and is apt to disappear in second and third generation crosses.

All in all, there was no lack of material for the nineteenth century rose garden; for it was only when this period was well advanced that a special department of the garden began to be set aside for rose growing alone. This was partly due to a theory current at the time (and much deprecated later by Robinson) that roses, and especially standard roses, did not associate well with other flowers, and were best segregated by themselves, and partly to a growing tendency to look on the shrub as a source of choice flowers for cutting or exhibition rather than as a decoration to the garden. The rose was not regarded as a florist's flower in the seventeenth or eighteenth centuries, when the tulip, carnation and auricula were at their zenith; it only arrived on the show bench in the mid nineteenth century. The first Grand National Rose Show was held in 1858, and was organized by the Rev. S. Reynolds Hole, subsequently Dean of Rochester, who even then had four hundred varieties in cultivation, and who became first President of the National Rose Society when it was founded by the Rev. H. Honeywood D'Ombrain in 1876. It now has a membership of over 72,000.

The enormous quantity of rose breeding done in the nineteenth century was at first rather to be called rose raising, as growers depended on chance and propinquity for their varieties. Artificial pollination was not practiced until the 1870s, and some of the more difficult species had to wait for their progeny until the twentieth century. *R. foetida*, the Austrian Briar, for example, had long been noted for its sterility. This was a very old garden rose, and highly valued because, with the exception of *R. hemispherica*, it was the only yellow rose known before 1800. It is an oriental flower, and was cultivated by the Moors in Spain in the thirteenth century; its misleading name arose because it was rediscovered in Austria, after a period of eclipse, by Clusius, who published its description in 1583. It reached England in time to be grown by Gerard, who was very scornful of the report that it had been created by "grafting a wilde Rose upon a Broome stalk." The variety *bicolor* with single red and yellow flowers is said to have been introduced about 1590, but the double form did not arrive till 1838, when it was brought from Persia by Sir Henry Willock, KLS, Envoy Extraordinary and Minister Plenipotentiary at Teheran. It was with this notoriously sterile and difficult rose that M. Perne-Ducher of Lyons elected to work, patiently cross-pollinating it with garden varieties from 1883 onwards; it was not until 1888 that he obtained a few seeds from "Antoine Ducher" with *R. foetida* pollen,

[1]Parkinson, 1629 [2]Langley [3]i.e., standards

and raised two hybrids, one of which proved sterile. The other—surely a rose "begotten by Despair, upon Impossibility"[1]—became the parent of the first Pernetti-ana Rose, "Soleil d'Or," exhibited in 1898 and distributed in 1900, which has had almost as great an influence in this century as the China roses did in the last, and has completely changed the character of our Hybrid Teas. By means of this rose the deep yellow, orange and flame shades were introduced—colors hitherto lacking, the only yellow rose previously used in breeding having been the pale and delicate Yellow Tea.

Meantime, new species were still being introduced, and used for new purposes and to suit new circumstances. Many came from China early in the present century, the most important so far being *R. moyesii*. Found by Mr. E. A. Pratt on the borders of Tibet in 1894, it was introduced by Wilson for Messrs. Veitch in 1903; the name commemorates the Rev. J. Moyes of the China Inland Mission, Wilson's host at the time the rose was collected. Seedlings raised from it rarely equal the bloomy, dusky, gipsy red of the original form, probably carefully selected by Veitch from the plants raised from Wilson's seeds; Bowles described it as the nearest approach to a carbuncle—presumably the lapidary sort—of any rose he knew. It is equally valuable for the autumn beauty of its hips; its hybrid, *R. × highdownensis*, raised in Col. F. C. Stern's garden at Highdown in Sussex, is even more spectacular in this respect, as it bears its flowers and fruit in clusters instead of singly. This was not the first rose to be grown for its hips; in 1629 Parkinson wrote of *R. villosa*, the Apple Rose, "The whole beauty of this plant consisteth more in the gracefull aspect of the red apples or fruit hanging upon the bushes, than in the flowers, or any other thing;" and up to the end of the eighteenth century at least, the large fruits were used to make "a sweetmeat greatly esteem'd." (This rose, now recognized as a native of Britain, received an Award of Merit for its decorative fruit in 1955.) The black hips of the Scotch Rose can be used for dyeing.

One rose, *R. rubrifolia* (Europe, 1814) is grown for the sake of its pretty, gray-mauve foliage; and there is even one cultivated entirely for its ornamental thorns—*R. sericea* var. *pteracantha*. Of this it might truly have been said, "She arayeth her thorn with fayr colour . . . ,"[2] for the scarlet, inch long, winged thorns that form a continuous double line up the young shoots, "glow out when lit by the setting sun like stained glass in old windows."[3] This

[1]*The Definition of Love*, Andrew Marvell, 1861 [2]*De Proprietatibus Rerum*, Bartholomeus Anglicus trans. Trevisa, 1398 [3]Bowles, *My Garden in Summer*

particular variety of *R. sericea* was another of Wilson's introductions for Veitch, before 1905; the variety *denudata* of the same rose is said to be as bald as a poached egg. It would seem to offer a good opportunity to test the theory of an eighteenth century nurseryman, John Cowell of Hoxton, who asserted that roses with few thorns loved water, while the prickly kinds would grow on drier soils. "And of the first sort, one may take of the Damask, and Monthly Roses, and with a large piece of Cork placed between the Root and the Branch, set them swimming in a Pond, and they will grow a long time."[1]

We have it on saintly authority that in Eden the rose was without thorns, which it acquired later owing to the wickedness of man.[2] But even today, no rose bears a true thorn, with a central woody core like that of the hawthorn or sloe, but only prickles, arising from the bark. One can see how useful these down-curving hooks must be to a plant scrambling up through other vegetation—quite apart from purposes of defense—and Bunyard suggests that when large hooked prickles are found on a dwarf rose, it signifies descent from a climbing ancestor. This thorniness makes the rose a useful hedge plant, and some of the more vigorous kinds have also been found to make effective snowbreaks. The *Daily Telegraph* for March 31, 1888 reported the success of rose hedges planted to protect a certain stretch of railway line in Hungary, previously snowed up every winter; they were found to "repel the fiercest onslaughts of their fleecy foe." (What a pity that "The Engineer" was not then available!) In October 1960 the same paper published an account of experiments in the use of rose hedges as anti-dazzle and crash barriers on American motorways, for which they were proving remarkably effective.

There seems to have been a very ancient belief in a connection between the thorniness of a rose and its scent, and Pliny was only following Aristotle and Theophrastus when he wrote, "If you would know a sweet-smelling Rose indeed, chuse that which hath the cup or knob underneath the flour, rough and prickile." The tradition lingered until the nineteenth century, for Victor Hugo said of the China rose "comme elle est sans épines, elle est sans odeur." Whether or not this is true, different species do vary widely both in character and degree of perfume, and keen-nosed experts find in scent a useful clue to the parentage of unidentified hybrids. In the old days, the scent was regarded as almost more important than the rose, and there is a large literature on the subject. The modern rose has been accused of having lost its fragrance; this may be true of varieties

[1]*The Curious and Profitable Gardener*, 1730 [2]The Persian Zoroaster (c. 630–553 B.C.) said much the same thing

that derive their brilliance of color from the faintly malodorous *R. foetida*, but the breeders are quite aware of the desirability of perfume, and although its inheritance is one of the most unpredictable factors in rose growing, some intensely fragrant varieties have recently been raised.

It should be noted, however, that there are many records of persons to whom the scent of the rose is not only disagreeable, but harmful—from the Ancient Greeks, who seem to have regarded it as fatal to beetles, to historic personages such as Marie de Medicis and the Duc de Guise. Robert Boyle quotes several examples in his *Essays of Effluviums* (1673), including that of an apothecary whose profession obliged him to handle great quantities of dried roses, "and the odor of Roses . . . makes such a colliquation of Humours in his Head, that it sets him coughing, and makes him run at the Nose, and gives him a sore Throat, and by an affluence of Humors makes his Eyes sore, in so much that during the season of Roses, when quantities of them are brought into his House, he is obliged for the most part to absent himself from home": obviously a case of an allergy similar to hay fever. As late as 1793 Dr. Woodville warns us that "where persons have been confined in a close room with a great heap of Roses, they have been in danger of immediate extinction of life." In a delightful essay on perfumes written about 1842, the author informs us that scents are out of fashion, except as a means of suicide; the blue-stocking will keep at hand a small vial of essence of roses, so that "whenever the cup of disenchantment shall be full, she will merely put the vial to her nose, and it will be all over with her."[1]

Boyle's apothecary friend would long be kept busy and miserable, because of the large number of preparations that were made from the rose. "What a pother have authors made with Roses!" said Culpeper; "What a racket they have kept! I should add, that Red Roses are under Jupiter, Damask under Venus, White under the Moon, and Provence under the King of France." He then proceeds to give more than three large pages of rose remedies. Besides the distilled water and the dried petals, there were Vinegar of Roses, Honey of Roses, Ointment of Roses, Syrup of Roses ("Pleasant and useful" as a laxative for children), Oil of Roses and Conserve of Roses (for colds and general purposes). It was usually the red Apothecary's Rose that was used for medical purposes; "amonge all floueres," wrote Bulleine, "none excelleth ye province Rose for his manifolde vertues; as against the trembling of the hart, dimnesse of sight, fransies, lack of sleape, corrupcion of the ayre,

heate above nature, flixes, etc . . ." but the Damask was used for the laxative preparations, and for the manufacture of rosewater and attar. The Damask Rose, moralized Parkinson, served "more for outward perfumes than for inward Physicke . . . and yet there is by many times much more of them spent and used then of Red Roses, so much hath pleasure outstripped necessary use."

Gerard mentioned that the hips of the wild rose[1] were used by cooks and gentlewomen for tarts, and there are many later references; in the eighteenth century a conserve made from them was "deemed good in consumptions and disorders of the breast"; but Dr. Woodville roundly stated that it had no medical virtue, and was useful only as a vehicle for "the more active articles of the Materia Medica." It was not till 1934 that dog-rose hips were found to contain more Vitamin C than any other fruit or vegetable, as well as small quantities of Vitamins A and P. The vitamin content was stronger in the north—hips from Scotland proved ten times richer than those from Cornwall—but on the average, rose hip syrup contains four times as much Vitamin C as blackcurrant juice and twenty times as much as orange juice. The quantity of the vitamin present also varies according to species, the Scotch Rose, for example, producing little or none; the richest of all is *R. cinnamomea*, the old Cinnamon Rose, whose range extends to the shores of the Arctic Ocean, where it may be buried in snow for eight months of the year, and whose fruit contains as much as 5 percent of Vitamin C. Dog-rose hips are collected in Britain for processing at the rate of four hundred to six hundred tons a year.

The universal name of "rose" is derived through the Greek and Latin from a Celtic word meaning "red"; so if we ever succeed in raising a blue rose it will have to be called by some other name, in order to avoid a contradiction in terms. So far, the pigment delphinidin, which alone could produce a true blue, has not appeared, though the geranium-scarlet pelargonidin turned up in 1930. According to Mr. Gordon Rowley, those who want a blue rose must try dipping a scarlet one in ammonia and detergent—the attempt to breed one from the lavender and mauve shades being like "going after rabbits with a dead ferret." The bewildered rosarian of today has reason to brood over Sir George Sitwell's apothegm—"A rose is neither red nor sweet, though we may think it so."

NOTES: The popularity of miniature roses in this country has not significantly diminished since Miss Coats' writing. In the United States, millions of rose plants are sold

---

[1]*The Flowers Personified.* Grandville, trans. Cleaveland, 1847

[1] Called "Nippernails" in some parts of the country

every year, almost without exception, modern hybrids. A great interest in the use of roses as garden shrubs has recently arisen as a result of new hybrids developed by the famous English breeder, David Austin. At the same time, a new group of landscape roses from France under the group name "Meidiland" are attracting considerable attention as easily cultivated, disease-resistant plants for hedge and border use. These departures from the stan-

Rubus odoratus
(FLOWERING RASPBERRY)   S. Edwards (1796)

dard hybrid roses will find a place in the garden, but the true rose enthusiast will most likely remain faithful to his lovely hybrid tea roses, floribundas, and polyanthas. Along with these, no rose garden would be complete without climbing roses. All told, rosarians have a brilliant palette of color to satisfy their every wish.

Rubus

The brambles, the little brothers of the roses, are usually banished from the flower garden on account of their rough ways and coarse manners; but among the very numerous species are some quite worth growing as ornamental plants. How many species there

are, nobody knows—probably three times as many as there are of roses, but the genus is even harder to disentangle. Augustine Henry used to say that in China he expected to find a new kind of Rubus with every ten miles of travel, and was rarely disappointed. We too are very rich in brambles, the five main species being the Blackberry, Raspberry, Dewberry, Cloudberry and Stone-Bramble—and not a true berry among them; botanically speaking the fruit of the bramble is an assemblage of drupes (it sounds like the definition of a particularly boring party), whereas a berry is a pulpy receptacle surrounding several seeds. Whatever we may call them—and "bumblekites" and "scaldberries" are among the local names—the fruits of the bramble have been welcome to many an English wayfarer, from the Babes in the Wood,[1] to Charles Wesley, who said of his tours in Cornwall: "We ought to be thankful that there are plenty of blackberries, for this is the best county I ever saw for getting a stomach, and the worst I ever saw for getting food."[2] Let us hope his travels were well timed, for no blackberry should be touched after Old Michaelmas Day—October 10—the day on which the Devil spits on them all.

Bean warns us that the advanced study of the British Rubi is "only suited to persons of abundant leisure." It is the blackberry that is the source of all the trouble in classification; it is a very variable plant, and botanists have now divided the old *R. fruiticosus* into no less than 389 "microspecies," all growing in Britain. It would be strange if, among so many, there were not a few worthy of garden cultivation; one or two varieties were grown for ornament in the eighteenth century, and are still available from nurserymen today. These are the double white, now classed as *R. falcatus* (syn. *R. thyrsoideus*) *fl.pl.*, and the double pink, *R. ulmifolius* var. *bellidiflorus*—useful for their ability to fend for themselves in rough places, and for the late season of their bloom. They were praised by Robinson, and Marshall in 1785 went so far as to call their flowers "beautiful beyond expression." The eighteenth-century gardener also cultivated the Cut-Leaved Bramble (*R. laciniatus*; handsome, but wickedly thorned) one or two variegated sorts, and one, var. *leucocarpus*, with white fruits, which "has often given occasion to a hearty laugh, by a bull which has been made by many on their first seeing this fruit, who have cried out with surprize, 'Here is is a Bramble that bears white blackberries.' "[3]

The various foreign brambles that have been brought to ornament our gardens are ornamental only; for even

[1]"Their pretty lips with Blackberries/ Were all besmeared and dyed . . ." [2]Quoted by Skinner [3]Quoted by Marshall from Hanbury

those that have the reputation of bearing good fruit in their native country, seem to have left either the fruit or the flavor behind when they left home.[1] The first to be cultivated here was the American *R. odoratus*, so named by Cornutus on account of the sweet balsamic odor of its leaves, which he compared to the scent of agrimony. Mr. Jason Hill says that the leaves smell of resin and cedarwood with a slight hint of pineapple; elsewhere the scent is likened to that of the mossy part of a moss rose. "The upright Pennsylvania bramble, or raspberry" figures in a catalog published by the Society of Gardeners in 1730, and at the beginning of the *Arboretum et Fruticetum Britannicum* Loudon attributes its introduction to Sir Hans Sloane in 1700, though later in the same work he follows Aiton in dating it from Miller, 1739. It was known as the "Flowering Raspberry," as it was for long the only species to be grown for ornament rather than for use; the crushed strawberry flowers with their pale cream-coloured stamens beings "so shewy, and so freely produced, that the plant has long been thought to merit a place in most shrubberies."[2]

Even in its native land the fruit of *R. odoratus* is said to be flat and flavorless; about that of *R. deliciosus* there seems to be a difference of opinion. This plant was discovered by Dr. Edwin James, when he accompanied Major Long's expedition to the Rocky Mountains in 1820. He described the fruit as being "of delicious sweetness and considerable size," and named the plant accordingly; but such fruit as it produces in Britain is completely lacking in flavor, and we grow it only for the sake of its arching thornless sprays thickly starred with flowers like white dog-roses. It was introduced to cultivation by Isaac Anderson Henry of Hay Lodge, Edinburgh, whose plants, raised from seed, flowered in 1870. A rather similar kind, *R. trilobus*, was introduced from the mountain of Citlaltepetl in central Mexico by Messrs. E. K. Balls and W. Balfour Gourlay in 1938; its handsome flowers are borne over a fairly long period, but too intermittently to make much of a show. In 1950 it was crossed with *R. deliciosus* by Mr. Collingwood Ingram, and a hybrid raised which has been named *R. × tridel* 'Benenden', and which is already coming to the fore as a garden shrub.

These two, *R. odoratus* and *R. deliciosus*, are the best of the flowering brambles, with *R. spectabilis* in the pink class and *R. parviflorus* in the white, as runners-up. (The two last were introduced by Douglas from western North America in 1827.) But there is another group of four or

[1] The miniature *R. arcticus* has both charming flowers and exquisitely flavored berries, but it rarely fruits well in gardens [2] *The Botanical Magazine*, 1796

Ruscus hypoglossum
(HORSE-TONGUE)   Ferdinand Bauer (1840)

Sambucus nigra
(ELDER)   J. Sowerby (1798)

five brambles which are grown not for their floral beauty but for the decorative value of their strikingly white stems in winter, which give the effect of being whitewashed. These "whitewashed brambles" are nearly all of Asiatic origin; perhaps the best—and the least invasive— is *R. biflorus*, found by Dr. Buchanan Hamilton near Chitlong in Nepal in 1802, and introduced in 1818. Besides its remarkable waxy-white young shoots, *R. biflorus* has a gentlemanly refinement of flower and leaf, and bears deep amber-colored fruit which is said to have a pleasant flavor and to make excellent jam.

Another bramble grown largely for its stems is *R. phoenicolasius*, the Japanese Wineberry—so clothed to its smallest twigs with reddish bristles that it seems to have grown a surrealist coat of red fur. Its large red berries are very ornamental, but seem to have left their wine in Japan; they are said to be "mawkish" in taste. Seed was sent to Paris by Maximowicz in 1864, and a plant came to Kew from the Jardin des Plantes in 1876. It is not quite so hardy as the other sorts described here.

This rufous species would seem to have the best right to the family name of Rubus, derived, like Rhus and Rosa, from a root meaning red; admittedly, many of the members are "red in tooth and claw." The young shoots of the blackberry were considered medicinal by the Romans, being so astringent that they served to fasten loose teeth[1] in the head. Anne Pratt repeats, rather doubtfully, a statement that badgers are said to be very fond of blackberries, and to thrive well upon them.

NOTES: *Rubus* 'Beneden' (syn. *R.* 'Tridel') is now a favorite among Southern gardeners for its large pure white flowers. Those a little further north might wish to try *R. odoratus*, a vigorous shrub with large, fragrant, rose-pink flowers with the thorns. For garden accent in winter, *R. biflorus* or *thibetanus* are prized for their striking white stems.

Ruscus

The Butcher's Broom, *R. aculeatus*, is one of the most curious of British natives. Although a shrub, it is a member of the Lily order; its stiff, scratchy "leaves" are actually flattened stems, known as "cladodes;" it bears these cladodes vertically, not horizontally, as if it dared not expose their surfaces to the torrid midday sun so frequent in English woodlands; the tiny true leaves and the small flowers are borne on the midribs of the cladodes, and the flowers are unisexual, so that unless both male and female plants are present no fruits follow, except in the rare hermaphrodite form; the large bright

berries do not seem to belong to the stiff, sombre branches, but look as though they had been artfully attached, in rather poor taste, to capture the Christmas market; and most of the seeds within them fail to mature, as "one of them commonly suffocates the rest."[1] Anne Pratt says that the Butcher's Broom is often found "skeletonized" in nature—more frequently than any other British plant; and this seems quite in keeping with its bizarre character.

Turner reported this botanical curiosity as growing wild in Kentish hedgerows—"but it beareth no fruite as it doeth in Italie"—and Gerard found it on "Hampsteede Heath, four miles from London." It was probably cultivated at first as a medicinal plant, for it had a reputation extending back to Dioscorides. Parkinson, who called it "one of the five opening diuretical rootes in the Apothecaries shoppes,"[2] said that it helped to expel gravel and stone, and Tournefort extolled it for dropsy. From very early times the young shoots were gathered in spring, to be cooked and eaten like its cousin, the asparagus; but they are bitter, and seem to have been taken more as a drug than as a delicacy. The plant got its name because its harsh and prickly branches were used by butchers to scour their blocks; Loudon says that in Brittany little scrubbing brushes were made of it to clean inside kitchen utensils, and one is reminded of old Adam Lambsbreath "clettering" the dishes at Cold Comfort Farm with a thorn twig. The branches were also used to protect foodstuffs from mice and rats; at Christmas, says Anne Pratt, "we often see them ornamenting the butchers' shops, or those in which bacon and cheese are sold, for they form an impenetrable and prickly hedge through which the mice cannot travel."

By 1659, at least, the plant was being grown for ornament, for Hanmer wrote that although a native, it was "preserved in some gardens as a fine greene and a rarity." According to Abercrombie in 1778, large numbers of young plants were dug up every year in the woods by "the herb people," and hawked in winter in the London streets, where "the citizens buy them to plant in pots of sand, to adorn rooms, where they will remain green a long time." Branches of Butcher's Broom in berry were also stuck in sand, with seed-heads of *Iris foetidissima* and *Paeonia corallina*, "which altogether made a show in rooms during the Winter."[3]

Two other species of Ruscus were early introductions from Europe, though seldom or never grown now; *R. hypoglossum* or Horse Tongue before 1597, and *R. hypophyllum*, about 1625. "What is the vertue of *Hippoglossum*,

[1] Natural ones

[1] Smith, in Rees' *Cyclopaedia* [2] The other four were parsley, fennel, "smallage" (i.e., celery) and asparagus [3] Martyn

whiche is called *Laurus*, or horse tong, as I have heard saie, by one that red me a pece of *Dioscorides*?" inquires Marcellus in *Bullein's Bulwarke of Defence* (1562), to which Hillarius the gardener replies "With this herbe, did that moste noble victorious Alexander, sometyme triumph withall, and his capitaines, he wearyng the diademe . . ." D'Alechamps (1582) also called this shrub the "true" Alexandrian Laurel, but later this name was transferred to another European and Near Eastern plant, known at first as *R. racemosus*, but set aside in 1787 as a monotypic genus and named *Danaë racemosa*. This is a much more elegant and pliable evergreen, and quite suitable for wreaths;[1] but it has been pointed out that most laurels shown on ancient classical monuments have fruits in the axils of the leaves, like those of *Laurus nobilis*, not *on* the leaves as in the Ruscus species or in terminal racemes like those of the *Danaë*. The latter was introduced to Britain in 1713, and was a valued inhabitant in eighteenth century wildernesses and ornamental plantations; in the 1820s nothing was more common in shrubberies and rustic gardens.

R. aculeatus has many English names besides that of Butcher's Broom, one of the oldest being Prickly Petigrue or Pettigree; it was also called Kneeholme or Kneehulver, from its height and its holly-like prickles. "Ruscus" is said to have started as "Bruscus," and to have been derived from the Celtic *beus-kelen*, meaning box-holly. "Danaë" was named after the lady imprisoned in a tower of brass, who bore Perseus after having been visited by Zeus in the form of a shower of gold; but the connection is obscure.

NOTES: The Butcher's Broom, while a very attractive ornamental, is very tender and rarely cultivated north of Louisiana.

Sambucus

When the elder blows, summer is established.
Gilbert White, *Journals*, 1772

I never saw the elder bushes so full of blossom, and some of the flowers, fore-shortened as they curve round, are extremely elegant; it is a favourite of mine, but 'tis melancholy; an emblem of death.
John Constable, letter to C. R. Leslie, 1833

S. *nigra*, the Common Elder, has been a native of Britain at least since the Atlantic period; and if its remains have been identified chiefly on occupied sites—Bronze Age, Iron Age and Roman—it is because it likes the "nitrophilous broken soil habitats found with human settlements."[1] The same preference explains its frequent occurrence on rabbit warrens, though our ancestors drew from this the conclusion that rabbits liked elders, rather than the other way around; and in the days when neither was regarded as a pest, recommended that elders should be planted about coney burrows, "for the shadowe of the conies."[2]

The elder has been called a weed among trees; but rank and coarse as it may be, no other non-foodbearing shrub has been so closely domesticated or so richly endowed with folklore and legend. That Judas hanged himself on an elder, we may read in *Piers Plowman*[3]—Sir John Mandeville, in his Travels, was shown the very tree—and the Jews' (or Judas') Ear Fungus that grows on it is the visible manifestation of the curse that has rested on the elder ever since.[4] The wood of an old specimen is hard, and burns and polishes well; but beware how you use it, for if a baby is laid in a cradle of elderwood it will pine, or the fairies will pinch it black and blue. Beating with a rod of elder will check a boy's growth, and if the wood is used in the building of a house, mysterious hands will be felt pulling at the legs. (Even to read of it conveys a faint leg pulling sensation.) To burn elder logs will bring the devil into the house. Women in Welsh farmhouses used to make green patterns on the newly scrubbed stone floors, with handfuls of bruised elder leaves, to keep away witches; a branch buried with a corpse served the same purpose, and "even now" wrote Dr. Fernie in 1895, "the driver of the hearse commonly has his whip handle made of elder wood." In the Tyrol, says Skinner, a cross of elder wood was planted on graves; if it grew and blossomed it signified the beatitude of the deceased; if not, "the relatives might draw their own conclusions." It was thought that the smell of the plant was narcotic, and that it was dangerous to sleep in the shade of an elder, or to plant it near the house. Evelyn tells of a house in Spain surrounded by these shrubs, that "diseased and killed almost all the inhabitants" until the elders were cut down. And these are only a few of the tales that are told.

The question arises whether this sinister shrub can properly be considered a suitable garden plant (though we are told that he who cultivates the elder will die in his own bed), but it has its beauties; Robinson said that a large elder with its branches sweeping the turf was "no mean object" on a lawn when in flower or fruit, and

[1] We pity any brow so unfortunate as to be wreathed by mistake in Butcher's Broom!

[1] Godwin [2] Gerard [3] Written between 1360 and 1399
[4] This fungus was gruesomely prescribed for the treatment of sore throats, quinsies and strangulations

there are a number of garden varieties, some of them cultivated for centuries. Gerard grew the cut-leaved and white-berried kinds (vars. *laciniata* and *alba*) and by the end of the eighteenth century the green-berried, gold-striped, silver-striped and silver-dusted had been added (vars. *viridis, auereo-variegata, albo-variegata* and *pulverulenta*). Marshall said that the elder should not be planted near walks or buildings, but would have an air of majesty at a distance, where it might "display its gaudy pride when in blow . . . while the sense of smelling is in no way discommoded by its strong disagreeable scent." Fourteen varieties are listed in the R.H.S. *Dictionary* including the Golden Elder—now perhaps over-planted, but a useful shrub to brighten drab parks in industrial areas—and one with double flowers, which must indeed resemble the "grand tree of cauliflowers" admired by a nineteenth-century Birmingham artisan on a country outing.

Some of the other species in the family are occasionally cultivated; the best of them is *S. racemosa*, the Mountain Elder, which would be a very handsome tree if it produced its scarlet berries as freely here, as it does in Switzerland or Spain. (It fruits well, however, in parts of Scotland.) Turner saw it growing in the Alps, and Gerard had it in his garden in 1596. It too sports into cut-leaved and gilt-edged varieties; and one that is both fine-leaved and golden, var. *serratifolia aurea*, is highly praised as a foliage shrub. Another foreign species, *S. canadensis*, was introduced from North America in 1761; Loudon said it was not uncommon in collections, but was "only fit for dug shrubberies in sheltered situations." Its variety *maxima* is said to be a handsome object where room can be found for a shrub with leaves 18 inches long, flower heads as much across, and the rest in proportion.

Parkinson in 1640 said that the common elder did not grow wild, but was "planted in all places, to serve for hedges, and partitions of grounds, vineyards etc. . . ." and Batty Langley recommended it to gardeners for wind breaks, which would "not only be a great Preservative by breaking off the cold Northern Winds from their tender Plants, but by its produce of Flowers and Berries, would turn to a very great Account also."[1] Peter Kalm saw "thick and beautiful" elder hedges in the market gardens about London in 1748, and it was still extensively planted for orchard and garden hedges in Kent a hundred years later, and the fruit brought to London "in immense quantities" for the making of elderberry wine. This seems to have been a popular brew both in

[1]Elder hedges are rapid in growth, but, says Martyn, "as their bottoms become naked in a few years, they are not so proper for that purpose"

town and country, especially as a hot drink in the winter, when Cobbett called it "a thing to be run for." Well made with raisins, sugar and spices, and three years old, it was known as "English Port;" and Andrew Young tells a story of a doctor who was puzzled because a patient insisted that port did his gout more good than medicine—until he found that the man was drinking a brand made from elderberries. (It was Douglas Jerrold of *Punch* who asked in a tavern for wine that was "old, but not elder.") But its flavor is not universally pleasing, and by the start of the present century its popularity had declined; far better the various products made from the flowers, which can be used to add bouquet to inferior wines, to impart a delicious muscat flavor to gooseberry jelly, or to make a sparkling and agreeable "champagne," which is said to resemble Frontignac. The distilled water of elder flowers was used to flavor confectionery, as a cosmetic to cleanse the skin of sunburn and freckles, and as the source of a drug still used by herbalists for colds and bronchial troubles; a tea made from the dried blossoms is described as fragrant and sudorific, but debilitating.

It is best not to embark on the medical uses of the elder, for there is no end to them. "If the medicinal properties of the leaves, bark, berries, etc. were thoroughly known," wrote Evelyn, "I cannot tell what our countryman could ail, for which he might not fetch a remedy from every hedge, either for sickness or wound." Every part was used; "Here's elder buds to purge your bloods" was a familiar street cry in eighteenth century London, and shoots, bark, leaves, flowers, berries and seeds were all pressed into service; it is a relief to find Culpeper remarking "I know no wonders the root will do." A Latin treatise on the plant, *Anatomia Sambuci*, by Dr. Martin Blockwitz, was translated into English in 1655; it prescribed the elder "for the Head-ach, for Ravings and Wakings, Hypocondriack and Mellancholly, the Falling-Sicknesse, Catarrhes, Deafenesse, Faintnesse and Feavours"[1]—and much else. No wonder the great Dutch physician Dr. Boerhaave always saluted the elder by raising his hat; it "deserves the Regard of Men for its many Galenical salubrious Uses,"[2] and requires to be treated with respect. It must not be gathered without first asking permission—"Elder, elder, may I pluck thy branches?" and if no rebuke follows, one should spit three times, and may then proceed.

"Sambucus" is derived from the Greek *sambuke*, a musical instrument said to have been made from the wood. "The Shepherds are thoroughly convinced" wrote Pliny "that the Elder tree growing in a by-place out of

[1]Coles, *Adam in Eden*, 1657   [2]Ellis, *Treatise on Foresty*

the way, and where the crowing of the cocks from any town cannot be heard, makes more shrill pipes and louder trumpets than any other." Children have made elder pipes and popguns from time immemorial, and the light pith that fills the hollow shoots is used today in electrical instruments.

NOTES: In the United States, our common *S. canadensis* occurs over the easternpart of the country and serves the same role as *S. racemosa* in Europe. Large shrubs with white flowers in flat clusters followed by bead-like black berries, the elderberry has been used for preserves and homemade wine since Colonial times. But it is not a garden plant and is best enjoyed in its native habitat or when used as a background plant for bird-feeding situations.

Sarothamnus, see under *Cytisus*

Sasa, see under *Arundinaria*

Schizophragma, see under *Hydrangea*

Sarcococca

The evergreen Sarcococcas had the misfortune of being introduced two centuries too late; if only they had been known at the end of the seventeenth century when "greens" of all kinds were so much in fashion, how popular they would have been! But they are natives of the Himalayas and of China, and most of them were only introduced in the present century. They belong to the Box family, and among the hosts of showier plants competing for our garden space, their quiet charms tend to be overlooked. Although some of the kinds will grow to a height of 5 or 6 feet, they are typically the "underlings" that Mr. G. S. Thomas calls them, excellent for covering the ground in shady places under trees. Their winter-borne flowers consist of styles and stamens only, without the protection of a single sepal or petal; yet with this inadequate equipment they manage unfailingly to produce a sweet and penetrating fragrance, which has earned them the name of Sweet Box. The berries of the previous year are often borne along with the inconspicuous flowers.

The two Himalayan kinds, *S. saligna* and *S. hookeriana*, were introduced in the nineteenth century. (Both are slightly tender, *saligna* is without fragrance, and *hookeri-*

Sarcococca pruniformis
(SWEET BOX)   M. Hart (1828)

Skimmia japonica
(JAPANESE SKIMMIA)   M. Smith (1905)

ana is said to be lost to cultivation.) All the rest were brought by Wilson from China—*confusa* (1900?), *hookeriana* var. *digyna* (1908), *humilis* (1907; sometimes classed as another variety of *hookeriana*), and *ruscifolia* (1901) with its narrow-leaved variety *chinensis*. Two of them were found on the sacred mountain of Wa Shan in Szechwan. *Humilis* is the shortest, *confusa* the tallest; *ruscifolia* is the only one to bear red berries (the others have purple or black fruits), but the pinkish flowers of *hookeriana digyna* have the sweetest scent; and if there is a hint of privet in their fragrance, even privet is welcome in November.

The name is derived from *sarkos*, fleshy, and *kokkos*, a berry. Some berries of *S. ruscifolia* grown in Midlothian were found to be polyembryonic, with up to seven embryos in one seed. This occurred before 1923, so cannot be attributed to the effect of atomic explosions.

NOTES: Where it can be grown reliably, probably Philadelphia southward and on the Pacific coast, *S. hookeriana* var. *humilis* is an exceptional evergreen shrub that should be used for diversity in the garden. In a choice shady spot, it can develop into a handsome planting, serving much the same purpose as pachysandra.

Skimmia

This is another family whose history was confused at first by a nomenclature muddle, owing to one mistaken identification by a learned botanist, aggravated by the unsuspected fact that one species was dioecious and had male and female forms. The first of the family was found by Kaempfer in Japan, and described by him in 1712 under its Japanese name of *Mijama Skimmi*—though he was mistaken in thinking it a "vast tree." It was officially named *Skimmia japonica* by Thunberg in 1784, and was introduced to Kew in 1838, but attracted little attention. It seems to have been the male form only that was brought, which bears no berries; and the flowers have few charms apart from their earliness and sweet scent. It was for the sake of this perfume that the plant was cultivated in China and Japan, and Siebold, who thought that it would not be hardy out of doors, suggested that it might be grown in a cool greenhouse among camellias, where it would "relieve with its fragrance the beauty of that scentless shrub."[1]

In 1848, Robert Fortune discovered in a Chinese garden a shrub that he was not immediately able to identify; the following year he sent plants of it to Messrs. Standish and Noble (labeled "Ilex sp., Honan") only one of which

[1]*Flora Japonica*

survived the journey. It flowered and fruited in 1852, and the nurseryman sent specimens to the celebrated botanist, Dr. John Lindley of the Horticultural Society, who unfortunately identified it as *S. japonica*, although it differed from that species in its habitat, in its smaller size, in its darker and oval berries, and the fact that, bearing flowers of both sexes, it fruited freely. So the new plant was put into commerce as *S. japonica*, while some of the best scented forms of the original species from Kew were distributed as *S. fragrans* or *S. fragrantissima*.

Thirteen years later Fortune sent home another Skimmia, this time from Japan, which was given the name of *S. oblata*. Complaints were made, however, that it would not fruit; until in 1864 it was realized that this was the female form of the plant of which "*S. fragrans*" was the male. So the happy pair were at last reunited, berries produced and seedlings raised. A fine form was grown by Mr. F. Foreman of Eskdale Nurseries, Dalkeith; it received an F.C.C. in 1888 as *S. × foremanii*, and was announced at first as a hybrid between *S. fragrans* and *S. oblata*, which aroused the comment from a contributor to the *Gardeners' Chronicle*. "Did you ever hear of a hybrid between a bull and a cow? I thought every tyro in gardening knew that these were two sexes of one plant." Actually this variety is a cross between the Japanese and Chinese species, and produces both round and oval berries, sometimes in the same cluster.

The confusion in nomenclature was eventually cleared up by Dr. Masters in an article in the *Gardeners' Chronicle* in 1889; he restored its proper name to *S. japonica*, with *S. fragrans* and *S. oblata* as synonyms, and the Chinese plant being left apparently nameless, he christened it *S. fortunei* after its discovery. However, it later transpired that at some stage Fortune himself had named it *S. reevesiana*, after John Reeves junior, who had been of great service to him while in China. So *S. reevesiana* it is today, and there the matters rests. This species makes a good pot plant and is popular in London for window boxes; its variety *rubella* (before 1900) is good for winter foliage color, but is sterile, and bears no fruit.

Sikimi in Japanese means a harmful fruit, and the plant is classed as poisonous both in Japan and China. In common with other members of its Order, the Rutaceae, skimmia seeds have a tendency to precocious germination, sometimes sprouting while still on the plant, as orange pips are occasionally found germinating inside the orange.

NOTES: This interesting broad-leaved evergreen shrub from Japan is seldom seen in gardens. It deserves more attention as a plant for shade and acid soil as a companion plant to azaleas and camellias. The bright red fruits are attractive. Since the sexes are on separate plants, both male and female plants are required for fruit.

Spartium

This is a monotypic genus, consisting of a single shrub (*S. junceum*) that would not fit comfortably either into the family of Cytisus or that of Genista. Its tiny leaves fall so early that many, including Dioscorides, have been deceived into thinking that it has no leaves at all. Turner said they were "so little as scarsely they deserve to be called leaves . . . or elles Dioscorides looked upon the braunches whych at that tyme had no leaves," but its evergreen stalks perform the leaf functions, and have the rush-like appearance to which the specific name refers.[1]

It was introduced to England before 1548, for Turner records that "it is found now in many gardines in Englande, in my Lordes gardine at Shene, and in my Lord Cobbam's garden a litle from Graves End." Both in the *Names of Herbes* and in the *New Herball* (1568) he calls it "French Broom," in spite of the fact that it "cam to us out of Spayn," but by the end of the century the name of "Spanish Broom" was firmly established. It may have been introduced as a medicinal plant; although dangerous if taken internally, "being applied outwardly, it is found to helpe the *Sciatica*, or paine of the hippes."[2] Before long it had won acceptance as a plant of ornament; London and Wise found it "so easie to be cultivated, that it must be a very ignorant Gard'ner indeed, that knows not how to order it . . . We make fine standards of them, which if kept well prun'd are a great Ornament to our Boccages[3] or little Woodworks." It is a tall lanky shrub, unless kept low by early pruning, and Phillips advised that it should be placed "so as to peep over the sombre evergreens, like the rays of the sun emerging from dark clouds."

"I first introduced the Spanish broom with double flowers," wrote Peter Collinson; "it was sent to me from Nuremburg, anno 1746, in a pot nicely wickered all over; it cost there a golden ducat; came thence down the Elbe to Hambro," and was brought by first ship to London, in good order. I soon inarched it on the single-flowered Broom, and gave it to Grey and Gordon, two famous nurserymen, and the public soon had it from them." This double form still exists, but is regarded as hardly worth Collinson's money and trouble, except as a

curiosity; perhaps it has deteriorated since the eighteenth century. A well-grown plant of the single sort will produce much finer blooms; E. A. Bowles said that "if one nips the telltale sharp-nosed keel out of a spike of extra large ones for a buttonhole, it is possible to excite a Sweet-Pea enthusiast into the belief that you are wearing a yellow variety of his favourite flower."

The Greeks made their first marine cordage out of rushes, and had the same word for a rush and a rope; and about 238 B.C. when they became acquainted with the Cartagena district of Spain, where this shrub grew, they used the broom for the same purposes and applied to it the same name of Spartum. Pliny had much to say on the subject; he described the barren places in which it grew ("Full well and properly it mought be called, the rush of a dry and leane ground . . ."), the difficulty with which the tough stems were gathered, and the processes by which the fiber was extracted, and said that while hemp made better ropes for use on land, nothing excelled "spart" for marine cordage, for it would "live and receive nourishment within the water, drinking now the full as it were to make amends for that thirst which it had in the native place where it first grew." (He insists, however, that "at what time as barges and vessels were sowed together with seames, it is well known, that the stitches were made with linnen thred and not with spart.") Up to the end of the nineteenth century at least, the fibers from the young shoots were still made into thread and cloth, on parts of the Continent where the soil was too poor to grow hemp or flax.

NOTES: The Spanish Broom is a Mediterranean plant, difficult to transplant and not frequently cultivated except in hot, dry climates.

Spiraea

Here is another Rosaceous family that is a bane to botanists, the species being many, similar, and promiscuous in their habits. The genus has been much diminished by separating from it eight groups of plants formerly classed as Spiraeas, and setting them up with households of their own—including the herbaceous Astilbes and Filipendulas, and the shrubby Holodiscus and Sorbaria. Even so, some eighty species still remain in the family, and many garden and natural hybrids. A contributor to Robinson's *English Flower Garden* complained that though no spiraea was worthless, there were too many of them, they were too similar, and too many flowered at the same time; there were few gardens, he thought, where more than a dozen kinds were needed.

[1]*Juncus* = a rush   [2]Parkinson, 1629   [3]French, *bocage* = grove or coppice

Spartium junceum

(SPANISH BROOM)   Ferdinand Bauer (1830)

Spiraea chamaedryfolia

(BRIDAL WREATH)   N. von Jacquin (1772)

For horticultural purposes, the family can be divided into two groups; those that produce umbels or corymbs of white flowers, like fairy parasols or bridal bouquets, on growths of the previous year, and those that flower on shoots of the current year, with terminal spikes or flat heads of fluffy white, pink or red flowers.

The first recorded species belonged to the second group—*S. salicifolia*, identified by Clusius as the Spiraea of Theophrastus, which he put among the trees that bear spikes. The plant was sent from Silesia to Clusius at Vienna, by one Sibesius, apothecary to the Duke of Briga; he thought it was a kind of lilac. It does not seem to have been grown here before 1640, for Parkinson took his description of "*Clusius* his Spiked Willow of *Theophrastus*" at second-hand; but by 1665 Ray could refer to it as "the common *Spiraea frutex*," and it has since become so thoroughly naturalized in parts of Britain (such as "Winandermere") that later botanists were deceived into thinking it indigenous. Its distribution ranges from east Europe and Asiatic Russia to Japan; Phillips says it "springs in the deserts to cheer the banished Muscovites, whom the tyranny of despotic rulers sends to waste their bloom of manhood in the dreary regions of Siberia." He considered it "singular and pretty," and compared its florets to Titania's pincushion. "They are of a pale red colour," said Marshall, and small, "but being produced in these thick spikes, four or five inches long, they have a good look. The vigorous shoots . . . that arise from the roots are very pliable and taper, and make good riding-switches."[1] The shrub acquired the English names of Bridewort and Queen's Needle-work, and at one time there were as many as twenty varieties; but it has been largely superseded by American species of similar growth, which have deeper coloured flowers—and are all the worse for that; for the pinks and reds in the spiraea family range infamously from blotting-paper to aniline.

The first of these American kinds to be introduced was *S. tomentosa*, the Steeple Bush, in 1736; but it is little grown now, being very similar to the better known *S. douglasii*, raised in Glasgow Botanic Garden from seeds sent home by Douglas in 1827. (*S. tomentosa* came from eastern North America, *S. douglasii* from the west.) Another very similar kind, *S. menziesii*, was introduced in 1838; Bean suggested that it might prove to be a hybrid between *salicifolia* and *douglasii*, or if itself a true species, its varieties *eximia* and *triumphans* might be of hybrid origin. The latter is the most frequent garden representative of this type of spiraea, and is rivaled in popularity

[1] One of the Chinese names for *S. prunifolia* means "Driving-Horse-Whip"

only by some of the varieties of *S. japonica*. The latter is a species with flattish terminal racemes of pink flowers, introduced from Japan about 1870. One of its numerous forms is the dwarf 'Bumalda', and it was from this plant that the well-known 'Anthony Waterer' was raised at the Knap Hill nurseries, before 1890. It resembles 'Bumalda' except in its stronger color, and the fact that the typical harsh crimson of the family is somewhat mitigated by foliage charmingly variegated with pink and cream. The red color is most tolerable in the very distinct and dwarf *S. bullata*—brought by Maximowicz from Japan in 1864, and grown here in 1879—where it is set against tiny leaves of so dark a green as to be almost black.

The white-flowered spireas began with *S. hypericifolia*—so called because its leaves are perforated like those of the St. John's Wort—which seems to have been grown here before 1640 and to have been much appreciated; Marshall called it "beautiful and elegant beyond description," and said that it had "the appearance of one continued flower, branching into as many different divisions as there are twigs." Parkinson thought that it came from America, and Miller from Canada; actually its range extends from south-east Europe to Siberia and northern Asia, but does not include the American continent, and its chief importance today is as a parent of one of the best hybrids. *S. crenata* and *S. chamaedryfolia* have a somewhat similar distribution, except that the first does not extend to Asia or the second to Europe. *S. crenata* was introduced in 1739, and *S. chamaedryfolia* some fifty years later. The flowers of the latter were described by Martyn as "biggish, white, having a weak virose scent, and fugaceous," and he praised the shrub for hedges, "being entirely covered with flowers in June." In Kamchatka the leaves of this spiraea were used as a substitute for tea.

Moving further east, and further down the centuries, we come to the Chinese *S. prunifolia fl. pl.*, which was introduced by Robert Fortune in 1844.[1] It was a favorite garden plant in China, with a name that meant "Smile-laugh-flowers," and was frequently planted on graves. The doubling of the flower is described as particularly regular and perfect, and the shrub has the additional advantage of good autumn color. East again, and another twenty years brought us *S. thunbergii* from Japan (about 1863)—the earliest to bloom of all the spiraeas, wreathing its bare twigs with small white flowers in April. It too has brilliant autumn color, and at one time was popular for forcing. Although cultivated in Japan, this species is a native of China, and so are three others, which were

[1]According to Bretschneider Bean says by Siebold, about 1845, and distributed by Van Houtte

Staphylea colchica
(BLADDER NUT)   M. Smith (1894)

Symphoricarpos racemosus
(now *S. albus*)
(SNOWBERRY)   W. Clark (1826)

introduced by E. H. Wilson for the firm of Veitch in 1900—*SS. henryi, veitchii* and *wilsonii*.

Many garden hybrids have been produced, among which *S. × arguta* is outstanding; its parentage is given as *S. thunbergii × crenata × hypericifolia*, the former predominating. It has been called 'Bridal Wreath' and 'Foam of May' and is the best, though not quite the first, of the early spireas; it was raised before 1897. *S. × van houttei*, raised about 1862 by a M. Billiard at Fontenay aux Roses, near Paris, had for one of its parents *S. cantoniensis*, which is not hardy out-of-doors in Britain; the hybrid is susceptible to damage by spring frosts, but is much used indoors for early forcing.

The name comes from the Greek *speira*, meaning a twist, spiral or coil. It may have been first applied to the meadowsweet—once *Spiraea*, but now *Filipendula ulmaria*—because of its use in garlands; but it is very well suited to the twigs of some of the white-flowered species, so densely wreathed with blossoms.

NOTES: Long used in gardens for their arching sprays which are often used in flower arrangements, some spiraeas have been overpopularized to the point where they have fallen into disfavor. But recently, the improved selections of *S. japonica* and *S. nipponica* as dwarf plants for rock gardens or other places where dense, mound-like plants are needed have brought them to the attention of nurserymen and gardeners. An extremely dwarf but charming species that I collected in the alpine region of Mt. Asama, Japan, is *S. betulaefolia*. Sad to say, it has never reached cultivation.

There are no cultural difficulties with the spiraeas. They propagate easily and can be grown in any good garden soil.

Staphylea

This is a small but interesting family, widely spread over the northern hemisphere; each species—there are only about ten of them—confined to its own area, with little or no overlapping. Botanists surmise that these more or less isolated groups are survivors of a very ancient family, now long past the zenith of its evolution, though each kind still manages to hold its own within the boundaries of its territory.

S. pinnata the Bladder Nut or St. Anthony's Nut, is the European representative of the genus. It is not a native of Britain, but was introduced before Gerard's day; he mentions several gardens in which it grew, including that of "the right honourable the Lord Treasurer, my very good Lord and Master, by his house in the Strand." In 1677 Ray reported that he had found it growing, apparently wild, in hedges near Pontefract in Yorkshire, "but not so copiously that I would venture to assert that it is spontaneous." It was cultivated for the sake of its panicles of white flowers, "in forme resembling a small Daffodill,"[1] and for its seeds in their large inflated bladders, which had "a very striking and singular look in the autumn."[2] It was probably owing to this seed-vessel that the kernels were given the name of *Nux Vesicaria*, Bladder Nuts, and prescribed for diseases of that organ; but they seem to have been early discredited. "Some Quacksalvers," said Parkinson, "have used these nuts as a medicine of rare vertue for the stone, but what good they have done, I never yet could learne." On the Continent the seeds were used—by "Friers," according to Gerard—as beads for rosaries.

S. colchica, from the Caucasus, is handsomer in flower but less striking in fruit than the European kind. No information about its introduction is to hand; it was not listed by Johnson in 1863, but in 1894 the *Botanical Magazine* said it had "lately" been much grown for forcing, "pot-plants of it being very ornamental," and it is still so used today. Robinson, in 1900, called it the only important member of the family, but at that date the Chinese *S. holocarpa* had not been introduced, though it had been discovered. Seed from Hupeh was sent by Wilson to the Arnold Arboretum in 1908, and a plant raised at Kew flowered for the first time in 1915. In the wild, Wilson thought this staphylea "most strikingly beautiful," but it is generally considered inferior to the other two species except in its pink variety, *rosea*, a lovely shrub which bloomed for the first time in two Cornish gardens (those of Canon Boscawen and Mr. J. C. Williams) in 1925. Another kind that is sometimes to be found in catalogues is *S. × coulombierii*. This is believed to be a chance hybrid between *pinnata* and *colchica*, and was noticed in the nursery of M. Coulombier at Vitry in 1887, among plants obtained from Alphonse Lavallée's arboretum at Segrez in 1872. Another hybrid of the same parentage has recently come to the fore, as *S. × elegans*.

"Staphylea" comes from the Greek *staphyle*, a cluster, from the arrangement of the flowers. According to Duhamel, one of its French names was *Nez-Coupé*; one might call it the shrub of a hundred spited faces. In the Language of Flowers, it stands for Prodigality.

NOTES: *S. colchica* is the only member of this genus of generally weedy shrubs that is cultivated as an ornamental to any extent. Growing easily in average soil, it produces a profusion of white flowers in the spring on plants that can grow to 12 feet tall.

[1]Parkinson, 1629 [2]Marshall

Symphoricarpos

The first of this family to be cultivated here was not the familiar snowberry, but the much less common *S. orbicularis*, the Coral Berry, Indian Currant or St. Peter's Wort. This native of North America was introduced in 1730, but has never been considered of much account as a garden shrub. Marshall praises its foliage, but admits that the flowers "are small, of an herbaceous color, and make no figure." He does not mention the purplish-red fruit, which it bears freely here only in very hot summers, but Miller describes it as being hollow, and shaped "like a pottage-pot." There is a variegated form that Robinson thought pretty. No one explains how the name of St. Peter's Wort, which properly belongs to a member of the Hypericum family, became transferred to this plant.

S. rivularis (syns. *S. racemosus*, *S. albus* var. *laevigatus*) the Common Snowberry, was among the seeds and plants which were brought home by Lewis and Clark from their expedition to the West in 1804–6, and given by President Jefferson to Bernard McMahon for propagation. Six years later McMahon sent some plants of it to the President, with the note—"This is a beautiful shrub brought by Captain Lewis from the River Columbia; the flower is small but neat, the berries hang in large clusters and are of a snow-white colour, and continue on the shrubs, retaining their beauty all the winter, especially if kept in a greenhouse. The shrub is perfectly hardy; I have given it the trivial English name of Snowberry-bush." Seed was sent to England in 1817, by a Mr. Robert Carr, and was raised by the nurseryman Conrad Loddiges, who unfortunately identified the plant as the *S. racemosus* described by Michaux in 1802—thus giving rise to a confusion of nomenclature that was not finally cleared up until 1927. (The true *S. racemosus* is a native of the eastern part of the American continent, not of the west, and is probably not in cultivation.) The rapid distribution of the snowberry in England was partly due to William Cobbett, who described it as "a very pretty dwarf shrub that comes into leaf more early in the Spring than any other that I know of, and has a leaf of singular beauty. I raised, the year before last (1827) great quantities from seed got from America . . . It blows in August a minute but pretty pendant rose-coloured flower, which is succeeded by a white berry about the size of a cherry, and which hangs on till the winter." "This elegant plant," says Maund, in the same year, "will properly assist in adorning the foreground of the shrubbery, or it may be still more advantageously shown . . . scattered over the lawn, or breaking the tameness of an open grassplot. Here its beautiful slender branches will spontaneously arise from the roots, and gracefully bending under the weight of their snowy-white berries, form an object of no little attraction."[1]

Unfortunately a closer acquaintance has revealed that the branches arise only too spontaneously, in the form of suckers, and the plant is now so common as to have become a nuisance, and is perhaps rather undervalued on that account. Loudon suggested that it might be grafted standard-high on some of the shrubby Lonicera species (for example, *L. xylosteum*), to make a "very elegant small tree"; he seemed to delight in converting good shrubs into unsuitable and short-lived trees, but in this case it might be advantageous to divorce the species from its own troublesome roots. A well-grown snowberry, attentively pruned, can be a handsome object; the late Mrs. Constance Spry appreciated the decorative value of its sprays of fruit in flower arrangements, and a specially fine form, named after her, is now in commerce. Nurserymen have recently been paying some attention to the family, and a race of hybrids with pink, white or red berries was exhibited at the R.H.S. in October 1955, by Mr. A. Doorenbos, Director of Parks at The Hague, Holland. They were the result of crosses between *S. rivularis*, *S. orbicularis* and *S.* × *chenaultii*, and have been collectively named *S.* × *doorenbosii*. *S.* × *chenaultii* was an earlier hybrid, between *S. orbiculatus* and the Mexican *S. microphyllus*; it was raised before 1912 by the firm of Léon Chenault et Cie of Orléans. Its white berries are so densely speckled with rose on the sunny side as to appear pink.

The hardiness of the common snowberry cannot be questioned, for its ranges as far north as Alaska. Birds are fond of the berries, and the small flowers are teacups of refreshment to the eager bees—so much so, that it has been suggested that it might be planted in waste places as a valuable source of late nectar. It is said that in its native country bees prefer its flowers even to those of the white clover, and an excellent honey is obtained from it.

The name is derived from *symphoreo*, to bear together, and *karpos*, a fruit, from the clustered berries; there are seventeen or eighteen species in the family, one Chinese and the rest American. The snowberries are closely allied to the honeysuckles, and at one time were classed under "Lonicera."

NOTES: *S. racemosus* is still generally incorrectly listed as a synonym for *S. albus*, but the true *S. racemosus* would probably not inspire much ornamental interest.

These hardy shrubs, which will thrive from Michigan

[1] *The Botanic Garden*, Vol II, 1827

to Texas, are native to northern America and valued for their interesting fruits. Most common is the Snowberry, *S. albus* which has white fruits; *S. paniculata* is grown for its blue berries.

Syringa

(LILAC)   W. Curtis (1800)

Syringa

> *There were few houses that did not contain in their gardens . . . lilac bushes . . . And so it came about that, all through the month of May, each small house found itself dowered with an unexpected magnificence, a whole, silent household staff of young lilacs gathered about the door and filling the interior with sweet air and fragrant smells, a staff which could have been supplied in an Eastern fairy-tale only by a fairy gifted with poetical powers.*
>
> *Jean Santeuil*, Marcel Proust[1]

The lilac belongs to the Oleaceae, and has a poor relation in the privet, and a rich one in the oil-bearing olive. About thirty species are known, and the family is divided into two groups: the Tree-lilacs (sometimes called *Ligustrina* from their resemblance to

[1]Posthumous: translated by Gerard Hopkins, 1955

the privets) whose flowers have protruding stamens, and the true lilacs, whose stamens are enclosed. These are again divided into two sections, the *vulgares* and the *villosae*, the former flowering on the previous year's wood, and the latter, rather later, on shoots of the current year; it is said that species belonging to these two sections will not cross. Some of the Tree-lilacs grow to a large size so they are not included here.

The common lilac was introduced to European gardens by way of Turkey in the sixteenth century. The first description came from Pierre Bélon, a French naturalist who travelled in Europe and the Near East under the patronage of the Cardinal de Tournon, and published his "*Observations*" in 1553. Among the flowers cultivated by the Turks he noticed "*un petit arbrisseau qui porte les feuilles de Lierre, qui est verd en tout temps, et fait sa fleur presque d'une coude de long, de couleur violette, entournant le rameau, gros comme une queue de Regnard.*" A specimen of this Fox's Tail, as the Turks called it, was among the plants brought home by Ogier Ghiselin de Busbecq, when he returned to Vienna in 1562 from his embassy to the court of Suleiman the Magnificent; it was much admired, and was illustrated in the fifth edition (1565) of the *Commentarii* of Matthiolus, where the name of "lilac" appears for the first time. Later, Clusius raised some plants from seed sent by Dr. von Ungnad, Busbecq's successor at Constantinople. It is uncertain when the plant reached Britain, but Gerard in 1597 had lilacs in his garden "in very great plenty," and described the leaves as "crumpled or turned up like the brimmes of a hat;" he correctly related it to the privet. A passage is often quoted from an inventory of the house and garden of Nonesuch, describing "a fountain of white marble with a lead cesterne, which fountaine is set round with six trees called lelack trees, which trees beare no fruite, but only a very pleasant flower."[1] This has been taken as a proof that the lilac was cultivated in England even earlier than on the Continent; but although the building of Nonesuch was begun by Henry VIII in 1538, the inventory in question dates only from 1650, by which time one would expect the lilac to be fairly common; so the passage is evidence only of the horticultural naïvety of Cromwell's commissioners. At whatever date, the fountain with its lilacs must have looked charming, and also the walk described by Liger half a century later, "of Lillachs planted at Twelve Foot distance from each other, the Stems of 'em about Ten Foot high, and between them a Palisade of Horn-beam; when they blow, there's nothing finer to look on, and whoever will give themselves the Trouble to raise them, will not repent of the

[1]*Archaeologia*, 1779

Labour bestow'd upon them." The shrub soon became naturalized in many parts of Europe (including Britain) but its native country remained for long unknown; it was vaguely set down as "The Orient" until 1828, when a naturalist called Anton Rocher found it growing truly wild at Banat, in what was then western Rumania; and later it was discovered elsewhere in the Balkans. Mr. Douglas Bartrum suggests that it may have been taken into cultivation by the Turks when they overran Greece in 1453.

The so-called "Persian" lilac, S. persica, was also an early introduction, and its true habitat long unsuspected. It was cultivated in Persian and Indian gardens from time immemorial; if the botanists of the sixteenth century were correct in identifying it with the "blue jasmine" of the Arab physician Serapio, it must have been known before 800 A.D. It reached Europe before 1614, this time through the Venetian ambassador to Constantinople; in 1629 Parkinson knew it only by repute, but by 1640 he was able to report that it was being grown by John Tradescant at South Lambeth; he thought it was "very like to have come first out of Persia, as the name importeth." Hanmer in 1659 still classed it apart from the other lilacs, as the Persian Jasmyn, though he thought it had little resemblance to a Jasmine except in leaf; he said it would grow "into a little tree strong enough to become a stander, and not to neede a wall or other support," and that although hardy, it was "often kept in tubbs or other vessells to bee hous'd in wynter." Up to this time the form grown was the one now known as var. laciniata, whose cut or lobed leaves have some remote resemblance to those of the jasmine, with which it was still classed as late as 1683.[1] The entire-leaved form was not distinguished until 1672, and unfortunately was the one selected by Linnaeus as the type of the species: unfortunately, because it is sterile, and known only in cultivation, whereas laciniatus bears fertile seed, and it was this variety that was at last found truly wild by Frank N. Meyer in 1915, near Kungchow in south-west Kansu, a region traversed by the ancient trade route by which other Chinese plants and products (peach, apricot, rhubarb, silk and musk) were taken to Persia. White or whitish forms of this lilac were first mentioned by du Monceau in 1755, and by 1785 the three kinds—blue, white and cut-leaved—were common. Marshall advised that their branches should be supported by sticks "disposed in such a manner to escape notice, except by the nicest examiner."

[1] Perfume was a factor considered in the early days of plant classification; and both jasmine and lilac belong to the same Order, the Oleaceae

The two species (vulgaris and persica), a hybrid between them, and a small number of varieties, were the only lilacs cultivated until the nineteenth century was well advanced; and the varieties were slow at first to appear. The first mention of a white lilac occurred in the Bavarian Hortus Eystettensis in 1613, and it was not a true white; Parkinson in 1640 described it as being of "a milky silver colour, which hath a shew of blue therein, comming somewhat neere unto an ash colour," and said that it had not yet been introduced to England. In 1659, however, Sir Thomas Hanmer cultivated three varieties of lilac, the common blue, the rare white, and the red, rarest of all.[1] A deep purple form—possibly Hanmer's "red," and probably almost identical with later varieties such as Rubra Major and Charles X—was listed in Sutherland's catalogue of the plants in the Edinburgh Physic Garden in 1683, and was subsequently known as the Scotch Lilac (var. purpurea). There is nothing to show whether Hanmer's "white" lilac was the greyish one already described, or a true albino; but the pure white form must have been well established by the early eighteenth century, for Collinson relates that Lord Petre was particularly fond of white lilacs and ordered his gardener to save seed of the white variety only; but when the 10,000 seedlings he raised flowered in 1741, only about twenty came white—the rest were blue. (The double white was first mentioned in Loddiges' catalogue for 1823.) About 1777 the two species (vulgaris and persica) were intentionally or accidentally crossed in the Botanic Garden of Rouen, and the resultant hybrid, named the Rouen Lilac, or S. × rothomagensis, reached England in 1795. By 1838, Loudon was able to list seven varieties of the common lilac, and large scale breeding was just beginning; six sorts, "tolerably distinct," but still unnamed, had been raised by Mr. Williams of Pitmaston, and the French nurserymen were "also in the possession of some new seedlings." A new species, S. josikae, had been introduced, and two others discovered (emodi and villosa)—all, Loudon thought, to be welcomed, for the shrub was so beautiful and so hardy "that the number of sorts, provided they are truly distinct, can hardly be too much increased."

Loudon could have had little conception to what degree, during the next ninety years, he was to be taken at his word; from that time onwards lilac varieties increased so rapidly that in 1928 Mrs. McKelvey was able to list and describe approximately 450 of them "as a starting-point for further study." Many countries con-

[1] His Garden Book remained in manuscript until discovered and published in 1933, so was not available to earlier historians of the genus

tributed, but the majority were raised in France, especially by the famous firm of Victor Lemoine et Fils, of Nancy, who between 1876 and 1927 produced 153 named sorts, many of them still firm favourites. The story goes that during the Franco-Prussian war, when Nancy was occupied by the Germans and business more or less at a standstill, Lemoine decided to occupy himself with the production of a good double lilac. He chose as a parent the variety *azurea plena*, which had double flowers, though they were sparse and small. These florets bore no stamens, and most of them had only abortive pistils; Lemoine's sight being defective, it fell to Madame Lemoine to stand on the top of a stepladder, search among the blossoms for the few with well formed pistils, and apply to them pollen taken from selected garden lilacs, and from the new Chinese species, *S. oblata*. In the first year, 1871, they were rewarded with seven seeds; in the next, with about thirty. Three of the plants raised from these seeds flowered for the first time in 1876. The cross with *S. oblata* was named *S. × hyacinthiflora*; among the other seedlings were several doubles, which were again crossed with the best garden varieties, the original *azurea plena* being discarded.

Up to this time most of the garden forms had been varieties of the common lilac, *S. vulgaris*; there were very few hybrids for the simple reason that there were very few species to work with. The species arrived late because nearly all of them were natives of distant China; the majority were introduced only in the present century, and the garden varieties being already so well established and, it must be admitted, so superior to many of the species in beauty, there was little inducement to experiment with further hybridization. The first new lilac to be introduced after the long hiatus was not, however, Chinese, but one of the few from eastern Europe—*S. josikae*, named after Rosalie, Baroness von Josika, who (about 1827) was the first to draw the attention of botanists to the fact that this species was distinct from *S. vulgaris*, which also grew wild in the same Hungarian locality.[1] A plant of it was sent from Hamburg to Dr. Graham of the Edinburgh Botanic Garden in 1833. *S. emodi* from the Himalayas was introduced by Dr. Royle in 1840, but the first of the true Chinese lilacs was *S. oblata*, a garden plant unknown in the wild, which was brought home by Fortune at the end of his third trip in 1856. It blooms very early and is vulnerable to spring frosts, but was used, as we have seen, by Lemoine as one of the parents of his × *hyacinthiflora*. A more important progenitor was *S. villosa*. Seed of this and two other lilacs (*oblata* and *pubescens*) was sent to several botanic gardens

by Dr. Emil Bretschneider when he was resident at Pekin from 1879–82 in the capacity of medical attaché to the Russian Embassy; Kew received its consignment about 1880, and the resultant plants bloomed for the first time in 1888. A botanist named Louis Henry of the Jardin des Plantes crossed this species with *S. josikae* in 1890, and produced the hybrid named after him *S. × henryii*.

But these were only the precursors of the great flood of lilacs from China that arrived with the new century. Half a dozen of the best were introduced during the first decade by E. H. Wilson, either for Messrs. Veitch of Exeter or for the Arnold Arboretum. They were *S.S. julianae, reflexa, swezingowii, tomentella, velutina* and *wolfii*, the first four of which received the Award of Merit. (*Julianae* was described by C. K. Sneider and named in honor of his wife; *wolfii* was described by the same botanist from a plant in the Forestry Institute at St. Petersburg, whose director was Egbert Wolf; and *swezingowii* was described from a specimen in the botanic garden at Roemershof in the province of Livonia, and named after the governor of the province.) Another Award of Merit lilac was contributed by Forrest within the same wonderful decade—*S. yunnanensis*, seed of which was sent to A. K. Bulley and to the Edinburgh Botanic Garden in 1906; while in 1910 Purdom, collecting for Veitch, sent home *S. microphylla*,[1] which won two awards.

The value of these Chinese species as possible parents of garden strains has hardly yet been explored. Most of them are extremely hardy. The lilac in general is a cold-climate plant; in Poland the common lilac is said to develop trunks 8 inches in diameter. It is one of the few flowering shrubs that thrives in the climate of Canada, and in 1833 the garden of every French-Canadian was said to have a lilac in the corner. From Canada, therefore, has come the most important development in lilac-breeding in the present century, thanks to the work of Miss Isabella Preston at the Central Experimental Farm, Division of Horticulture, at Ottawa. In 1920 she crossed *S. villosa* with *S. reflexa*, using the former as the seed, the latter as the pollen-parent, and raised nearly 300 seedlings, which were planted out in 1922 and bloomed in 1924. The best of them were then selected for naming and propagation, and form the very beautiful and distinct race known as *S. × prestonae*, many of which are now available. Miss Preston also crossed *S. reflexa* (pollen-parent) with *S. josikae*; only one plant resulted, which was not considered of much value, but four of its seedlings

[1] *S. josikae* belongs to the other section of the genus, the *villosae*

[1] The true identity of the dwarf or 'Doll's House' lilac sometimes listed as *S. microphylla* or *S. palibiniana*, has been investigated by Mr. Roy Elliott in the *Alpine Garden Society Bulletin*, vol. 29, no. I; it appears to be a form of *S. meyeri*

of the second generation proved outstanding. This group is known as *S.* × *josiflexa*.

These Canadian shrubs are quite unlike the *vulgaris* varieties which have so long ruled the lilac roost; but they are not so strange as the crosses suggested by Loudon, who thought that some "curious hybrids" might be raised between lilac and privet, privet and olive, or lilac and ash (all members of the Oleaceae) most of which, he asserts, may be grafted on each other. The grafting of olive on lilac and of lilac on ash seems to have been tried with little success; but grafting on privet was frequently practiced, usually with deleterious results—"these shrubs," says Robinson briefly, "are often grafted on the privet, but die on it." The practice has come to be regarded as "an unpardonable offence," but some varieties are difficult to propagate except by grafting, and when stocks of the common lilac are used the suckers are a nuisance. If a graft is worked low down on a privet stock and the point of union buried, the scion will eventually form its own roots.

Forcing lilacs for out-of-season bloom was practiced as early as 1774 by a M. Matthieu of Belleville, who used for the purpose the old red 'Lilas de Marly'. (Colored lilacs forced in heat and darkness come out white; and the darker forms usually give a better white than light ones.) The flower has always been particularly popular in France, and in the 1880s forced lilac could be bought in Paris in every month of the year except July and August, when there was no demand. In 1890 the method of forcing by the use of chloroform or ether was invented in Denmark, and was subsequently much used in Germany, in spite of the risk of explosions. This is still among the chemical methods practiced; but most of the forced lilac seen in this country is not produced here, but imported from Holland.

Loudon remarked that when forced, the flowers of *S. persica* lost their scent; this fortunately is not true of all the kinds, for a lilac without perfume has lost practically its whole *raison d'être*. Not all the species are fragrant—some show only too clearly their affinity with the privet, and others have little or no scent. *S. swezingowii* is said to be the most fragrant of them all, and by any name would doubtless smell sweeter still. Few can surpass our common lilac for scent; in this respect, some of the very large garden varieties are decidedly inferior to the original type. The perfume is extracted, like that of jasmine, by means of *enfleurage*. Lilac-wood is said to be fragrant when burned.

" 'Tis said that Lilach is an Arabick word, tho' some pretend 'tis derived from *Lilium*, because its Flower is like a little Lilly, and that it came originally from the *East-Indies*."[1] Reliance is not to be placed on any of the foregoing statements; the word actually comes from the Persian, and appears in English in every possible form, from laylock to lily-oak; Hanmer said it was "corruptly called . . . the LELAPS by many." Nor should one put much faith in Loudon's statement that "Syringa" comes from *sirinx*, "native name in Barbary." It is a "poetical name" derived from the Greek *syrinx*, the pipes of Pan, and was applied at first both to the lilac and the philadelphus, because their pithy shoots were used in Turkey to make pipes. This name too appears in various forms in various languages, and is found masquerading as "Scringa" or "Syringe."

The lilac is one of several purple flowers—heliotrope, lavender, violet and "mauve"[2]—to have given its name to a particular tint; and it may be this association with the color of mourning that accounts for some of the traditions associated with the flower, which are mostly of a somewhat melancholy nature. In several countries it is considered unlucky to bring lilac into the house, and it is frequently associated with death:

> *Here, coffin that slowly passes*
> *I give you my spray of lilac.*
> *O death, I cover you with roses and early lilies*
> *But mostly and now the lilac, that blooms the first . . .*[3]

In Persia, it is said to stand for "the forsaken," and to be given by lovers to their mistresses when they quitted them; in England and America it was said that the girl who wore lilac—except on May Day, when it was permitted—would never wear a wedding ring, and that to send a spray to a fiancé signified a wish to break off the engagement. There is a useful tradition in Germany, said to be widely held, that when the lilac blooms people become especially tired and indolent.

NOTES: Our forebears brought the common lilac *S. vulgaris* with them to the colonies. Throughout New England, the descendants of these popular fragrant shrubs can still be found around old farmhouse sites. Over the years, extensive collections have been established at the Arnold Arboretum, Jamaica Plain, Massachusetts; Highland Park, Rochester, New York; Swarthmore College, Swarthmore, Pennsylvania and the Morton Arboretum, Lisle, Illinois. Of the many common lilacs named prior to the Second World War, it appears that the majority came from the famous Lemoine nursery in France. These are still among the best offered in nurseries.

There are a few worthy exceptions, such as the popular dwarf lilac from Korea, *S. patula* 'Miss Kim', more

[1]Liger   [2]French for "mallow"   [3]Walt Whitman

Tamarix gallica
(FRENCH TAMARISK)   Ferdinand Bauer (1819)

Tricuspidaria lanceolata
(now *Crinodendron hookerianum*)
(LANTERN TREE)   M. Smith (1891)

suited to small gardens than the others. Derived from an obscure species, 'Miss Kim' has a history of serendipity. Professor Elwyn Meader of the University of New Hampshire was serving with the American Red Cross in South Korea. Like any good horticulturist, he took advantage of the opportunity to climb the mountains north of Seoul where he encountered a rare species, *S. patula*, in November 1948. Meader sent back thirteen seeds to the plant breeder Professor A. F. Yeager, who successfully germinated them. 'Miss Kim' was selected as the best seedling and named in 1954.

This suggests that in the United States, the professional plant breeders in public institutions play a far greater role in the enhancement of ornamental plants than in Europe, where nurserymen and astute amateurs make significant contributions. Perhaps the closest American examples of the European system occur in the large genera, such as Rhododendron, where enthusiasts play an equally important role.

Tamarix

The introduction of the tamarisk to England was attributed by a contemporary to Bishop Grindal, a prominent cleric of the reign of the first Elizabeth. "When this archbishop returned out of Germany" says Hackluyt, "he brought into this realm the plant of tamariske from thence, and this plant he hath so increased, that there be here thousands of them, and many people have received great health by this plant." That would make the date of introduction 1558, when Grindal returned from his exile during the persecutions of the reign of Mary, to become shortly afterwards Bishop of London and eventually Archbishop of Canterbury. Tamarisk was considered sovereign for the spleen; but it must be remembered that there were at this period two "tamarisks," the French or Italian (*T. gallica*) and the German; the latter has now been put into a different genus, under the name of *Myricaria germanica*. It seems probable that it was this German plant that Grindal introduced, and that this was also the kind mentioned in Turner's *Names of Herbes* some ten years earlier: "Myrica, otherwise named tamarix . . . I dyd never see thys tree in Englande, but ofte in high Germany, and in Italy. The Potecaries of Colon[1] before I gave them warning used for thys, the bowes of ughe."[2] Bulleine in 1562 says that tamarisk grew plentifully "in one part of Germanie, within a certaine Ilande, belonging to one of the Germaine Bishops," and that it was made known in England

[1]Cologne   [2]Yew

principally by "the famous learned man William Turner Phisicion, . . . whiche doctor is a iewell among us English men . . . for his synguler learning, knowledge and iudgement."

The French Tamarisk, *T. gallica*, was introduced at some unknown date between Turner's time and Gerard's; so English apothecaries of the seventeenth century had no need to resort to the dangerous yew for the treatment of what were called "splenetick" diseases—for which the tamarisk was considered so effective that it was sufficient in many cases for the patient merely to drink from a cup made from the wood, "as though the drincke which was given them out of such cups should doe them good."[1] (Pliny also recommended an ointment made from the leaves for the treatment of "nightfoes, or chilblanes.") Henry Phillips makes a mysterious reference to some magical powers; he says "they ascribed qualities to this plant too much against common reason and decency to mention." Perhaps he was thinking of the mystic *baresma* or bundle of tamarisk twigs used by the Magian priests of Persia for divinations. According to Pliny, it was commonly called "the unluckie tree," because it bore no fruit and was never set or planted; it was accursed, and was used to garland malefactors. In the Language of Flowers it is made to stand for "crime."

Perhaps the most interesting product of the tamarisk is the Arabian manna. This is obtained from a variety of *T. gallica* when the weather conditions are suitable, and is caused by the punctures of a scale-insect, *coccus manniparus*. After a season of heavy rainfall, the sap exudes from these punctures in resin-like beads, at first clear and transparent, then crystallizing to resemble rough barley-sugar; it falls or is shaken from the branches, but melts in the sun, so that it must be collected before sunrise. It is said to have a sweet, slightly aromatic taste, and to be considered a great delicacy among the Arabs. It answers in some respects, though not in all, to the manna of the Israelites, which was "like coriander seed, white; and the taste of it was like wafers made with honey . . . when the sun waxed hot it melted."[2] The shrub is not common in Palestine except on the saline soil of the Wadi-el-Sheik, not far from Mount Sinai, where it is abundant; the manna was collected there by the Arabs and sold to the monks of St. Catherine's, who disposed of it to pilgrims, some of whom valued it greatly and would purchase it at high prices.

Miller says that the common tamarisk (*T. gallica*) has small leaves that "lie over each other like scales of fish" and flower-spikes "about an inch long, and as thick as a large earthworm"—a description that may be vivid, but

[1] Dioscorides [2] Exodus XVI, 21, 31

is hardly attractive. (Nevertheless, by 1675 the shrub was "usually planted by those that respect variety and pleasure.")[1] The botanical differences among the various species in the family are small; for garden purposes, the best are the early *T. tetandra* (1821) whose flowers have four stamens and are borne on the previous year's shoots, and the later-flowering *T. pentandra*, whose five-stamened flowers are borne on shoots of the current year. (Both are natives of south-east Europe.) Some botanists consider that *T. anglica* and several other "species" are only local forms of the widespread *T. gallica*, whose range may extend to China, were the tamarisk goes by the charming name of Hear-Prayers Willow. *T. gallica* and *T. anglica* both grow wild in southern coastal districts of the British Isles, but neither is truly indigenous; they have been extensively planted as seaside shrubs, and are so easy to propagate, that a tamarisk-hedge can be formed by merely driving in a series of cuttings the size of a small walking-stick.

The tamarisk, said Dioscorides, "is a known tree, growing by marishie grounds and standing waters;" and the ancient Latin name is said by some authorities to be derived from the river Tamaris (now the Tambra) in Spain. But is it not equally possible that the river may have been named from the abundance of the plant? NOTES: The various species of Tamarisk are occasionally listed by nurserymen because of their interesting scale-like leaves and small pinkish flowers. These are plants of the Near East and, while long cultivated in Europe, do not have the same appeal in the United States.

Thelycrania, see under *Cornus*

Tricuspidaria

T. lanceolata, Lantern Tree. This is another plant for those unfairly favored gardeners of the south and west, and one of the most beautiful evergreen shrubs it is possible to grow—where it is possible to grow it. It is one of the red-flowered forest shrubs of the Andes, and like the Embothrium, was introduced by William Lobb for Messrs Veitch in 1848. Exactly what had been introduced was for long obscured by one of the usual nomenclature muddles. The first of the family was discovered by one Juan Ignacio Molina in 1782, and the Spanish botanist Cavanilles, describing it from a drawing that Molina provided, named it *Crinodendron*, lily-tree. The French botanist de Candolle rejected this name in

[1] Worlidge, *Systema Agriculturae*

favor of *Tricuspidaria* (three-pointed—i.e., the petals) given to the present species by Ruiz and Pavon in 1784, because he found Cavanilles' description "so imperfect and misleading"—not realizing, perhaps, that it applied to a different member of the genus, which comprises, besides the present subject, *T. dependens* and *T. patagua*, both white-flowered trees of a very different aspect. The rules of botanical nomenclature decree that the oldest name must be regarded as the authentic one, and the correct, though unfamiliar, name for the Lantern Tree is actually *Crinodendron hookerianum*.

This shrub has the unusual habit of producing its flower buds in the autumn, and slowly expanding them to the pendant bells of soft but glowing red of the following summer. It was established here too late to be incorporated into the Language of Flowers; otherwise it might have been made to symbolize the Pleasures of Anticipation.

NOTES: These shrubs are not cultivated as ornamentals here.

Ulex

. . . Ulex, a herb like rosemary, with a quality of attracting gold.
Christopher Smart, *Rejoice in the Lamb, c.* 1756

Like many other things, the Common Gorse (*U. europaeus*) is not so very common, although it is very abundant—which is not at all the same thing; it has, one might say, many points of interest. It is a plant of the Atlantic coasts, and is not entirely hardy, occasionally suffering severely from frosts in inland districts, though seldom or never killed outright. It is thought not to be indigenous to Scotland, at least in the Highlands—a theory partly based on its habit of winter flowering; "no truly Scotch plant would be so rash."[1] Its range has been considerably extended by cultivation, for it has been planted on poor soils in Cornwall, Wales and Scotland for use as shelter, fodder and firing. Its employment for the latter purpose is ancient indeed; records of gorse wood and gorse charcoal are common from Neolithic sites, and are taken to indicate that by then the great forests had been cleared, and gorse had become the handiest fuel. Gerard said that about Exeter, where this shrub grew particularly tall, "the great stalks are dearely bought for the better sort of people, and the small thorny spraies for the poorer sort." (Gorse-bundles were commonly used for heating bakers' ovens.) If crushed in a mill to break the spines, the green growths made good

[1]Dr. Walker, quoted by Loudon

fodder for cattle and horses; I remember being told as a child that horses fed on gorse developed those long whiskers on the lip that made certain cart-horses look so venerable.

Foreigners from lands where the gorse will not grow have always been highly appreciative of its beauties. "In the colder countries to the east," says Gerard "as Danzicke, Brunswicke, and Polande, there is not any branch hereof growing, except some fewe plants and seedes that myselfe have sent to Elbing, otherwise called Meluin, where they are most curiously kept in their fairest gardens, as also the common Broome, the which I have sent thither likewise, being first desired by divers earnest letters." Linnaeus lamented that in Sweden he could hardly keep it alive in a greenhouse. In the eighteenth century our gorse-covered commons were much admired by visiting botanists, one of whom is said to have fallen on his knees in ecstasy at the sight—but which? The story is told by Sir. J. E. Smith about Linnaeus, during his visit to England in 1736, while the same anecdote was related by Martyn about Dillenius, who came to England from Germany in 1721. Linnaeus is supposed to have experienced his raptures on Putney Heath, Dillenius at Hounslow; and a more sober but better authenticated admiration was expressed by Peter Kalm at Fulham Common in 1748. The greatest compliment to the plant, however, was paid by an Englishman; the late Captain Kingdon-Ward, familiar though he was with the marvelous flora of western China and Upper Burma, said that gorse was the one plant he would have chosen to have introduced to Britain, if it had not been already a native.

If gorse and broom were less common, Batty Langley thought, "they would be valued and cultivated with as much Eagerness as any Evergreen or Shrub whatsoever; for in Fact, the Furze is capable of any Form required, but much finer when let grow in its rural Manner as Nature directs. If the beauty of this Shrub . . . was compar'd with that of the long-esteem'd Yew, whose Aspect is melancholy, the true Image of Sadness, 'twould not only be found to be much the finer Plant of the two, but more deserving the Gardener's Care." Nevertheless, the common gorse cannot really be considered a good garden plant; it is hard to keep within bounds, and gorse hedges "wherein many Birds take great Pleasure to build their Nests" are dangerously inflammable. But there is a double variety which is more compact in growth, and is a useful plant on very poor soils, where it flowers best. Loudon says it was first found in Devonshire, about 1825; in the *Transactions of the Horticultural Society, vol. VII* (1852) it is mentioned as a rarity, to be obtained from John Miller of Bristol "in whose nursery it was first raised," but Cobbett knew it well in 1833, and praised it as being "sweet-smelling, an abundant flowerer, and evergreen. It should be in every shrubbery, and it does not disgrace a border even." It received the highest award of the Royal Horticultural Society, the A.G.M., in 1929. Both Gerard and Parkinson speak of a white-flowered gorse, supposed to grow "on barren grounds in the North," and this variety was still listed in 1778, although nearly two centuries had gone by without anybody having actually seen it. (It is unknown today.) Two other native species of dwarf growth, *U. minor* (syn. *U. nanus*) and *U. galli*, are recommended for the rock or heather garden; the latter in particular has been called "a glorious native neglected." It frequently grows and blooms along with the bell-heather, vying with the Assyrian cohorts in purple and gold.

Culpeper called the gorse a plant of Mars, and prescribed a decoction of its flowers for the cure of jaundice and the stone—"Mars doth also this by sympathy." Yellow was Mars' color; but the shrub may also have been associated with him on account of its war-like prickles. "A plant altogether a thorne" Gerard justly called it; for leaves appear only on young seedlings, and as the plant matures every would-be leaf-stalk is converted into a spine with a tip of steel. "Ulex" is said to be derived from the Celtic *ec* or *oc*, meaning a prickle or sharp point. Of the English names, Furze is said to be used in the south, Gorse in the north and Whin in the east. The domination of our eyes over our ears is proved by the incomprehensibility of versions such as "Furres" or "Furs"; we are also taken aback when Turner speaks of the plant in the singular, as "a Fur or Whine." "Whin" is also current in Scotland, and in the ancient *Lyke Wake Dirge* the soul of the departed encounters a gorse-covered waste called Whinny-muir on the outskirts of Purgatory:

> *If ever thou gavest hosen and shoon,*
> *Sit thee down and put them on. . .*
> *If hosen and shoon thou ne'er gavest nane,*
> *The whinnes sall prick thee to the bare bane . . .*

Nevertheless the plant has its gentler side; "when the gorse is out of bloom, kissing's out of season"—and there is always bloom on the gorse. Its early flowers afford welcome bee-pasture, and have a rich honey-and-almond scent:

> *Sweet as the breath of the whin*
> *Is the thought of my love*
> *Sweet as the breath of the whin*
> *In the noonday sun*
> *Sweet as the breath of the whin*
> *In the sun after rain.*

> *Glad as the gold of the whin*
> *Is the thought of my love*
> *Glad as the gold of the whin*
> *When wandering's done*
> *Glad as the gold of the whin*
> *Is my heart, home again.*[1]

NOTES: This weedy shrub is rarely cultivated in America.

Veronica

The shrubby Veronicas are as much a product of New Zealand as the Kiwi: only one, *V. elliptica*, having strayed as far as Chile, by way of the Falkland Islands and Tierra del Fuego. About 200 out of 250 species of Veronica are found in New Zealand, and between seventy and eighty of them are shrubby. Hooker said that they formed a conspicuous feature of the New Zealand vegetation "both from their number, beauty and ubiquity, from so many forming large bushes, and from the remarkable forms the genus presents."[2] *V. gigantea* can grow into a tree 40 feet in height, and the sizes range all the way to pygmy shrublets suitable for the rock-garden. Both in gardens and in the wild they interbreed very freely; in their native home "hybrid swarms" frequently occur, and a story is told of a New Zealand gardener, learned in his native flora, being shown a collection of veronicas in a British west-country garden—not one of which had he ever seen before. As early as 1789, the shrubby members of the family were put into a group by themselves under the name of "Hebe," and this classification was followed by many later botanists; but the subject is still controversial, and the R.H.S. *Dictionary of Gardening* authorizes the retention of the more familiar name.

Unfortunately in this country the shrubby veronicas, like their fellow-countrymen the olearias, are only on the borderline of hardiness. Several of them, maritime shrubs in their own land, will grow near the coast, and will endure any amount of wind, but are vulnerable to frost. The broad-leaved kinds are on the whole less hardy than the narrow-leaved, and needless to say the most beautiful are also the most tender. The lavender *V. hulkeana*, for example, is universally praised, and has been called one of the most beautiful of flowering shrubs; but it needs all the protection a warm wall can afford. Seed of it was sent from New Zealand by "the indefatigable Dr. Mueller"[3] and the plant flowered for

[1] Wilfred Wilson Gibson [2] *The Botanical Magazine*, 1864
[3] F. von Mueller, prolific writer on botany and authority on the Australian flora

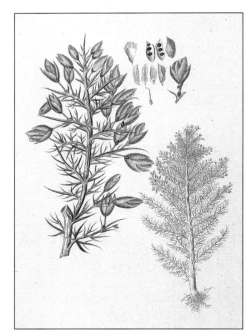

Ulex europaeus

(GORSE)   *Flora Danica* (1775)

Veronica speciosa
(now *Hebe speciosa*)

(SPEEDWELL)   W. H. Fitch (1843)

the first time in the cool greenhouse at Kew in 1864. The showy *V. speciosa*, the most brightly colored member of the family, is but little hardier. Great expectations were aroused prior to its introduction by the descriptions and specimens sent home by its discoverer, Richard Cunningham. "Since the country around its locality, at the mouth of the Hokianga River, is now occupied by Europeans" wrote Cunningham's brother Alan in the *Botanical Magazine* (1836) "let us hope soon to receive the seeds, which we may reasonably expect will succeed . . . in the open borders; the elevated grounds occupied by our VERONICA, being greatly exposed to the prevalent tempests of its weather-beaten coasts." Ultimately live plants were brought from the locality mentioned, by a Mr. Edgerley, and sold to Knight's nursery at Fulham, where they flowered in 1843; a specimen was presented to Kew, and was solicitously kept in a cool greenhouse, being too rare to risk out-of-doors. There are now a number of garden varieties of this species in rather lurid purple and crimson shades; but they can safely be grown only in the south and west.

The hardier veronicas all have white or whitish flowers, which compensate for their monotony by their profusion, and in some cases, by their fragrance. The principal kinds are *VV. anomala, brachysiphon, elliptica* and *salicifolia*. *V. elliptica* is the far-flung species, and was introduced from the Falkland Islands by John Fothergill in 1776, considerably earlier than any other member of this Antipodean family. It is hardy anywhere on the coast of Britain, and reasonably hardy inland; its purple-marked flowers have "a most delicious fragrance . . . not mentioned by authors, and we believe scarcely known, having never heard it spoken of by those who have cultivated the plant."[1] It is chiefly important as the probable parent of the popular variety or hybrid 'Autumn Glory', unkindly described as "a shrub of low stature and uncertain parentage." Its origin is obscure; Messrs. A. R. and K. M. Goodwin record in their catalogue that their nursery received it from Miss Gertrude Jekyll in 1916 *V. brachysiphon* (syn. *V. traversii*) introduced in 1868, is hardier still, being rarely injured by cold; it is the best-known and most reliable species, and is said to make a good hedge, as does *V. anomala*, another of the hardier sorts, which was found by a Mr. Armstrong at 4,000 feet in the valley of the Rakaia River, Ashburton Province, South Island, and which flowered at Kew in 1886. *V. salicifolia*, though rather more tender, has managed to become naturalized in suitable parts of the country. In New Zealand it is one of the most widespread and variable species, and also one of the most fragrant; a dwarf

[1] *The Botanical Magazine*, 1794

and hardy garden form known as 'Spender's Variety' has received an Award of Merit.

Many of these maritime veronicas have their leaves neatly arranged one above another in four ranks, like those of our own common ling; and some have a double layer of protective epidermal cells in their leathery leaf-surfaces—both devices to reduce transpiration and enable the plant to endure long droughts or desiccating sea winds. But there is one group that has adopted even more extreme measures—the "whipcord" veronicas, whose leaves are reduced to scales closely clasping the stem, and giving by their regularity the impression of a plaited thong. In general appearance these shrubs resemble a tamarisk or a dwarf cypress, until they startle us by the production of their pale lavender flowers. *V. cupressoides*, the best of them, adds to the conifer effect by the possession of an aromatic scent like that of cedar-wood, which in warm moist weather is perceptible at a considerable distance from the plant. This remarkable plant-mimic was found by Dr. Andrew Sinclair[1] in the upper valley of the Wairoa River, and introduced to Kew in the 1880s.

The two ladies, Veronica and Hebe—Christian saint, and pagan immortal—appear to have little in common, and the connection of either of them with this or any other group of plants is far to seek. Hebe, a daughter of Juno and Jupiter, acted as cup-bearer to the gods, until banished from Olympus by her father for falling and spilling nectar at a grand festival; she subsequently married Hercules and lived happily ever after. (Possibly the shrubs, too, have a reputation as nectar-bearers?) According to Charlotte M. Young, "Veronica" comes from the Latin *veras*, true, and the Greek *eikon*, an image; the name of *veraiconica* was given to a portrait on a piece of linen, shown at St. Peter's at Rome and said to be a "true image" of Our Lord—and "superstition, forgetting the true meaning of the name, called the relic St. Veronica's handkerchief, accounting for it by inventing a woman who had lent our Blessed Savior a handkerchief to wipe His face during the passage of the Via Dolorosa, and had found the likeness imprinted on it."[2] The markings on the flowers of some of the herbaceous veronicas are said to resemble those on the sacred veil; but the derivation is uncertain.

NOTES: We tend to use these handsome blue-flowered plants as perennials for borders or rock gardens. The new species of shrubs in cultivation are simply not hardy, being natives of maritime New Zealand.

[1]Dr. Sinclair was Colonial Secretary for New Zealand from 1844–56, and was drowned when crossing the Rangitata River in 1861
[2]*A History of Christian Names*, 1884

Viburnum

The large family of the Viburnums contains many valuable garden shrubs, and some that are indispensable; their versatile beauties adorn every season. Most of the species in the family are shrubs of the northern temperate zone; two are natives of Britain, including *V. lantana* (thought to be the original "Vibur-

Viburnum tinus
(LAURUSTINUS)   J. Sowerby (1796)

num" mentioned by Virgil) which still retains Gerard's English name of Wayfaring Tree, in spite of Parkinson's protest that "no travailer doth take either pleasure or profit by it, more then by any other of the hedge trees." This species was formerly planted by cowshed doors to protect the cattle from witchcraft; but it cannot be considered a garden plant, though a variegated form was occasionally cultivated.

V. opulus is the other British native, its specific name being derived from the fact that it was identified by the French botanist Jean Ruel (1474–1537) as the "Opulus" of Dioscorides. At first it was classed as a kind of elder, on account of its pithy wood, and appeared under such names as *Sambucus aquatica* (Water-elder) or *Sambucus rosea* (Rose-elder) and in English as the Marris (Marsh)

Elder, Ople or Dwarffe Plane-Tree. The misleading name of Guelder-Rose did not appear in print till 1597, when Gerard referred to it as the Elder-Rose, "not rightly called the Gelder Rose," and gave its Dutch name as *Gheldersche Roosen*.[1] The sterile garden form we call the Snowball Tree was known on the Continent as early as 1554; it may have reached England by way of Holland, and brought the Dutch name with it:

> Sambucus *too from* Gueldria's *Plains will come,*
> *Drest in white robes she shows a Rose-like Bloom,*
> *Be kind, and give the lovely Stranger room . . .*[2]

Gerard had two varieties (one with purplish flowers) and did not speak of them as novelties. It has been a favorite garden plant ever since, as can be seen from the number of its pet names, which include Tisty-tosty, Whitsun-boss, Love-roses and Pincushion-tree. Its commonest name was considered by Miss Kent to be so appropriate that she had "more than once heard it remarked by persons who knew it only by its more general title of Guelder-rose, that it *should* have been called the Snowball-tree," but it seems to date only from the eighteenth century, and makes its first appearance in Miller's *Gardener's Dictionary* (1759). Hanbury (1770) said that the flowers had "the appearance of balls of snow, lodged in a pleasing manner all over its head," and Phillips compared them to "the work of the finest chisel upon the purest alabaster." These

> . . . *flower-globes, light as the foamy surf*
> *That the wind severs from the broken wave*[3]

are composed, like those of the garden hydrangeas, of sterile flowers only, and produce no fruit; and many purists claim to prefer the wild species with its flat corymbs of sterile and fertile flowers, which are followed by brilliant berries. A variety with yellow fruit has been cultivated since the late nineteenth century, and there is said to be a kind that bears berries of a beautiful and uncommon shade of salmon- or coral-pink. Miller said that the wild Guelder-rose was "seldom suffered to stand very long in Gardens; but I have seen one in an old Garden whose Stem was more than two Feet and a Half round."

V. tinus,[4] the evergreen laurustinus or "Wilde Baie" from the Mediterranean region, has been cultivated for as long as the Snowball-tree,[5] though it was centuries before it was realized that there was any connection between them. As its name denotes, this species was at first classed as a variety of laurel; and the two forms grown before 1597 had increased to six main kinds "and some other trifling varieties" by the end of the eighteenth century, including those with gold- or silver-striped leaves. "The Laurus-Tinus . . . is all the Winter the most beautifullest Plant of any in the Garden," wrote Batty Langley enthusiastically in 1728, "Also Hedges of this Plant is wonderful fine, if their beauty is not destroyed by the unskilful Hand of the Gardener, in clipping them at the same Season as he does Yews, Hollies, etc." Marshall, half a century later, praises it for its winter bloom "in spite of all the weather that may happen; and the boldness of these buds, at a time when other flowers and trees shrink under oppressive cold, is a matter of wonder and pleasure." In cold districts, specimens were often grown in pots and tubs, and treated like oranges or myrtles; Langley suggested a "small Hedge of Laurus-Tinus Plants, planted in large Flower-Pots" as the best ornament for little-used rooms in the winter. The variety *lucidem* is handsomer than the type, but more tender; it is an old favorite, and might even be identical with "the Fulham, or great blowing Laurus Tinus" grown by London and Wise in 1706. The laurustinus tolerates urban conditions better than most evergreens, and Loudon suggested it might be used "for varying the iron palisades, pales or brick walls, which separate the front gardens of street and suburban houses"—with a warning that the fallen leaves must be swept up, as they acquire "a remarkably foetid odour."

The supremacy of these two ancient viburnums, the laurustinus and the snowball-tree, remained for long unchallenged. During the eighteenth century several American kinds were introduced, but none was of great horticultural importance. Then, as so often happened, the influx of new species from the Orient in the nineteenth and twentieth centuries doubled and trebled the number of viburnums in general cultivation. First came the Chinese and Japanese snowball-trees, both introduced by Robert Fortune in 1844. The Chinese Snowball, *V. macrocephalum*, has the largest flowers in the genus, but is rather tender, and is safest against a wall; it is often forced under glass, for the flowerheads in their small and early stages are a bright apple-green, and very popular with flower-arrangers. The hardier Japanese snowball (*V. tomentosum* var. *plicatum*) is remarkable for its ribbed leaves and its horizontal habit of growth; some authorities put it among the dozen best deciduous shrubs. Both of these viburnums were long-cultivated garden forms; the wild type of *V. tomentosum* was introduced about 1865, but is not so hardy as its variety *plica-*

[1]After the district of Gelderland, where it was abundant [2]Rapin, *Of Gardens*, trans. James Gardiner, 1728 [3]Cowper [4]"Tinus" was the old Latin name, used for this species by Pliny [5]The date is given as 1560

tum, or as the handsome large-flowered var. *mariesii*, introduced by Messrs. Veitch about 1879.

Hitherto no garden viburnum had been remarkable for fragrance; but the start of the present century brought two species especially valuable for their delicious scent—*V. carlesii* and *V. fragrans*. William Richard Carles (1849–1929) was the British Vice-Consul in Korea from 1883 to 1885, and during that time made three exploratory journeys from his base at Chemulpo into the almost unknown interior. It was on one of these trips that he discovered this viburnum, dried specimens of which he sent to Kew in 1885; it was not until nine years later that he learned it had been named in his honor. (Carles was afterwards awarded a C.M.G. for the part he played in the Boxer Rising of 1901, but he is remembered today for his bush rather than his bravery.) In the same year of 1885, a living plant of this species was given by a lady in Korea to Mr. Alfred Unger of the nursery firm of L. Boehmer and Co. of Yokohama; after growing and propagating it for a number of years, he sent some leaves and flowers to Kew for identification, followed by the presentation of a plant in 1901, which bloomed for the first time in England in 1906. Mr. Unger then sold his whole stock of this viburnum to the French firm of Lemoine, who were responsible for its distribution.

It is surprising that *V. fragrans* should have been so late an introduction, as it was a favorite garden plant in China, and was known to European botanists in the mid-eighteenth century. But it is a plant of the north, more often recorded from Pekin (where it seems to have been a special feature of the Imperial gardens) than from Shanghai or Canton, where the earlier botanists made their collections. (It is grown in Chinese cottage gardens at an altitude where even corn only ripens one year in three.) It is said to have been first introduced by William Purdom when collecting for Messrs. Veitch in 1909, but was not noticed or identified until rediscovered in 1914 by Reginald Farrer, growing wild in Shi-hoi in Kansu. Purdom was also present on this second occasion, but the credit for the introduction is always given exclusively to the more voluble Farrer. He sent home abundant seed, and would have sent more, but for an unfortunate falling-out with his Highness Yang Tusa, Prince of Joni, who had promised to save the kernels when his crop of viburnum fruits came to be gathered, but who in a fit of pique "set to, and sedulously ate up all the Viburnum fruits in his palace garden, and threw away the seed."[1] Unfortunately these edible and highly ornamental berries are rarely produced here.

Both *V. fragrans* and *V. carlesii* received several

awards; but once the first flush of excitement was over, it was realized that both had defects as garden plants, and breeders set to work to improve them. The first British hybrid was *V.* × *burkwoodii*, raised about 1924 by Messrs. Burkwood and Skipwith of Kingston-on-Thames, from *V. carlesii* crossed with *V. utile*—the latter a Chinese species introduced by Wilson for Veitch in 1901, and given the name of "useful" for the rather insufficient reason that the shoots were used for pipe-stems by the Chinese. *V. carlesii* was also crossed with *V. macrocephalum* to produce *V.* × *carlcephalum* (*c.* 1932)—clumsy in name and nature; but the best of the *carlesii* hybrids is said to be the American *V.* × *juddii*. In this case the other parent was the rather similar *V. bitchiuense*, which was introduced in 1911 and received its unfortunate name from the province in Japan in which it originated; the cross was made in 1920 at the Arnold Arboretum in Massachusetts, and named after William Henry Judd, the arboretum's propagator. The most notable descendant of *V. fragrans* is *V.* × *bodnantense*, raised by Lord Aberconway at Bodnant and exhibited for the first time in 1947; it was the result of a cross with *V. grandiflorum*, (another winter-flowering Chinese kind, collected by R. E. Cooper in 1914) and is said to inherit the best points of both parents.

All the Viburnums so far mentioned are grown chiefly for their flowers; but there are others whose value lies in the brilliance of their autumn color, both in leaf and fruit. Our own wild Guelder-rose has been said to come "from that garden of Aladdin, where all the fruits of the trees were jewels."[1] Perhaps the best of the berrying bunch is *V. betulifolium*, whose branches in autumn bend and sway with the weight of their shining fruits; it was among the treasures introduced from China by E. H. Wilson in 1901, for the firm of Veitch. Farrer says that the berries of *V. fragrans* "can be eaten with avidity and good result" so long as the kernels are discarded; those of the guelder-rose are said to be sometimes fermented and eaten—"a statement" says the Rev. C. A. Johns "which will seem scarcely credible to anyone who has chanced to smell them."[2] But the berries of the laurustinus, which are "of an excellent pale bright blew colour," were reported by Parkinson so to burn and inflame the mouth and throat "that it is almost unsupportable, and not but in a long time, and by drinking milke, and holding it in the mouth, to be taken away or eased."

"Viburna" (in the plural) seems to have been a name anciently applied to any shrubs used for binding or tying; the shoots of the Wayfaring-tree, "easie to be

[1]Farrer, *The Journal of the Royal Horticultural Society*, October 1916

[1]Quoted by Nuttall, *Beautiful Flowering Shrubs*, 1944 [2]*Flowers of the Field*, 1853

bowed and hard to be broken," were used for basket-making and for tying faggots until quite recent times. The leaves of *V. coriaceum* (syn. *V. cylindricum*) are waxy on the underside, and were used long ago by leisurely Chinese gentlemen to send rural messages to their friends in town; characters scratched on the back of the leaf with a sharp point turn white, and remain visible for several days. The other sorts carry only dumb mes-

Vitex agnes-castus
(CHASTE TREE)   Ferdinand Bauer (1830)

sages; the laurustinus signifies "I die, if neglected," and the Snowball-tree, the "Winter of Age."

NOTES: Most of the viburnums in the nursery trade today are the old standards discussed by Miss Coats. But the cultivars developed by the Dr. D. Egolf of the National Arboretum are marked improvements. Egolf began his viburnum work as a graduate student and devoted himself to this important genus, emphasizing larger flowers, improved habit and better fruiting. His selection of *V. plicatum* var. *tomentosum* 'Shasta' is perhaps his outstanding contribution to the genus to date. All told, Dr. Egolf released seventeen selections of various Viburnum species.

Viburnums are easy to cultivate in good garden soil and respond well to a light application of garden fertil-

izer in early spring. They prefer a sunny position in the shrub border. Given adequate care, viburnums will reward the gardener with handsome displays of large creamy white flowers in late spring and red or black fruits in autumn to attract birds. Viburnums are among our most valuable garden shrubs.

Vitex

"There is a kind of tree named Vitex, not much different from the Willow," wrote Pliny; "the Greeks, some call it Lygos, others Agnos, i.e., chaste; for that the dames of Athens, during the feasts of the goddesse *Ceres*, which were named Thesmophoria, made their pallets and beds with the leaves thereof, to coole the heat of lust, and to keep themselves chaste for the time." All parts of the plant, *V. agnus-castus*, were believed to have "peculiar sedative properties"; Turner said that the seed "both fried and not fried, stayeth the desyre to the pleasure of the bodie," and that the flower and the leaf had the same effect, either eaten or drunk, or when "strowed all about wher folke trede." Other improbable medical virtues were also attributed to it by Greek and Roman physicians.

With this reputation, it is not surprising that the Chaste Tree—a member of the Verbena order and a native of Sicily—should have been an early introduction to Britain. Chaucer refers to it, as an attribute of the virginal Diana; and we know from de l'Obel that it was grown in English gardens before 1570. Broad and narrow-leaved kinds were recorded in the eighteenth century, but on the whole there is little variation. Useful, aromatic, ornamental in leaf and flower, the shrub nevertheless is little grown today; not so much because it is not completely hardy, as because it blooms so late in the year that in some seasons the flowers fail to open. In good years and in favorable situations it well deserves the Awards of Merit that have been granted both to the blue- and the white-flowered forms. A few other species in the family have been cultivated, such as the Chinese *V. negundo*, which was grown by the Duchess of Beaufort in 1697; but most of them are only suitable for the greenhouse.

The ancient Latin of "Vitex" is said to be derived from *vieo*, to plait or weave, because of the pliability of the shoots, which are still used by Greeks and Cretans for basketmaking. The "absurd officinal name" of *agnus-castus* is a repetition, the Greek *hagnos* having the same meaning as the Latin *castus*—a double chastity; it has nothing to do with the Latin *agnus*, a lamb. By a curious misapprehension, this shrub has given its name to the totally unrelated medicinal Castor Oil. It seems that when the true Castor-oil plant, *Ricinus communis*, was introduced to cultivation in Jamaica in the eighteenth century, it was confused by the Portuguese and Spaniards with the Vitex (their recollection of which must have been remarkably hazy, for the resemblance is slight) and called by them "*agno-casto*"; from which in turn the English planters and London traders got the name of "castor."

NOTES: The Chaste-tree is not a common garden shrub, probably because its hardiness range limits it to the mid-Atlantic states where it must compete with numerous other desirable plants. Only this one species has been cultivated in our gardens to any extent.

A newly introduced species from Japan and Korea has recently come to our attention. This is *V. rotundifolia*, a procumbent, sandbinding, woody plant that may have some use as a wall plant in coastal gardens. I first introduced it in 1955 and again with S. March and F. Meyer during our 1978 National Arboretum exploration of the Japan Sea coastal region of Japan. More recently, the National Arboretum collecting team brought material from Korea and it is now secure in cultivation but still rare in nurseries.

Vitis

As a plant of utility the vine has been grown for some sixty centuries; as a plant of ornament, barely for one. Grapes were cultivated and wine made in Egypt about 4,000 B.C., and by the Romans in Britain at the end of the third century A.D.; but it was only when English vineyards had been finally abandoned (except by a few devoted amateurs) and the cultivation of dessert grapes had long been relegated to the greenhouse, that the vine began to be appreciated as a decorative plant. In 1863 Johnson wrote that few were worth growing except the grapevine (*V. vinifera*) and its varieties—"the other species are valued chiefly in this country as botanical curiosities"; and most of the handsome Oriental kinds have only been introduced since that date.

Among the "curiosities" cultivated at a relatively early period were one or two forms of *V. vinifera*, the most ornamental being the Purple-leaved Vine (var. *purpurea*), believed to be identical with the "Teinturier" grape, which is one of the oldest kinds recorded in Britain. Parkinson in 1629 had a "Teint" grape, "whose juice is so deepe a colour, that it serveth to colour other wine." The Parsley-leaved Vine (var. *apiifolia*) followed shortly

after (1640) with "leaves more jagged and deeply-cutt" and a large white berry like the White Muscadine, "and of as good a rellish." Another very distinct variety was Miller's Grape (var. *incana*) sometimes listed as the "Meunier" or Dusty Miller, on account of its downy appearance, but, according to Loudon, actually so called because it was raised by Philip Miller, about 1720. Loudon also mentioned two variegated forms which were

Vitis vinifera
(WINE GRAPE)   J. Abbott (1797)

"more ornamental than useful."

The beauty of the autumn coloring of certain varieties is occasionally mentioned, and sometimes lists are given of the most decorative kinds. William Speechly, in *A Treatise on the Cultivation of the Vine* (1779) makes the interesting point that "whenever the least tint of red, purple or scarlet appears on the leaves of the Vine at the time of their maturation, it is a certain sign that the grape will be grizzly, red or black"—thus making it possible to distinguish between white grapes and dark ones, when still in the seedling stage. Generally speaking, the darker the grape, the more spectacular the autumn color of the leaf.

Several American species were early introductions, *V. aestivalis*, *V. labrusca* and *V. vulpina* (syn. *V. riparia*) all being grown here by 1656; the last, the Riverside Grape, was particularly renowned for the sweet mignonette-like scent of its flowers. But on the whole, little use was made of the vine as a plant of ornament until the Asiatic kinds began to arrive; *V. amurensis* from Manchuria in 1854, *V. davidii* from China in 1885 and *V. thunbergii* from Japan towards the end of the century. The most spectacular both for size of leaf and for flaming autumn color was *V. coignetiae*. Seeds of this vine were brought to France from Japan in 1875 by the Mme. Coignet after whom the species was named; but owing to the difficulty with which it was propagated, and the restrictions imposed on account of the vine-disease of phylloxera, it did not spread beyond France. A plant procured from Japan by the firm of Anthony Waterer through the East India merchants Jardine and Matheson was presently identified as belonging to this species, but again difficulties of propagation prevented any wide distribution. At last, in 1893, a consignment of seed was imported, and the plant successfully launched in cultivation; it is said, however, that the seedlings raised were not so handsome as Waterer's original form, which was for long the glory of the Knap Hill nursery.

The ornamental vines, so late to arrive, were also early to depart; they are little grown now, because most of them require a great deal of space, and they refuse to adapt themselves to walls and gardens that get smaller and smaller. (Even the common grapevine, it is well known, will in time grow to a great size, and send its roots to a corresponding depth.) One species, however, is more restricted in its demands—*V. flexuosa*, a neat and elegant Chinese vine, introduced in 1880, which might well be more often grown, especially in its variety *parvifolia*.

The name Vitis is said to come from *viere*, to twist, for obvious reasons:

> Sweet Ivy, bend thy boughs, and intertwine
> With blushing Roses, and the clust'ring Vine—

but some authorities give it a much older origin, through the Greek *voinos* from the Semitic *yain* or *wain*. The Virginia Creepers, once classed as vines, are now to be found under Parthenocissus.

NOTES: Only one grape species commands ornamental garden use and even then it is not common. *V. coignetiae*, a hardy grape from northern Japan, received considerable attention from the famous landscape architect Beatrix Farrand for its brilliant red autumn color. Despite its high praise as the best grape for ornamental use, it rarely has gotten beyond the confines of aboretums.

Wiegela, see under *Diervilla*

Wisteria

This is another of the families whose members occur only on the eastern side of North America, and in China or Japan. As usual, the first to be discovered was one of the American kinds—*W. fruticosa*, which was introduced by Mark Catesby in 1724 under the name of "Carolina Kidney Bean." It never seems to have become a very popular plant; it proved hardy, but not very floriferous. The nurseryman Conrad Loddiges had a fine specimen trained on his house at Hackney, which bore profusely in some years, and was probably much admired—until the oriental species were introduced; after which, the American wisteria is no more heard of.

The Chinese wisteria, *W. sinensis*, was described in a letter from a French Jesuit missionary[1] in 1723; but little more was heard of it until about 1812, when it was seen by John Reeves growing in the garden of a Canton merchant called "Consequa"—an anglicized version of his Chinese name. He was one of the eleven "hong" agents permitted to trade with the foreigners, and seems to have been a generous, easygoing character who was shockingly swindled by his English and American clients, and died about 1823, reduced to penury and misery.

It seems to have been Robert Fortune who first saw this wisteria growing truly wild; he found it on the island of Chusan, during his first expedition in 1843–6, growing on trees and hedges by the sides of the narrow mountain roads. He sent home from the same expedition the first plant of the beautiful white variety, which as late as 1898 was regarded by Hibberd as "still in the nature of a little stranger," though by that time the rather unsatisfactory double blue and double white kinds were also in cultivation.

The Chinese wisteria can be trained as a standard; and the Japanese wisteria (*W. floribunda*) should always be grown in this way, or on a trellis, for it neither flowers nor displays its beauty so well, when trained against a wall. The plant so popular in Japan (where it has an enormous literature) is a cultivated form, *W. floribunda* var. *macrobotrys* (syn. *W. multijuga*) with very long racemes of flowers—said in some cases to reach 7 feet in length[2]—but with smaller and more widely spaced individual blooms than those born by the Chinese plant. Mr. Collingwood Ingram has described the famous specimen at Ushijima near Kasukaba, supposed to have been

[1] *Dominicus Parennin*, quoted by Bretschneider [2] Wilson saw none that exceeded 5 feet 6 inches

planted over a thousand years ago by the priest Kobo Daishi; in 1920 this venerable plant had a trunk 32 feet in circumference, and covered a trellis approximately 400 square yards in extent with upwards of 80,000 close-hung trusses of flowers. Nothing like this is yet to be seen in Europe, where this species was only introduced by Siebold at the comparatively recent date of 1830; how and when it reached England has not apparently been recorded. This wisteria too has white, pink and double forms.

When the first species was introduced from America, the botanists of the day allotted it to the Glycine family (whose most notable member is *G. soja*, the Soya Bean) and named it *Glycine fruticosa*. This family, however, was large and heterogeneous, and in 1818 Thomas Nuttall created a new genus for the plant, and named it after Professor Caspar Wistar, M.D., of the University of Pennsylvania—"a philanthropist of simple manners and modest pretensions, but an active promoter of science"—who had died earlier the same year. Wistar was of German descent, and his grandfather's name was Wüster; when it became anglicized it was spelled Wistar by one branch of the family and Wister by another. Nuttall himself wrote the plant's name with an "e" and the Professor's name with an "a." "Glycine" came from the Greek *glykys*, sweet—the source of our word "glycerine." In France the wisteria still keeps its original name, and it was a pardonable confusion that led the two ladies in *Northbridge Rectory*,[1] with happy memories of a Mentone villa called *Les Glycines*, to name their residence "Glycerine Cottage." Darwin noted that wisteria shoots twine counterclockwise; and that while his own plant, grown in a pot, "tried in vain for weeks to get round a post between five and six inches in thickness," a plant at Kew easily managed a trunk more than 6 inches in diameter. NOTES: Famed in Japan since ancient times, the wisteria is symbolic of the pleasures of the Japanese strolling garden. In addition, the wisteria is important as a bonsai, as standard plants in tree form, and for flower arrangements. There are numerous selections in Japan. Curiously, the Chinese and American species climb by twisting from left to right while the Japanese species twists from right to left. Our American species, *W. frutescens*, grows along the borders of streams and lowland woods along the coast from Virginia to Florida but is not cultivated, illustrating again that at present it is the oriental counterpart species that are the garden plants of choice, a result of their long history of use in Chinese and Japanese gardens.

Perhaps the main problem with wisteria is encourag-

[1]By Angela Thirkell

ing the plant to bloom. Recommended treatments include severe root pruning, girdling the trunk, withholding fertilizer, severe pruning and any other manner of intimidating the plant. One specimen in the Brooklyn Botanic Garden has defied attempts for several decades. In the South, wisteria has escaped into the roadside canopy and makes a handsome display. However, it can eventually strangle a tree with its thick twining stems.

Yucca

Like the Butcher's Broom, but perhaps more recognizably, the Yuccas are members of the noble family of the Lilies; and all the species—supposed to number about thirty—are confined to the southern United States, Central America and Mexico. It is natural that representatives of so striking and conspicuous a group of plants should have been among the earliest introductions from America to Europe; what is strange is that any of the species should tolerate our climate. A good half-dozen, however, have proved hardy or nearly so; before 1836 Messrs. Backhouse were able to list at least five main kinds and some minor varieties which they had grown successfully in their nursery at York.

Y. gloriosa, the Adam's Needle, Spanish Bayonet, Mound Lily or Palmetto Royal of the Florida keys, was the first kind to be introduced. A plant of it was brought to Gerard in 1593 "by a servant of a learned and skillful Apothecarie of Excester, named Master *Thomas Edwards*"; it grew and prospered, but had not yet flowered at the time of the publication of the *Herball* in 1597, and he could only describe the leaf, which "with advised eie viewed, is like a little Wherrie or such like bote," tough, prick-pointed, and green summer and winter, "notwithstanding the injurie of our colde climate, without any coverture at all." Gerard died in 1612, and his plant "perished with him that got it from his widow, intending to send it to his Country house"[1]; but during his lifetime he had given a slip to Jean Robin at Paris whose son, Vespasien Robin, gave a plant to John de Franqueville, who in turn gave one to Parkinson. Meantime other specimens had reached the country, for we hear of a plant (origin unknown) in flower for the first time in 1604, in the garden of William Coys at North Okington in Essex. Yucca flowers were then so rare as to cause some excitement in the horticultural world; in 1633 Gerard's editor Johnson recorded that another had bloomed "in the garden of Mr. Wilmot at Bow, but never since, though it hath been kept for sundry yeares in

[1]Parkinson

many other gardens, as with Mr. Parkinson, and Mr. Tuggy."

When at last it flowers, this yucca is truly *gloriosa* (though Andrews regarded the word as a "metaphysical hyperbole, very inapplicable to any plant, however beautiful"),[1] but some of the later introductions have proved more generally useful—the Silk Grass, *Y. filamentosa*, for example, which is even hardier, and blooms more regularly and at an earlier age. This species is reported to have flowered in 1675 in the garden of George Crook of Waterstock near Oxford; it was over and in seed when

Wisteria sinensis
(CHINESE WISTERIA)   E. D. Smith (1827)

seen by Robert Morison of the Oxford Botanic Garden, who mentioned that the same kind had also been cultivated for many years by a Mr. Walker in "his suburban garden in the village of St. James . . . but I never saw it flower there." It received its name because of the threads that hang from the leaf-margins. Another early introduction was the tender *Y. aloifolia* which was grown in the gardens of Hampton Court in 1696, and its variety *draconis*, grown in James Sherard's garden at Eltham, before 1732. The latter was also grown by Miller, from seed that was sent to him "by the Title of Oil Seed" and

was one of the kinds that Backhouse found hardy at York.

There was a long pause before any other hardy yucca of importance was introduced, and then three came close together; *Y. recurvifolia* in 1794, *Y. glauca* in 1813, and *Y. flaccida* about a year later. The first, found in Georgia by Lecomte, resembles *Y. gloriosa* in its tree-like habit of growth, with a stem that lengthens yearly; the other two, like *filamentosa*, form no stem, and keep their rosettes firmly on the ground.

Evelyn pointed out that the yucca was hardier than was supposed, and it was unnecessary to bring it into the conservatory for "hyemation"; and he suggests that being easily propagated, it might be used as the Aloe was used in Languedoc, to make "one of the best and most ornamental fences in the world for our gardens, with its natural palisadoes . . . But we believe nothing improvable, save what our grandfathers taught us." Both Loudon and Robinson prized it for producing exotic effects; Loudon thought it might be planted in vases, as a substitute for the Agave, "in imitating Italian scenery round an Italian villa." In 1836 Andrews said that though eminently suitable, the yuccas were never seen in seaside gardens—an omission that has since been rectified.

A nineteenth-century American botanist declared that the flowers of *Y. filamentosa* not only looked their best by moonlight, but actually delayed their blooming until the moon was full. "This flower is made for the moon, as the Heliotrope is for the sun, and refuses other influences, or to display her beauty in any other light." At night the drooping bells erect themselves and expand, and the "transparent, greenish-white leaves, which look dull by day, are melted by the moon to glistening silver."[1] All this only signifies that the special family of moths on which the yucca depends for its fertilization is a night-flying one. *Tegeticula yuccasella* can no more exist without the yucca, than the yucca can set fertile seed without the services of this interesting insect—or of man; for in countries like our own, where the moth does not exist, the flowers can be successfully pollinated by hand, and many hybrids have been raised by this means. The principal yucca cultivators were Continental; first J. B. Deleuil of Marseilles (from 1874) and then Carl Sprenger, a nurseryman of Vomero near Naples, who raised more than a hundred varieties between 1897 and 1907. Unfortunately, few of their products survived—at least, in this country. A more recent arrival is Jackman's *Y. flaccida* variety, 'Ivory', whose flowers maintain their erect night-time position during the day.

---

[1] *The Botanist's Repository*, vol.7

[1] *The Autobiography and Memoirs of Margaret Fuller Ossoli*, 1852

The plant was named "Yucca" because Gerard wrongly believed it to be the Manihot or Iucca, from the root of which cassava (and, incidentally, tapioca) is made. The error was corrected by his editor Johnson, in 1633, and by Parkinson in 1640, but by that time the name was already established. Parkinson said that the natives of Virginia used yucca roots for bread in much the same way as the natives of "Hispaniola" used the roots of the true Hyjucca or Manihot, but later authors say that this was done only in times of extreme scarcity, "and when more grateful roots fail them."[1] The fibers of *Y. filamentosa*, the Silk Grass, were used by the Indians for the making of clothes; but in 1748 it was reported that they had become accustomed to getting textiles from Europe, and the method of preparation of the yucca fabrics had been forgotten. Our sailors' beds, on the other hand, are probably derived from the "hamacks" which the Indians wove from yucca-cordage. The fruit of some sorts is edible, that of *Y. gloriosa* being "of the form and size of a slender cucumber, which, when ripe, is of a deep purple colour . . . and of an agreeable aromatic flavour, but bitter to the taste"[2] and purgative if eaten to excess. The hard spines in which the leaves of this species terminate are formidable defences; "Aldinous[3] relateth," said Parkinson, "that the wound made by the sharp point ende of one of these leaves in his own hand, wrought such intolerable paines, that he was almost beside himselfe, untill by applying some of his owne *Balsamum* thereto, it miraculously eased him of the anguish, and all other trouble thereof."

The most romantic species of yucca is not in general cultivation. This is the Joshua Tree, *Y. brevifolia*, which grows in Nevada, California and Arizona, and was first discovered in the Mojave Desert in 1844. It is said to have been given its name by the Mormons in Utah, in the erroneous belief that its leaves all pointed one way, and would serve as guides across the waste, as Joshua was guided: "Thou shalt follow the way pointed for thee by the Trees." It grows to a height of about 30 feet, and considerable forests of it still exist in some areas, looking from a distance like nightmare apple-orchards. It is thought to be the oldest species in the genus, and looks it, growing in the most fantastic and demented shapes, so that it is no surprise to learn that it is proved to have formed part of the food of the extinct *Nothoterium* or Giant Ground-Sloth. The panicles of flowers can weigh as much as fifty pounds, and the creamy, leathery petals are a quarter of an inch thick; they have a curious earthy

smell which some find distasteful, and which persists even after the flower has been immersed for months in formalin solution.

NOTES: Only *Y. filimentosa*, native along the eastern seaboard and into the mountains in the Southeast, is cultivated to any extent. Known as Adam's needle, or beargrass, numerous cultivars appeared in Europe around the turn of the century and, as Miss Coats points out, have become aspects of "the vanishing garden." Today, other than escaped plants dug from abandoned fields or roadsides, only the variegated form is offered

Yucca filamentosa

(SILK-GRASS)　　　　　　P. J. Redouté (1802)

in the nursery trade. The cultural range of Yucca far exceeds its natural distribution, making it a plant for the sunny location.

The variegated form 'Gold Sword', popular in the nursery trade, has yellow leaves with green edges. This is far more dramatic in the garden than the common green type. It is a rugged plant and tolerant of poor soil and water conditions.

[1] Bryant [2] *Travels through North and South Carolina, etc.*, William Bartram, 1792 [3] Physician and Controller of Gardens to Cardinal Edvardo Farnese at Rome

Some Brief Biographies

*. . . Such great names never fail to excite lively and pleasing emotions; it is the
ignorant and unfeeling alone who can ridicule the associations of the names of travel-
lers and naturalists with those of animals and plants.*
J. D. Hooker, *Himalayan Journals*, 1854

THE PIONEERS

The personalities associated with the early history of the flowering shrubs
are the same as those associated with herbaceous plants; we depend for our
information on the same authorities and such introductions as are recorded
were made through similar channels. But toward the end of the seventeenth
century the specialists began to appear—botanists and gardeners who, though
wide-ranging in their tastes, are particularly remembered for their collections
of foreign trees and shrubs. Reference to some of the more outstanding of
these personages is frequently made in the text, and it was thought that some
further information about them might be of interest.

HENRY COMPTON (1632–1713) was among the pioneers. Born at
Compton Wynyates in Warwickshire of a noble Royalist family, he became
Bishop of Oxford in 1674, and Dean of the Chapels Royal in 1675; and was
promoted to the bishopric of London on the death of its previous holder, in
the same year. As Dean of the Chapels Royal one of his functions was to
supervise the education of the young princesses, Mary and Anne; he took
up the task with enthusiasm, and inspired them both with a love of protes-
tantism and of gardening that they were never to lose. He was not on the
best of terms with their father, the Duke of York (later James II), and in
the Revolution of 1688 supported William of Orange, then the husband of
his pupil Mary. His other pupil Anne, afraid (with reason) that the King was
preparing to whisk her off to France, appealed to Compton for help; to
which he responded in the best cloak-and-dagger tradition, carrying her off
at midnight in a hackney coach, accompanied by Sarah, Duchess of Marlbor-
ough and their attendants. Later, as their flight to the north turned into a
triumphal progress, he himself preceded the cavalcade, "in a buff coat and
jackboots, with a sword at his side and pistols in his holsters"[1]—remembering,
perhaps, a short period before he entered the church, when he had been a
cornet in the Guards. In this adventure he was assisted by his ex-gardener,
George London, who had left him seven years before to become the founder
of the famous nursery firm of London and Wise, and who was rewarded for
his share in the escape by the post of Page of the Back-stairs to Queen Mary.

This picturesque episode was of minor importance in Compton's long and
notable career; he played a large part in political and ecclesiastical history,
but it is only his gardening activities that concern us here. "This Reverend
Father was one of the first that encouraged the Importation, Raising, and
Increase of Exotics, in which he was the most curious Man in that Time";[2]
and there were few days in the year that he was not in his garden at Fulham,
"ordering and directing the Removal and Replacing of his Trees and Plants"[3]
As a collector he had two great advantages: he resided for thirty-eight years
in one see, which gave time for the collection to grow, and as Head of
the Church for the American colonies, he had special opportunities for the
importation of American plants. He is said to have grown over 1,000 species
of exotics under glass, and about half as many hardy trees and shrubs, which
were always accessible to interested botanists—"As himself takes great delight
in observing the same, so doth he freely admit others, curious in Botanism,
to do the like."[4] The upkeep of all this was costly, and Compton seems to
have been troubled in conscience on account of his garden expenditure, and
to have contemplated the sale of all his curious and exotic plants, in order
to have more money for his numerous charities. (At his death his plants and
his library were almost his only assets.) His style of gardening was not
appreciated by his successor, Bishop Robinson, who cut down many of Compton's choice trees and cleared away his rarities, to make room for "the more

ordinary productions of the kitchen garden,"[1] The gardener was allowed to
sell what he pleased, and fortunately some items in the collection were saved
through being purchased by the enterprising nursery firms of Furber and
Gray.

JOHN BAPTIST BANISTER (1654–92) was one of the chaplains sent
out by Compton, first to the West Indies, and then to Virginia, where he
became one of the Bishop's most enthusiastic collectors. He compiled a catalogue of American plants, which appeared in the second volume of Ray's
Historia Plantarum (1680) and was the first study ever to be published of the
American flora. He was interested in all aspects of science, and contemplated
writing a book on the natural history of Virginia; some beautiful drawings
which he sent (together with dried specimens) to the botanist Petiver, were
probably intended for this work. It was never accomplished; Banister was
killed by a fall from a height when collecting on one of his excursions, at
the comparatively early age of thirty-eight. Among the seeds and plants he
sent home were the first of the American azaleas, the Swamp Magnolia (*M.
virginiana*), *Echinacea purpurea* and *Dodecatheon meadia*.

MARY SOMERSET (1630?–1714) first Duchess of Beaufort, was one of
the earliest of Britain's distinguished lady gardeners. Very little is known
about her personal life; she was widowed in 1699,[2] but did not lose her love
of gardening, for in 1702 she was particularly anxious to obtain the services
of William Sherard as a tutor for her grandson, "hee loving my diversion so
well," and Sherard added more than 1,500 plants to her collection. (She
would probably have been annoyed had she known that Aiton[3] was to date
her introductions from the year of her husband's death, not of her own).
Switzer tells us that two-thirds of her time, "excepting for the times of her
Devotions," was spent in the gardens at Badminton or Chelsea, tending her
collection of thousands of exotics, drawn from all over the world, which she
"kept in a wonderful Deal of Health, Order and Decency." Contemporaries
praised her skill with tender or difficult subjects, "Her Grace having what
she called an Infirmary or small greenhouse, to which she removed sickly or
unthriving plants, and with proper culture by the care of an old woman
under her Grace's direction, brought them to greater perfection than at
Hampton Court or anywhere."[4] Her herbarium in twelve volumes,
bequeathed to Sir Hans Sloane, is preserved in the British Museum, and a
two volume florilegium containing paintings of some of her treasures is in
the library at Badminton. Unfortunately the majority of her introductions
were tender greenhouse plants; but we owe to her the first zonal pelargo-
nium, the ageratum and the blue passion-flower.

All three pioneers are commemorated in plant names—Comptonia, Banis-
teria and Beaufortia. It is curious that the shrub named after the Bishop—
the American Sweet Fern, *Comptonia asplenifolia*—was first grown not by him,
but by the Duchess of Beaufort; for so far as is known, there was no connection between them. One would expect that two such keen gardeners, almost
exact contemporaries in age, and both having gardens in or near London,
would be sure to be friends, or at least rivals; but no record of any acquain-
tanceship seems to have survived.

THE GENTLEMEN OF SCIENCE

The eighteenth century was a wonderful age for scientific exploration and
discovery, and it brought to the fore two new figures—the professional full-
time collector and naturalist, and the noble patron or patrons who supported
him. The large gardens and parks then being laid out gave ample scope for
the planting of new tree and shrub introductions, which came pouring in,
particularly from America, in response to the demand.

MARK CATESBY (1679–1749) was one of those men, Pulteney assures
us, "whom a passion for natural history very early allured from the interesting pursuits of life"[5]—meaning such occupations as might have led to prefer-
ment, riches and honor. He migrated to London from his native Essex, but

---

[1] Carpenter, *The Protestant Bishop*, 1956   [2] Switzer, *Ichnographia Rustica*,
1742 edition   [3] Quoted by Carpenter   [4] Newton, quoted by Britten,
The Sloane Herbarium

[1] Watson, *Phil. Soc. Trans.*, 47   [2] According to Rees and Aiton; 1700
in the D.N.B.   [3] In *Hortus Kewensis*   [4] Sloane, quoted by Britten,
The Sloane Herbarium, 1958   [5] Pulteney, *Sketches*, 1790

little is recorded of him before 1712, when he embarked for his first visit to Virginia, where his sister was married to Dr. William Cocke, the Secretary of State. Catesby stayed in America for seven years, and during this time sent dried specimens and living plants to friends at home, including the nurseryman Fairchild and the botanist Samuel Dale. His drawings and specimens were brought to the notice of William Sherard and Sir Hans Sloane, with the result that after his return to England in 1719 he was induced by these and other scientists to go back to America "with the professed design of describing delineating and painting the more curious objects of nature";[1] and a subscription was raised to defray his expenses. This time he was away for four years (1722–6) visiting Carolina, Georgia and Florida, and making a prolonged stay in the Bahamas on the way home. He then settled down, at first in Hoxton and afterwards at Fulham, to the production of his great work *The Natural History of Carolina, etc.*, which occupied—and supported—him for the remainder of his life. Finding that it would be very costly to get his drawings engraved in Paris or Amsterdam, he took lessons in etching and prepared the plates himself. The drawings were made from specimens "fresh and just gather'd" and the pictures colored by hand, either by himself or under his own supervision. They were published in sets of twenty, at two guineas a set; the first appeared in 1730, and the last—a twenty-plate appendix to the two previous volumes, each containing a hundred plates—in 1748.[2] In order to economize in space and labor, birds and insects—sometimes even fishes—were included on the same plates as the plants; and if the results were sometimes a little naïve—as Catesby himself admitted, he was not bred a painter—they set the pattern that a much greater draughtsman and naturalist, John James Audubon, was to follow a hundred years later. The admiration of Catesby's contemporaries was unqualified; Kalm thought the plants and animals "incomparably well represented . . . so that no-one can say they are not living, where they stand with their natural colours on the paper."[3] One can see that he coveted a copy, but the two volumes at that time cost from 22 to 24 guineas—"therefore, not for a poor man to buy." Kalm visited Catesby in May 1748, and found him still at work on his masterpiece, though now very short-sighted and nearly seventy years of age. He died in December 1749; a letter from Dr. Mitchell dated May 1750 informed Linnaeus that Catesby was dead, "after having completed his book, which is already fallen in price, being now sold at seventeen guineas and a half."[4]

In 1733 Catesby became a Fellow of the Royal Society, among whose members he was "greatly esteemed for his modesty, ingenuity and upright behaviour,"[5] and to whose *Transactions* he contributed a paper, much in advance of his time, on the migrations of birds.

Money to cover the production costs of his great work was lent to Catesby without interest by PETER COLLINSON (1694–1768), "to enable him to publish it for the benefit of himself and family, else through necessity it must have fallen a prey to the bookseller."[6] Collinson was a Quaker, the son of a linen-draper of Gracechurch Street. At the age of two, he was sent to be brought up by relations at Peckham, who imparted to him their own love of gardens, and it was in Peckham that his first garden was established; he began keeping a record of the plants he grew there, in 1722. Kalm, who saw it in 1748, described it as "full of all kinds of the rarest plants, especially American ones, which can endure the English climate and stand out the whole winter. However neat and small this garden was, there was, nevertheless, scarce a garden in England, in which there were as many kinds of trees and plants, especially the rarest, as this." In 1724, having some time before inheriting the family business, Collinson married Mary Russell, whose father owned a considerable property at Mill Hill near Hendon; as early as 1737 he obtained permission to plant some of his new introductions in his father-in-law's garden, but even so, when he ultimately inherited Ridgeway House in 1749, it took him two years to transfer the treasures from his Peckham garden to his new home.

Collinson made use of the foreign connections established by his prosperous business to obtain seeds and plants from abroad; he corresponded with many foreign scientists, including Benjamin Franklin, Bouffon, Gronovius

and Linnaeus, but his most fruitful association was his long partnership with John Bartram. He had so pestered his American correspondents to send him seeds and plants—"what was common with them (but Rare with us) they did not think worth sending"—that at last, as he said himself, "some more artfull Man than the Rest contrived to get rid of my importunities By recommending a Person whose Business is should be to gather seeds and send over plants. Accordingly John Bartram was recommended as a very proper Person for that purpose, being a native of Pennsylvania with a numerous Family—the profits arising from Gathering Seeds would Enable Him to support It." The correspondence began about 1730, and continued until Collinson's death. It was arranged that Bartram should send over some twenty boxes a year, each containing a hundred different species of seeds and plants, at a price of five guineas apiece, for which Collinson should find purchasers; he in return sent clothes, books and European seeds and plants. Lord Petre was the first to patronize the scheme, and was followed by the Dukes of Richmond, Norfolk and Bedford, and others; so that Bartram became, in effect, the first professional plant collector to work for a syndicate of garden owners. It involved Collinson in a great deal of trouble and expense—correspondence, accounts, delayed payments, visits to the Customs, dispatch of cases and losses at sea—for which he was repaid only by the delight of himself and his friends in the new acquisitions.

Collinson was "of amiable manners, very communicative, and of a very benevolent heart,"[1] and "Persons in a superior station to his own, treated him with familiarity and respect."[2] Apart from his plant introductions, he conferred many benefits on the nation, both through his generosity to individuals such as Mark Catesby, and through more public works—he was instrumental, for example, in the establishment of Cambridge Botanic Garden and the founding of the British Museum. But his last years were troubled; his business almost totally declined (perhaps owing to neglect?) and at the age of seventy-four he was obliged to solicit a small pension from the Government—which was refused. Between 1765 and 1768 his garden sustained three major robberies, many of his most valuable plants being stolen and much damage done. After his death the property fell into "the most barbarous and tasteless hands"[3] until taken over in 1797 by R. A. Salisbury, who solicitously rescued such trees and plants as remained; it subsequently became a school.

JOHN BARTRAM (1699–1777) was also a Quaker; he was born, as we have seen, in Pennsylvania, of a Derbyshire family which had emigrated seventeen years before. Having inherited some land from a bachelor uncle, he began life as a farmer; the story goes that he was struck one day by the beauty and complexity of a flower torn up by his plough, and could not rest until he had learned Latin (which he mastered in three months) in order to take up the study of botany. In 1728 he bought a property on the Schuylkill River, enlarging the house that stood on it with his own hands, to accommodate his already numerous family,[4] and began the planting of a botanic garden. By 1740 his consignments of seeds and plants to Collinson had become a "settled Trade and Business" and he made long collecting trips into the interior of the country, at that time very little known to Europeans. He also corresponded with other botanists, including Phillip Miller and Sir Hans Sloane, and was persuaded by Dillenius to take an interest in mosses, which he had hitherto regarded "as a cow looks upon a pair of new barn doors; yet now . . . I have made good progress in that branch of Botany, which really is a very curious part of vegetation."[5] In 1765 Collinson was able to secure for his friend the appointment of King's Botanist—rather a barren honor, for the utmost salary that his influence could obtain was £50 a year.[6] "Thou knows the length of a chain of fifty links," he wrote to Bartram, "Go as far as that goes, and when that's at an end, cease to go any further."[7] Bartram managed to make it stretch as far as Georgia and Florida, on a much wished-for journey hitherto beyond his reach; it was in the course of this, his last and longest expedition, that he found the rare and beautiful

---

[1] Smith, in Rees' *Cyclopaedia*   [2] Two of the plates were contributed by Ehret   [3] *Kalm's Account of his Visit to England*   [4] Smith, *Linnaean Correspondence*, ii   [5] Pulteney, *Sketches*   [6] Brett-James, *Peter Collinson*

[1] Lemprière, *Universal Biography*, 1808   [2] Smith, in Rees' *Cyclopaedia*   [3] Smith, *Linnaean Correspondence*, i   [4] He had in all nine children, two of whom died in infancy   [5] Quoted by Britten in *The Sloane Herbarium*   [6] A rival collector of very inferior merit, a protégé of the Queen, had £300; but George III had little interest in botany   [7] Brett-James, *Peter Collinson*

tree, *Franklinia altamaha*, even then confined only to one small locality, and now believed to be extinct in the wild. After Collinson's death in 1768 Bartram continued to send occasional consignments to English friends, but shipments and pension were alike cut short by the outbreak of the Revolutionary War in 1776; and Bartram's own death was hastened by suspense at the approach of troops after the battle of Brandywine—he feared they would ravage his garden. The garden, however, survived this and other vicissitudes, and though long engulfed by the city of Philadelphia, is preserved as part of a public park, with Bartram's house and some of his trees still standing.

Bartram was accompanied on his Florida trip by his fifth son WILLIAM BARTRAM (1739–1823) the only one to follow in his father's footsteps. William was a talented draughtsman, and Collinson procured for him some commissions for drawings of objects of natural history. Among his patrons was the Quaker doctor and keen plantsman John Fothergill, who in 1772 agreed to finance William for another trip to the south; he was to receive £50 a year and his expenses, and drawings were to be paid for separately. He "sat off," in his own phrase, in 1773, and was away until 1778. His account of this expedition, *Travels through North and South Carolina, Georgia, etc.*, was published in 1791, and was widely read; Bartram's descriptions of plants and scenery were vivid and his outlook romantic, and the book had a considerable influence on writers such as Coleridge, Wordsworth and Southey. He seems to have got on well with the Indians, and was proud of the title they gave him—Puc Puggy, the Flower-hunter;[1] but in the remaining forty-five years of his life he never made another journey, nor wrote another book.

Perhaps none of the foregoing—Collinson the merchant, Bartram the farmer, Catesby, his own artist and editor—would have laid claim to the title at the head of this section; but there is no doubt about the standing of JOSEPH BANKS (1743–1820) who has been called "the aristocrat of the philosophers." He began to take an interest in botany while still a schoolboy, and having inherited £6,000 a year from his father at the age of eighteen, devoted the rest of his life with single-minded ardor to the advancement of science. In 1766 he went on a trip to Newfoundland and Labrador, and two years later he accompanied Captain Cook on his voyage around the world, in the capacity of botanist—an adventurous journey during which he showed great courage, initiative and endurance. In 1772 he went to Iceland, and brought back from there the blocks of lava that were afterwards incorporated in the rock garden of Chelsea Physic Garden. This was his last voyage; he then settled down to become the dominating figure in the botanical world for the next fifty years. He was made Honorary Director of Kew in 1772 and President of the Royal Society in 1778, both of which posts he held until his death, and was knighted for his services to science in 1781. Except for two small species from Newfoundland (the rest of his collection having been swept overboard in a gale), he made no direct introductions—he knew where plants were to be found, but sent others to fetch them; like the Duchess of Hertfordshire, he "sat aloft, and beckoned desirable specimens up."[2] He was responsible for sending Masson to South Africa and Kerr to China; Menzies in California and Goode in Australia worked under his instructions, and collectors trained by him at Kew traveled all over the world. It was Banks who in 1787 dispatched Bligh in the ill-fated "Bounty," to take the breadfruit tree from the South Sea Islands to the West Indies; after the famous mutiny, a second and successful attempt was made in 1790 under the same commander, but the ungrateful West Indians subsequently showed little appreciation of the fruit that had been procured for them with so much trouble.

For thirty-eight years of Bank's lifetime England was at war, usually with France; but in those civilized days science was universally regarded as having no concern with political squabbles, and the movements and affairs of botanists were little affected, except for the additional risk of loss at sea. Banks was always most punctilious about extending help to foreign botanists in difficulties, and returning captured natural history collections to their proper owners; he is said to have been instrumental in getting no less than ten such collections restored to the Jardin des Plantes, "which had fallen prey to our naval superiority."[3] He had a finger in every botanical pie in Europe, and it

is remarkable how many valuable Continental herbariums found their way to Britain through his agency, after the death of their owners—including that of the great Linnaeus himself. His own collection and library at Soho Square, looked after by his assistant Solander and his librarian and taxonomist Robert Brown, were available to all visiting botanists and were eventually transferred to the British Museum; his garden at Spring Hill was also a place of pilgrimage for all interested in plants.

Solander, formerly a pupil of Linnaeus, had been Banks' companion on his voyage with Captain Cook. Linnaeus was so impressed by the quantity of new species brought back from this trip (many of them from the place named on that account Botany Bay) that he suggested that the newfound continent should be named Banksia. But Captain Flinder's name of Australia being preferred, the name of Banksia was given instead to an important genus of Australian plants—unfortunately not hardy.

THE NURSEYMEN

In the eighteenth century, the nurserymen began to play an important part, not only as propagators, but as introducers of exotic plants. At first it was only a matter of the skillful cultivation of material sent home by various travelers, but as the century waned firms began to arise of sufficient size and importance to send out their own collectors, and to embark on extensive programs of breeding and hybridization. The nurserymen also had the important function of training gardeners and collectors, and clients frequently applied to them for persons qualified to look after the plants they purchased.

CHRISTOPHER GRAY (c. 1694–c.1768) was the first to grow foreign exotics on a large scale; he had a nursery on both sides of the King's Road at Fulham, where he specialized in American plants. Information about his life is scanty, and comes mostly from Collinson, who tells us that Gray purchased part of Bishop Compton's collection from his successor. Collinson estimated his age at "about seventy" in 1764; if so, Gray must have been a very young man at the time of the Bishop's death in 1713. Catesby, Collinson, Miller and Dr. Garden were among those who contributed plants to his nursery. The catalogue of trees and shrubs for sale that Gray published in 1740 is said to have been the first nurseryman's list; and he had one of the first specimens of *Magnolia grandiflora* in the country. The date of his death is not known. In 1763, a book called *Hortus Brittano-Americanus* made its appearance, nominally by Mark Catesby (who had died thirteen years before) but actually a compilation based on his *Natural History of Carolina*, which according to some authorities was produced by Gray. A second edition of this work was published in 1767; and it seems as though Gray was not only still living at that date, but was very much alive to the value of advertisement, for it is mentioned in the Introduction "that Mr. GRAY at Fulham[1] has for many years made it his business to raise and cultivate the plants of America (from whence he has annually fresh supplies) in order to furnish the Curious with what they want; and that through his industry and skill a greater variety of American forest-trees and shrubs may be seen in his gardens, than in any other place in England." Reference is made to "Christopher Gray's nursery" as late as 1777.

JAMES GORDON (1708?–81). All that is known of Gordon's early life is that he was a Scot who worked as a gardener for Dr. James Sherard and Lord Petre. After the latter's death in 1743, Gordon started a nursery at Mile End, with a seed-shop in Fenchurch Street. He very soon became renowned for his skill in raising foreign plants from seed, especially the more difficult subjects. "Before him" wrote Collinson in 1763, "I never knew or heard of any man that could raise the dusty seeds of the Kalmia's, Rhododendrons or Azalea's. These charming shrubs . . . he furnishes to every curious garden; all the nurserymen and gardeners come to him for them; and this year, after more than twenty years' trial, he shewed me the Loblolly Bay[2] of Carolina coming up from seed in a way not to be expected . . . and his sagacity in raising all sorts of Plants from cuttings, roots and layers surpasses

---

[1] Properly speaking, the full title should have been *Puc Puggy Faya*—Puc Puggy meaning "a flower" only   [2] Max Beerbohm, *Seven Men*, 1919
[3] Edward Smith, *Life of Sir Joseph Banks*, 1911

[1] This might, however, refer to a son; a "William Gray, junior" also of Fulham, is mentioned as one of the original subscribers to Miller's *Gardener's Dictionary*   [2] Subsequently named *Gordonia* in his honor

all others by which our gardens are enriched . . ." It was not only American plants that were successfully propagated; the Chinese maidenhair-tree was first grown in Gordon's nursery, and it was he who introduced the camellia[1] and the gardenia to British commerce. After his death in 1781, the firm continued as Gordon, Dermer, Thompson and Co., then as Gordon, Forsyth and Co., and finally as Thompson and Co., until the land was requisitioned for building, about 1839.

About the time that Gordon was starting his nursery at Mile End, another Scot was founding an even more famous establishment at Hammersmith. This was JAMES LEE (1715–95). He was born in Selkirkshire—perhaps of gentle kin, for he brought with him to England a fine Ferrara sword, which is still treasured by his descendants. He walked south about 1732, and after a break in the journey caused by an attack of smallpox at Lichfield, took service in London as a gardener, first with the Duke of Somerset at Syon, and afterwards under the Duke of Argyll at Whitton. In 1745 he went into partnership with a rather shadowy Lewis Kennedy, and started a nursery on ground that had formerly been a vineyard. The firm seems to have been slower to become established and to have received less patronage than Gordon's; Collinson, after speaking of the help he had given to the latter, refers briefly to "the ingenious *Mr. Lee* of Hammersmith, who, had he the like assistance, would be little behind him."

In Scotland Lee had received a sound basic education, and during his service with the Duke of Argyll is said to have been allowed the run of his library. He made such good use of his opportunities that in 1760 he was able (with some gratefully acknowledged assistance) to bring out his *Introduction to Botany*. This was largely a translation of Linnaeus' *Philosophia Botanica*, and was the first exposition of the new Linnaean system to be published in English. It became very popular, running eventually into ten editions, and brought Lee and his nursery a great deal of publicity and renown. The business steadily expanded; a seventy-six page catalogue issued in 1774 offered seeds and plants for every department of gardening, from kitchen-garden produce to hothouse exotics. (Lee is credited with having been the first to practice artificial fertilization, to obtain seeds from his greenhouse plants; but it is not recorded that he made any intentional crosses.) He was among the first to receive seeds from Botany Bay, and thereafter specialized in "New Holland" plants, but he also grew many South African species, and seems to have had a special fondness for mesembryanthemums.

Lewis Kennedy died in 1783, and was succeeded by his son John, a much stronger personality, who played a large part in the history of the firm. He was called in by the Empress Josephine to advise on the planting of her garden at Malmaison, and thereafter traveled frequently to France, in spite of the fact that the Napoleonic Wars were then in progress. Meantime Lee, too, had died; but under his son, a second James, the Vineyard Nursery went from strength to strength. Plants worth up to £2,600 were purchased by Josephine in 1803 alone, while the Marquis of Blandford's bill the following year came to more than £15,000. Collectors were sent to North and South America, and one (shared by Josephine) to South Africa; the original nursery-ground in Hammersmith was augmented by departments in Fulham, Kensington, Feltham, Stanwell and Bedfont, and later in Ealing, Isleworth and Hounslow. Kennedy retired in 1813, but James Lee II carried on with undiminished vigor; Loudon in 1822 said that the Vineyard was "unquestionably the first nursery in Britain, or rather the world," in spite of considerable competition, by this time, from the younger firm of Loddiges.

James Lee II had four sons, whom he intended for four departments of the firm—the seed business, the counting house, the greenhouses and the hardy trees and plants. But he died in 1824, when only his first son, John (b. 1806), was old enough to enter the business; after that, the importance of the firm gradually declined, though its impetus carried it on until John's death in 1899. Long before then, the various branch nurseries had been surrendered one by one to builders, the original Vineyard plot in Hammersmith becoming the site of Olympia.

CONRAD LODDIGES (1743?–1826) was a Dutchman who settled about 1761 among the Nonconformists of Hackney. Ten years later he took over a nursery whose previous owner, a German called John Busch, had just been appointed gardener to Catherine the Great of Russia. Loddiges of Hackney soon became famous for the cultivation of new and rare plants; but the firm rose to its greatest heights early in the nineteenth century, after Conrad's son George (1784–1846) had entered the business. George Loddiges devised a system of heating glasshouses by steam, and also a method of producing "artificial rain" which won him a medal from the Horticultural Society in 1817. Both heating and watering systems were installed in the "largest greenhouse in the world" (80 × 60 × 40 feet) which Loddiges erected about this time to house their collection of palms, of which they had more than eighty species. Camellias and greenhouse exotics were their specialties, but they also had a display nursery of hardy trees and shrubs arranged in alphabetical order, in beds ingeniously devised to show them to the best advantage. "Such a collection of Plants as is in possession of Messrs. Loddiges does not exist elsewhere in the world" wrote Johnson in 1829; "The stock if sold at retail price is worth £200,000. The gardens and houses contain 8,000 species, exclusive of varieties. . . . Of Hardy Trees and Shrubs they have 2,664 species."

In 1826 Conrad Loddiges died, but the firm continued to prosper for another twenty years under George's able and enterprising management. In 1833 and 1834 it assisted in making the first tests of the new Wardian Case for the transport of plants by sea, and it supplied much of the material for Loudon's *Arboretum et Fruticetum Britannicum* (1838). George Loddiges died in 1846, and the firm was closed about ten years later.

The success of Lee's *Introduction to Botany* encouraged other nurserymen to turn to authorship as a form of advertising. Loddiges' contribution took the form of a serial production called *The Botanical Cabinet*, which appeared in twenty volumes between 1818 and 1833. Many of the 2,000 illustrations were done by George Loddiges; most of the rest were by George Cooke, a connection by marriage. The botanical information given in this work is scanty, but the pictures are charming and the moral tone impeccable. "The astonishing productions of plants" were expressly designed to afford Man "an inexhaustible source of the purest and most innocent pleasure"; and George Loddiges naturally wished that they should be "diffused throughout the world, that all may participate that have a mind capable of delighting in them."

It was a dynasty rather than a family that was founded by JOHN VEITCH (1752–1839) when he came from Jedburgh in Roxburghshire to Killerton in Devon, at the end of the eighteenth century. After working for some time as land steward to Sir Robert Ackland, he rented a piece of land near Killerton in 1808, and started the nursery business that was to make the name and fortunes of five generations of Veitchs, and last more than a hundred years. Finding Killerton too remote, John Veitch and his son James and his grandsons James II and Robert, moved in 1832 to Mount Radford near Exeter, henceforth to be known as the Exeter Nursery. In 1853 the firm, now James Veitch and Sons, purchased the business of Knight and Perry of the King's Road, Chelsea; James Veitch II moved to London to look after the new branch, while the Exeter nursery was kept up by his father until his death, and afterwards by his younger brother Robert and his descendants. The London business thrived,[1] and further grounds were added at Feltham, Langley and Coombe Wood, near Kingston. When James Veitch II died rather suddenly in 1869, the business was carried on by two of his sons, John Gould and Harry James; but John died only a year later, and it fell to Harry to keep the firm going until John's two boys, James Hubert and John Gould II, were old enough to take charge. James Hubert died in 1907, and this time there were no more sons to inherit; the aging Harry returned for a time, but in 1914 the lease of the Coombe Wood nursery expired, and the business was dissolved shortly before the outbreak of the First World War.

The great importance of this dynastic firm lay in the fact that they maintained almost from the start a constant flow of collectors; no other firm has combed the world for contributions on so large a scale. In the period between 1840 and 1905 they sent out twenty-two plant hunters, including three members of the family. The most outstanding contributions of hardy plants were made by the first and the last—William Lobb, who worked in Chile and California between 1840 and 1857, and E. H. Wilson, who made two trips

---

[1] A camellia variety afterwards raised in this nursery was named 'Mile-Endii'

[1] Its connection with the Exeter firm was served in 1863.

to China, 1899–1902 and 1903–5. Some valuable introductions were also made by John Gould Veitch, who visited Japan in 1860, when it had only recently been opened to Europeans, and by Charles Maries, in Japan and China from 1877–80. Messrs. Veitch were also notable for the large-scale raising of hybrids, especially of begonias and orchids; their success both as introducers and hybridists can be gauged by the frequency with which *"veit-chii"* and *"exoniensis"*[1] still appear in horticultural works and on trade lists.

DOCTORS IN JAPAN

Japan was rigorously closed to Europeans from 1639 to 1858, the only key-hole being the entire flowery kingdom being the tiny island of Deshima in the harbor of Nagasaki, where the Dutch East India Company was granted a trading station. During the whole of that period, the only information that could be gained about the Japanese flora came through three or four doctors in the service of the company, whose profession gave them special advantages. Even those botanists who got as far as Deshima had to depend almost entirely on such plants as they could get brought over from the mainland, and these were usually cultivated garden flowers, many of which were afterwards found to be natives of China, not of Japan. Some of the earliest information came from Dr. Andreas Cleyer, a colorful character who was on Deshima from 1682 until 1686, when he was banished by the Japanese for smuggling. He purchased a collection of 739 drawings of Japanese plants and was the first to describe the "Japanese" anemone.

ENGELBERT KAEMPFER (1651–1728) was born at Lemgo in Lippe, and received his first schooling at Hameln, though more than three centuries too late to be beguiled by the Pied Piper. Later he went to the universities of Cracow, where he took a degree in languages and philosophy, and Königsberg, where he studied natural history and physic. In 1683, in the course of a visit to Sweden, he obtained the post of secretary to the Swedish Ambassador to Persia, and accompanied him through Russia to Isfahan, where he stayed for two years. He then became Chief Surgeon to the Dutch East India Company, and sailed with their fleet to Ceylon, Sumatra and Batavia. From there he went to Japan, where he lived on Deshima for nearly two years, and twice accompanied the annual Dutch embassy to the Court at Tokyo. (On Deshima he obtained the friendship of Japanese officers and interpreters by giving them medical advice, instructions in mathematics and astronomy, and "a cordial and plentiful supply of European liquors.") In 1692 he returned to Batavia, and in 1693 to Amsterdam; he took his degree in physic at Leyden in 1694, and then went back to his native Lemgo. There he became physician to the Count of Lippe, got married, and set up in practice, and was so much in demand, that he had very little time to prepare his notes on his travels for publication. In 1712 he produced the *Amoenitates Exoticae* in five books, the fifth of which was devoted to the plants of Japan; it was intended only as a preliminary to later works that never materialized. Many plants now familiar in our gardens were described for the first time by Kaempfer—under their Japanese names, for they had no others until classified by Linnaeus about half a century later. Kaempfer's only other book was a *History of Japan,* which was not published until after his death; a copy of it belonging to Sir Hans Sloane was translated into English and published in two volumes in 1728. In it Kaempfer asserts that the fruits of Japan lack flavor, and the flowers "fall as short of others of their kind, growing in other countries, in strength and agreeableness of smell, as they exceed them in the exquisite beauty of their colors."

CARL PETER THUNBERG (1743–1828) was the next botanist to penetrate Japan's defences. The son of a Swedish clergyman, he studied botany under Linnaeus, and graduated from the University of Upsala in 1770. The following year, "by the interference of his friend Burmann,"[2] he obtained a post with the Dutch East India Company—nominally as a surgeon, but his medical duties never seem to have hampered his botanical researches. He went first to Cape Town, where he spent three years, learning Dutch in readiness for his assault on Japan, and studying the rich local flora, to such good effect that he is frequently called the "Father of South African Botany." In 1775

he proceeded to Java and thence to Japan. On Deshima "the jealousy of the inhabitants and the mistrusting conduct of the government limited his excursions,"[1] but he was able to make one strictly supervised two-month visit to Tokyo, though he was not allowed to penetrate further inland. It is said that much of his knowledge of Japan's wild flora was gained from studying the weekly hay supply that was brought over to feed Deshima's animals. He left Japan in 1776, and after a prolonged visit to Ceylon, returned to Europe in 1779. Linnaeus had died during his absence, and in 1781 Thunberg succeeded him as Professor of Botany at Upsala. His *Flora Japonica* was published in 1784, and contained descriptions of over 300 plants not previously known.

PHILLIP FRANZ von SIEBOLD (1796–1866) was a more forceful and perhaps less scrupulous character than either of the foregoing; he not only succeeded in entering the forbidden country, but was instrumental in opening its closed doors to others. He was born in Wurzburg, of a scientific family, and graduated as a physician specializing in eye diseases in 1820. Two years later he entered the employment of the Dutch East India Company, and accompanied their embassy to Japan in 1823. His skill as an eye specialist was greatly in demand among the Japanese, and he used the influence this gave him to obtain many unusual privileges. (He may also have been helped by the fact that he married a Japanese girl.) He was permitted to visit Tokyo, where he made a large collection of Japanese plants; he also collected other things, including a map of Japan, the possession of which by a foreigner was a treasonable offence. His Japanese confederates were beheaded, and he was confined to Deshima for two years; but when he returned to Europe in 1830, he managed to bring with him a collection of 485 plants—the first considerable influx of living Japanese plants to reach the west. Half of them were intended for the gardens of his patron, the Duke Ursul, near Brussels; but unfortunately Siebold disembarked at Antwerp just at the time that war was breaking out between Holland and Belgium. His collections were confiscated, and though some of the plants were eventually restored to the botanic and nursery garden he established at Leyden, others fell into the hands of the horticulturists of Ghent. The final effect, however, was the same—the new flowers became available to European gardens.

At Leyden, Siebold built up a considerable plant importation business, with the help of his Japanese connections; he also wrote several books on the language, customs and natural history of the country, including (in collaboration with J. G. Zuccarini) a two-volume *Flora Japonica* (1835) with handsome plates in color, European in style but executed by unnamed Japanese artists. Meanwhile the march of events was rendering it impossible for Japan to maintain her isolation, and in 1859 Siebold was asked by the Dutch government to return to Japan and use his influence to mediate between that country and Europe. He rapidly regained and even increased his former privileges, but grew arrogant, abused his position, and was again obliged to leave. He returned to his native Germany in 1864, and died two years later. One of his more amiable actions (in 1826, during his first visit) was to erect a monument in Deshima Botanic Garden to the memory of Kaempfer and Thunberg.

COLLECTORS IN CHINA

China was almost as strictly closed to Europeans as Japan, though not for so long a time. From 1755 to 1842 foreign trade was limited to the port of Canton, and the township and environs of Macao, which had been a Portuguese settlement since 1537. The political difficulties of plant importation were equalled by the difficulty of keeping the subjects alive during the long sail home. In 1819 it was estimated that only one Chinese plant in a thousand survived the journey, and the average price being 6s 8d, the survivor had cost more than £300. Nevertheless the camellia, hydrangea, japonica, Yulan magnolia, tree-paeony, wintersweet and several roses were introduced by various hands before the end of the eighteenth century.

Collecting on a large scale began immediately after the opening of the treaty-ports in 1842, when anti-European feeling was still high; it never became one of the safest of occupations, and the life of a collector of Chinese plants was seldom devoid of incident. Fortune beat off attacks by pirates with

---

[1] = "of Exeter"   [2] Lemprière, *Universal Biography,* 1808. Johannes Burmann was for many years Professor of Botany in Amsterdam

[1] Lemprière

the gun that the Horticultural Society had reluctantly allowed him; Wilson hunted for his Handkerchief Tree in country disturbed by the Boxer uprising; Forrest escaped with difficulty from murderous Tibetans, and Kingdon-Ward survived one of the worst earthquakes ever recorded. Their books qualify for classification under "Travel and Adventure" rather than under "Botany"; yet none of these collectors made less than four trips to the scene of their dangers and triumphs.

JOHN REEVES (1774–1856) belonged to an earlier period, and was not in any sense an adventurer—a modest tea broker, born in West Ham, who sailed to China in 1812 as Assistant Tea-Inspector to the East India Company, and stayed there for nearly twenty years, with only two short holidays at home. He lived at Canton during the tea season, and for the rest of the year at his house and garden at Macao. He was an enthusiastic garden lover and was soon in correspondence with Sir Joseph Banks and officials of the Horticultural Society, and sending home seeds and plants. By a mixture of tact and influence he persuaded the reluctant captains of the Company's tea clippers to take an interest in the consignments he put into their charge, and even to compete with each other in getting them safely home. He established the plants in pots in his own garden months beforehand, and hardly a ship left without some of his "little portable greenhouses" on deck. Even so, there were many losses. When Reeves himself came home on leave in 1816, he brought with him a hundred plants, ninety of which survived the journey; but when the plants were sent in the charge of others, the percentage of casualties was much higher.

It was taken as a matter of course that these imports would consist of Chinese garden plants—the only kinds available. "I have tried in vain," Reeves later reported, "to get the gardeners at Fa-te[1] to collect their own wild plants, of which they have so many beautiful ones . . . nothing will drive them out of old custom." Apparently he succeeded in getting a few, for in 1823 it was noted that he intended to send "a case of wild plants of the hardiest kind."

In 1815 the Horticultural Society decided to form a collection of plant drawings, and from about 1819 Reeves commissioned Chinese artists to make a series of drawings under his own supervision, which were sent to the Society and eventually became known as the Reeves Collection. These drawings were among items sold during a period of near-bankruptcy in 1859; but five volumes of them have been recovered and are now in the Lindley Library.

In 1831 Reeves left China and retired to Clapham; but one of his greatest services to horticulture was still to be rendered. In August 1842, at the end of the Opium War, the treaty was signed that ceded Hong Kong to Britain, and opened four more ports to trade—Amoy, Fuchow, Ningpo and Shanghai. Reeves realized that this afforded a golden opportunity to obtain new plants, and he urged the Horticultural Society to send out a collector. A Chinese Committee was formed, of which he was the moving spirit; and before the end of the year (when the Government Department concerned was still advising caution and delay) the collector was chosen and his equipment settled.

ROBERT FORTUNE (1812–1880) was the man whose application was accepted, and seldom can the choice of an untried candidate have been so amply justified. He was a Scot, born at Kelloe in Berwickshire, and had worked for more than two years in the Edinburgh Botanic Garden before being recommended to the post of superintendent of the hothouses at the Horticultural Society's gardens at Chiswick, only a few months before the Chinese opportunity arose. He set sail in the *Emu* on February 26, 1843 (being then thirty years of age), armed with a long list of instructions, equipment that included a fowling piece and pistols and a Chinese vocabulary, and a salary of £100 a year plus (limited) expenses. He also took with him for trial a number of the new Wardian cases—an invention which was to revolutionize the transport of plants for long distances, and which helped to bring about the success of his introductions. He made his first base at Hong Kong, and later went on to Amoy, Chusan and Shanghai. The country was still too unsettled for him to venture far from a port—except in disguise—and most of the introductions from this first journey were of garden plants; but he reaped a wonderful harvest from the hitherto untouched riches of the more northerly regions.

It seems to have been expected that when Fortune returned to England (in 1846) he would go quietly back to his former job in the glasshouses; but he broke with the Horticultural Society, and we next hear of him employed as the Curator of Chelsea Physic Garden—but not for long. In 1848 he sailed again for China, this time under the auspices of the East India Company and with the object of procuring the Chinese teaplant and introducing it to India. This was no easy task, as the plant and its method of preparation had always been a jealously guarded Chinese monopoly; but Fortune achieved it, and it was largely thanks to him that the tea industry of India was founded.[1] In 1853–6 he made another trip, also on behalf of the East India Company and in 1860–2 a fourth, this time independently, or perhaps in collaboration with the nurseryman Standish of Bagshot, to whom many of his plant consignments were sent. This was his last voyage, and it included a visit to Japan.

After each of his journeys Fortune wrote a book giving an account of his adventures, and also much information about China and the Chinese, valuable at a time when so little about the country was known. He is said to have introduced nearly 190 species or varieties of plants, 120 of which were new to botany or horticulture, and brought home herbarium specimens which included a further twenty-five new species.

During the latter half of the nineteenth century, botanists and collectors of various nationalities made inroads into Chinese territory, following up with great rapidity every opportunity afforded by the shifting political scene; but none of them equalled Fortune's contributions to horticulture. An important influence, however, was exerted by AUGUSTINE HENRY (1857–1930) a Customs and Medical Officer in the Chinese Customs Service, who was stationed in 1882 at Ichang, a hundred miles up the Yang-tse Kiang, and later at outposts still more remote. He started to collect plants in order to relieve the boredom of his leisure hours, and eventually became the greatest botanical authority of the day on the Chinese flora. The dried specimens he sent to Kew aroused great interest among horticulturists. When some inquiry was made to Henry about the possibility of procuring plants and seeds, he replied "Don't waste money on postage—send a man"; and this advice was followed by Harry James Veitch, who sent out a collector in 1899, with special instructions to obtain seeds of the Davidia or Handkerchief Tree.

ERNEST HENRY WILSON (1876–1930) was the person chosen for this exploit. He was born at Chipping Campden, and served his apprenticeship in Hewitt's nursery at Solihull in Warwickshire. For a time he worked at Birmingham Botanic Garden, and then went on to Kew, whose Director, Sir W. T. Thistleton Dyer, recommended him to Veitch. Wilson was seven years younger than Fortune when he first set sail for China, but he was much more carefully trained for the work he was to do. First of all he spent six months in Veitch's Coombe Wood nursery, learning their requirements; then he was sent to China by way of America, so that he might visit the Arnold Arboretum near Boston, and study the latest methods of collection, packing and transport; finally, he was instructed to seek out Henry at Szemao in southwest Yunnan, in order to obtain local information, including the whereabouts of the only known Davidia tree. During this journey Wilson underwent many dangers and disappointments, but came triumphantly home after three years with a rich cargo of plants and seeds; and after only a few months at Coombe Wood, set off again for another trip, this time to Szechwan and the border of Tibet, which lasted from 1903–5. He was a much more wholesale collector than Fortune—Dr. Fairchild speaks of his "systematic dragnet methods"—and these two trips yielded seeds of about 1,800 species of plants, besides 3,000 bulbs, an unspecified quantity of living roots and rhizomes, and some thousands of herbarium specimens.

During his visit to the Arnold Arboretum, Wilson became friendly with its Keeper, Professor Charles Sprague Sargent—himself a traveler and collector, and an authority on woody plants; and when Wilson again went to China, in 1907–9, it was as an employee of Harvard University, to which the Arboretum belongs. Henceforward he lived in America, and collected for the Arboretum and for American subscribers; but such was the generosity of Professor Sargent in distributing the seeds Wilson sent home, that England received almost as much benefit from his later expeditions as his adopted

[1] The nursery-garden area, about three miles from Canton

[1] Another variety of the tea-plant was afterwards introduced from Assam

country. He made a fourth trip to China in 1910–11, and it was on this occasion that his leg was broken in two places by a fall of rock, when he was returning from collecting bulbs of the magnificent Regal Lily in the Min River Valley, three days' journey from the nearest mission post. At one time it was feared that the leg would have to be amputated; fortunately it was saved, and he was left with only a slight limp.

After this, Wilson went no more to China, but he made two fruitful journeys to Japan, Korea and Formosa (in 1914 and 1917–8). In 1919 he became Assistant Director to the Arnold Arboretum, and when Professor Sargent died in 1927, Wilson succeeded to his post of Keeper. He published several books, of which *A Naturalist in Western China* (1913) is probably the best known. After surviving all the hazards of travel in the Orient, he was killed in a car accident in Massachusetts.

Wilson introduced more trees and shrubs than any other collector, and nearly all his finds were characterized by great hardiness. He was very successful in getting them safely home, though some of his early introductions were afterwards lost through the dissolution of the firm of Veitch. No less than a hundred of the plants he introduced won awards from the R.H.S.

GEORGE FORREST (1873–1932) takes us back again to the gardeners' homeland; he was born at Falkirk, and educated at Kilmarnock Academy. He went to work in a chemist's shop, and received some pharmaceutical training, which included botany, elementary medicine and first aid—all of which proved very useful later on. But feeling, perhaps, the need for wider horizons, he left the shop and emigrated to Australia, where he led a roving life for some years. Returning by way of South Africa in 1902, he applied for a job at Edinburgh Botanic Garden, and accepted the only post then available, that of an assistant in the herbarium, where he had to sort and classify thousands of dried specimens of plants from all parts of the world.

About this time Mr. Arthur Kilpin Bulley, a wealthy Liverpool cottonbroker and founder of the firm of Bees Seeds Limited, was laying out a fine garden at Ness on the Wirral peninsula, and was anxious to obtain foreign plants. He wrote to missionaries in distant places and asked them to collect the seeds of wayside flowers, but the chief result was "the best international collection of dandelions to be seen anywhere,"[1] so he decided to send out his own collector. He inquired of his friend Sir Isaac Bailey-Balfour, the Director of the Edinburgh Botanic Garden, for a suitable person; and Sir Isaac recommended Forrest.

In 1904, therefore, with very little training or experience, George Forrest set out on his first Chinese journey; and it was very nearly his last. His destination was in the borderland between Yunnan, Burma and Tibet; Chinese and Tibetans were then in arms over frontier disputes, and both were hostile to foreigners. In the summer of 1905 a party of Tibetans descended on the French mission station where Forrest was staying; the inhabitants fled—but too late; they were overtaken, and of eighty persons, only about a dozen survived. Forrest escaped with his life, after a nine-day nightmare of starvation and pursuit, but lost all his possessions, including the large collections he had already made. Only three months later, however, he was exploring the neighboring Salween valley—even more remote and unknown, and penetrating still farther into Tibet; and he stayed another year before returning home in 1907. He made six more Chinese journeys, but no other was quite so perilous. The second (1910) was again undertaken for Bulley, but the third (1912–14) was largely financed by Mr. J. C. Williams of Caerhays, Cornwall and a syndicate of other garden owners; and the later expeditions[2] were partly subsidized by the Rhododendron Society (founded in 1915).

Forrest was the first to train and organize a team of Chinese collectors to work under his direction, which enabled him to cover greater areas and to make his introductions on a very large scale. From his last trip, for example, he brought back "approximately two mule-loads of good clean seed, repressing some 400–500 species; and a mule-load means 130–150 lb." Altogether he made 31,015 seed-collectings, including 5,375 of rhododendrons, over 300 of which were classed as new species by his old chief, Balfour of Edinburgh, to whom he sent beautiful herbarium specimens of his new finds. Forrest brought us some of our finest plants; but owing to the more southerly region in which he worked, not all of his introductions are as hardy as those of Wilson.

Forrest was not voluble, either about his finds or his adventures; he wrote no books and few articles. Perhaps he would have turned author in his retirement; but he collapsed and died suddenly of heart failure, at the end of what he intended as his last collecting journey. He had been satisfied with the results of this trip, and had remarked in a letter, "I have made a rather glorious and satisfactory finish to all my past years of labour."

In 1911, Forrest having been engaged by Williams and his syndicate, Bulley again applied to Balfour for a collector for his own use, and this time Balfour recommended the son of a friend of his, Harry Marshall Ward, a Professor of Botany at Cambridge. The young man, FRANK KINGDON-WARD (1885–1958) was already in China; after taking an honors degree at Christ's College he had accepted the first post that offered travel in the Orient, and was teaching at a school in Shanghai. He had already accompanied one expedition (albeit a zoological one) and had sent home some botanical specimens. He joyfully took advantage of the opportunity that Bulley provided, and embarked on a career of plant hunting that was to continue, except for periods of Army service during two world wars, for forty-seven years.

Kingdon-Ward did not, like Forrest, employ native labor to help with the work; he preferred to see every plant for himself, and if possible to mark especially fine forms from which to collect seed in due season. He was also a keen geographer, and mapped, surveyed and photographed many of the remote places where he traveled. (He has acknowledged receiving help from the reports of a secret agent, who provided a sketch map of a river in an area otherwise completely blank.) His first two expeditions were to north Yunnan and the borders of Tibet; but as Forrest and the American Dr. J. F. Rock were also working in that area, Kingdon-Ward moved farther west, and except for two trips in 1921 and 1922, when he returned to Yunnan, made all his journeys after 1914 to Assam, Upper Burma and south Tibet.

Twice during his long career Kingdon-Ward refused a permanent official post as a botanist. After 1948, when he was over sixty, he was offered retirement and a small Government pension; but he could not give up his nomadic life (he had no settled home) and after that made five more journeys, on four of which he was accompanied by his intrepid second wife Jean Macklin,[1] whom he married in 1947. (She was with him in the Lohit valley during the devastating earthquake of 1950.) He was still planning further excursions when he died, rather suddenly, in 1958, at the age of seventy-three. In all, this indefatigable man went on twenty botanical or geographical expeditions, during which he made 23,068 collectings of seeds. He introduced another hundred new rhododendrons, some collected "blind"—that is, from plants he had never seen in bloom, and did not recognize when encountered afterwards in an English garden. He was also a fluent writer, and published twenty-three very readable books, from *The Land of the Blue Poppy* (1913) to the posthumous *Pilgrimage for Plants* of 1960.

Kingdon-Ward was the latest—let us not say the last—of the great collectors. Sir Frederick Stern[2] found the perfect text to serve as his epitaph:

"The range of the mountains is his pasture, and he searcheth after every green thing." Job, XXXIX, 8.

[1] Mrs. Bulley, quoted in the *Journel of the Royal Horticultural Society*, 1960, p. 218 [2] Their dates were 1917–19, 1921–2, 1924–5, 1930–2

[1] After whom *Lilium mackiniae* is named   [2] In *A Chalk Garden*, 1960

Recent Ornamental Plant Explorations

The following is a list of ornamental explorations since the Second World War. The U.S. Department of Agriculture by itself and with collaborating institutions has produced the most impressive record of sustained effort collecting and introducing ornamental plants into cultivation. While other institutions have sent collectors into the field on an occasional basis and a number of private individuals have made collecting trips, none have provided the continuity engendered in the collecting trips listed here.

These expeditions are collected here in two separate lists: the U.S.D.A./

Longwood Gardens Cooperative program and then those under auspices of the National Arboretum in collaboration with other gardens. Under the U.S.D.A./Longwood Gardens Program, thirteen explorations were completed between 1956 and 1971. One collecting trip in 1955, which lasted eight months, was a prelude to the Longwood cooperation. After this cooperative program was terminated in 1971, similar efforts were initiated at the National Arboretum five years later. Through 1989, the Arboretum led or participated in thirteen explorations to China, Japan and South Korea.

U.S.D.A. and U.S.D.A./Longwood Gardens Sponsored Explorations

Name of Collectors	Date	Locality
J. L. Creech	1955	Okinawa to Hokkaido, Japan
J. L. Creech	1956	Southern Japan, Yakushima
F. G. Meyer	1957	Southern Europe
G. H. Spaulding	1958–9	Australia
L. O. Williams	1958	Brazil
F. G. Meyer	1959	Northern Europe
J. L. Creech	1961	Japan, Hong Kong
J. L. Creech and F. de Vos	1962	Nepal
J. L. Creech and D. H. Scott	1963	USSR (fruits/ornamentals)
F. de Vos and E. G. Corbett	1964	India, Sikkim
E. G. Corbett and R. W. Lighty	1966	South Korea
J. L. Creech	1967	Taiwan, Hong Kong
H. F. Winters and J. J. Higgins	1970	New Guinea
J. L. Creech	1971	USSR, Siberia

National Aboretum Cooperative Explorations

J. L. Creech and S. G. March	1976	Japan, Yakushima
W. L. Ackerman	1977	Japan, Okinawa
J. L. Creech, F. G. Meyer, and S. G. March	1978	Japan Sea coast
T. R. Dudley	1980	China
R. Jefferson	1981	Japan
M. Kawase, S. G. March, F. N. Meyer, D. Nielsen	1982	Hokkaido, northern Honshu
R. Jefferson	1982–3	Kyushu-Hokkaido, Japan
B. Yinger, S. G. March, D. Apps, P. Bristol, P. Meyer	1984	South Korea
B. Yinger, T. Dudley, J. C. Raulston, A. P. Wharton	1985	South Korea
R. Jefferson	1986	Japan, Taiwan
T. Dudley	1987	China
L. Lee	1988	China
S. G. March and P. Meyer	1989	South Korea

Selected Bibliography

A Man will turn over half a library to make one book
Dr. Johnson

Abel, Dr. Clarke   *Narrative of a Journey in the Interior of China, etc.*   1819
Abercrombie, John   *The Universal Gardener and Botanist*   1778
Anderson, A. W.   *The Coming of the Flowers*   1950
Arber, Agnes   *Herbals*   1938

Bartram, William   *Travels Through North and South Carolina, etc.*   1792
Bean, W. J.   *Trees and Shrubs Hardy in the British Isles*   1914–33
Bélon, Pierre   *Les Observations de Plusieurs Singularitez, etc.*   1554
Bowles, E. A.   *My Garden in Spring*   1914   *My Garden in Summer*   1914   *My Garden in Autumn and Winter*   N.D.
Bretschneider, E.   *History of European Botanical Discoveries in China*   1898
Brett-James, Norman G.   *The Life of Peter Collinson*   N.D.
Britten, J., ed. Dandy, J. E.   *The Sloane Herbarium*   1958
Brook, Richard   *A New Family Herbal*   N.D.
Bryant, Charles   *Flora Diaetetica*   1783
Bulleyn, William   *Bullein's Bulwarke of Defence, etc.*   1562

Carpenter, E.   *The Protestant Bishop*   1956
Catesby, Mark   *Hortus Europae-Americanus* (2nd edn)   1767   *The Natural History of Carolina, etc.* (2nd edn)   1754
Cobbett, William   *The English Gardener*   1833
Cowan, J. MacQueen   *The Journeys and Plant Introductions of George Forrest*   1952

Cox, E. H. M.   *The Introductions to England of Reginald Farrer*   1930
Cox, E. H. M. and Cox, P. A.   *Modern Shrubs*   1958
Culpeper, N.   *The Complete Herbal and English Physician Enlarged*   1653 edn

Dawson, Warren R.   *The Banks Letters*   1958
Dillwyn, L. W. ed.   *Hortus Collinsonianus*   1843
Duhamel du Monceau   *Traité des Arbres*   1760

Evelyn, J., ed. Hunter   *Silva, or a Discourse on Forest Trees*   1776 edn

Fairchild, David   *Exploring for Plants*   1930   *The World was my Garden*   1938
Farrer, Reginald   *On the Eaves of the World*   1917   *The Rainbow Bridge*   1921
Fernie, W. T.   *Herbal Simples* (1895)   1914 edn
Folkard, Richard   *Plant Lore, Legends and Lyrics* (2nd edn)   1892
Fortune, R.   *Three Years' Wanderings in the Northern Provinces of China*   1847

Gerard, John   *The Herball, or Historie of Plants*   1597
Gerard, ed. Johnson   *The Herball, or Historie of Plants*   1633
Godwin, H.   *History of the British Flora*   1956
Grandville, A., trans. Cleaveland   *The Flowers Personified (Les Fleurs Animés)*   1847
Gunter, R. T.   *The Greek Herbal of Dioscorides*   1934

Hadfield, M. *Gardening in Britain* 1961 *Pioneers of Gardening* 1955
Hanmer, Sir Thomas, ed. Elstob *The Garden Book* (1659) 1933
Harvey, A. G. *Douglas of the Fir* 1947
Haworth-Booth, M. *The Flowering Shrub Garden* (2nd edn) 1947
Hibberd, S. *Familiar Garden Flowers* 1898
Hill, Jason *The Curious Gardener* 1932 *The Contemplative Gardener* 1940

Johnson, G. W. *A History of English Gardening* *c.* 1829

Kaempfer, E. *Amoenitatum Exoticarum, etc.* 1712
Kaempfer, trans. Scheuzer *History of Japan* 1728
Kalm, P., trans. Foster, J. R. *Travels in North America* (2nd edn) 1772
Kent, Elizabeth *Flora Domestica* 1823
Kingdon-Ward, F. *Berried Treasure* 1954 *Pilgrimage for Plants* 1960

Langley, Batty *New Principles of Gardening* 1728
Le Texnier, F. *Essaies sur l'Histoire de Quelques Fleurs d'Ornement* 1906–11
Li, H. L. *The Garden Flowers of China* 1959
Liger, L., trans. London and Wise *The Retir'd Gard'ner* 1706
Loudon, J. C. *Arboretum et Fruticetum Britannicum* 1838
Lucas, J. (trans.) *Kalm's Account of his Visit to England* 1892
Lyte, Henry *A Nievve Herball* 1578

Marshall, William *Planting and Ornamental Gardening* 1785
Miller, Philip *The Gardener's Dictionary* (7th edn) 1759
Miller, ed. Martyn *The Gardener's Dictionary* (9th edn) 1805

Parkinson, John *Paradisi in Sole, Paradisus Terrestris* 1629 *Theatrum Botanicum* 1640

Phillips, Henry *Floral Emblems* 1825 *Sylva Florifera* 1823
Pliny, trans. Philemon Holland *Historie of the Worlde* (*c.* A.D. 77) 1601
Pratt, Anne *Flowering Plants of Great Britain* 1855
Pulteney, R. *Historical and Biographical Sketches of the Progress of Botany in England* 1790

Ray, John *Catalogus Plantarum Angliae* 1677
Robinson, W. *The English Flower-Garden* 1900 edn

Siebold, P. F. and Zuccarini, J. G. *Flora Japonica* 1835–42
Skinner, C. M. *Myths and Legends of Flowers, Trees, Fruits and Plants* 1911
Smith, Edward *The Life of Sir Joseph Banks* 1911
Smith, John *The Records of the Royal Botanic Gardens, Kew* 1880
Swem, E. G. *Brothers of the Spade* 1949

Thomas, Graham Stuart *Colour in the Winter Garden* N.D.
Threlkeld, Caleb *Synopsis Stirpium Hibernicarum* 1727
Tournefort, J. P. de *The Compleat Herbal, etc.* 1719–30
Turner, Robert *Botanologia; the Brittish Physician* 1687
Turner, William *Names of Herbes* 1548 *A New Herball* 1551–68

Veitch, Sir J. H. *Hortus Veitchii* 1906

Willson, E. J. *James Lee and the Vineyard Nursery, Hammersmith* 1962
Wilson, E. H. *A Naturalist in Western China* 1913
Wood, Samuel *A Plain Guide to Good Gardening* 1891
Woodville, William *Medical Botany* 1790

Monographs

Ahrendt, L. W. A. *Berberis and Mahonia* 1961

Bartrum, Douglas *Hydrangeas and Viburnums* 1958 *Lilac and Laburnum* 1959 *Rhododendrons and Magnolias* 1957
Bunyard, E. A. *Old Garden Roses* 1938

Chapple, Fred J. *The Heather Garden* 1952
Cox, E. H. M. and Cox, P. A. *Modern Rhododendrons* 1956
Curtis, S. *Monograph on the Genus Camellia* 1819

Ebel, Marcel *Hydrangea et Hortensia* (2nd edn) 1948

Freeman-Mitford, A. B. (Lord Redesdale) *The Bamboo Garden* 1896

Harvey, N. P. *The Rose in Britain* 1953
Haworth-Booth, M. *The Hydrangeas* 1950
Hibberd, S. *The Ivy* 1872
Hooker, Sir J. Dalton *Rhododendrons of the Sikkim Himalayas* 1894
Hume, H. Harold *Camellias in America* (2nd edn) 1955
Hyams, E. *The Grape Vine in England* 1949

Johnson, A. T. *Hardy Heaths, and some of their nearer Allies* (2nd edn) 1956
Johnstone, G. H. *Asiatic Magnolias in Cultivation* 1955
Jones, G. Neville *A Monograph of the Genus Symphoricarpos* 1940

Kingdon-Ward, F. *Rhododendrons* 1946

Lindley, John *Rosarum Monographia* 1820

McKelvey, Susan Delano *Yuccas of the South-Western United States* 1938–47
Millais, J. G. *Magnolias* 1927 *Rhododendrons . . . and the Various Hybrids* 1917
Monro, Col. *A Monograph of the Bambusaceae* 1868
Moore, T. and Jackman, G. *The Clematis as a Garden Flower* 1872

Paul, W. *The Rose Garden* (9th edn) 1888
Porcher, Felix *Histoire et Culture du Fuchsia* (4th edn) 1874

Rehder, A. *A Synopsis of the Genus Lonicera* 1903

Shepherd, Roy L. *History of the Rose* 1954
Stern, F. C. *The Genus Paeonia* 1946
Street, Frederick *Azaleas* 1959 *Hardy Rhododendrons* 1954
Sweet, Robert *The Cistinae* 1829–30

Thomas, Graham Stuart *The Old Shrub Roses* 1955 *Shrub Roses of To-day* 1962

Urquehart, Beryl Leslie *The Camellia* 1956 *The Rhododendron* 1958

Van Rensselaer, M. and McMinn, H. E. *Ceaonthus* 1942

Watson, W. *Rhododendrons and Azaleas* 1911
Whitehead, Stanley B. *Garden Clematis* 1959
Willmott, Ellen *The Genus Rosa* 1914
Wood, W. P. *A Fuchsia Survey* 1950

Periodicals and Works of Reference

Aiton, W. *Hortus Kewensis* (2nd edn) 1810
Andrews, H. C. *The Botanist's Repository* 1797–1812

Backer, C. A. *Verklarend Woordenboek* 1936
Britten, J. and Boulger, G. S. *Biographical Index of Deceased British and Irish Botanists* 1931

Chittenden, F. J. ed. *The R.H.S. Dictionary of Gardening* (2nd edn) 1956

Edwards, Sydenham *The Botanical Register* 1815–49

Johnson, G. W. *The Cottage Gardener's Dictionary* 1863

Loddiges, G. *The Botanical Cabinet* 1818–33
Loudon, J. C. *The Encyclopaedia of Gardening* 1822

Rees, Abraham *The Cyclopaedia, or Universal Dictionary* 1819–20

Wijk, Gerth van *A Dictionary of Plant Names* 1911

The Gardener's Chronicle 1841–
Journal of the Royal Horticultural Society 1846–
Kew Bulletin 1887–
The New Flora and Sylva 1928–40

Selected Bibliography to the New American Edition

Bean, W. J., ed. Clark, D. L. and Taylor, Sir G. *Trees and Shrubs Hardy in the British Isles* (8th edn) 1970–1980
Brickell, C. and Sharman, Fay *The Vanishing Garden* 1986

Cunningham, I. S. *Frank N. Meyer Plant Hunter In Asia* 1984

Meyer, F. G. and Walker, E. H. *Flora Of Japan* 1965

Radford, A. E., Ahles, H. E. and Bell, C. R. *Manual of the Vascular Flora of the Carolinas* (8th edn) 1981

Spongberg, S. A. *A Reunion of Trees* 1990

Index of English and American Names

Aaron's Beard *Hypericum calycinum*
Adam's Needle *Yucca* spp.
Alaternus *Rhammus alaternus*
Alexandrian Laurel *Ruscus hypoglossum*
Alexandrian Laurel *Danae racemosa*
Althaea *Hibiscus syriacus*
Alpine Mespilus *Amelanchier ovalis*
Alpine Rose *Rhododendron hirsutum*
American Witch-hazel *Hamamelis virginiana*
Amur River-North *Ligustrum amurense*
Apothecary's Rose *Rosa gallica* var. *officinalis*
Apple Rose *Rosa villosa*
Arabian Jasmine *Jasminum sambac*
Architect's Friend *Cotoneaster microphyllus*
Aromatic Vine *Aristolochia macrophylla*
Austrian Briar *Rosa foetida*
Austrian Heath *Erica carnea*
Avignon Berry *Rhamnus infectoria*

Baked Apples and Pears *Cornus canadensis*
Bamboo *Arundinaria* spp.
Barberry *Berberis* spp.
Bay *Laurus nobilis*
Bear Grass *Yucca filimentosa*
Beauty-Bush *Kolkwitzia amabilis*
Bedwind *Clematis vitalba*
Bell-heather *Erica cinerea*
Belly-wind *Clematis vitalba*
Birthwort *Aristolochia hirta*
Biting Clematis *Clematis flammula*
Blackberry *Rubus fruticosus*

Blackthorn *Prunus spinosa*
Black Locust *Robina pseudoacacia*
Bladder Nut *Staphylea pinnata*
Bladder Senna *Colutea arborescens*
Bloody Choak-berry *Amelanchier confusa*
Bloudy Rod *Cornus sanguinea*
Blue Blossom *Ceanothus thyrsiflorus*
Blue Pipe-Tree *Syringa vulgaris*
Blue Spiraea *Caryopteris mastacanthus*
Boston Ivy *Parthenocissus tricuspicata*
Bottlebrush Buckeye *Aesculus parviflorus*
Box *Buxus sempervirens*
Boxberry *Gaultheria procumbens*
Box Thorn *Lycium* spp.
Bramble *Rubus* spp.
Bramble-Rose *Rosa multiflora*
Bridal Wreath *Spiraea chamaedryfolia*
Bridewort *Spiraea salicifolia*
Broad-leaved Birthwort *Aristolochia macrophylla*
Broom *Cytisus* spp.
Brush-bush *Eucryphia* spp.
Buckthorn *Rhamnus catharticus*
Buffalo Currant *Ribes aureum*
Burning Bush *Euonymus americanus*
Butcher's Broom *Ruscus aculeatus*
Butterfly Bush *Buddleia davidii*

Cabbage Rose *Rosa centifolia*
Calico Bush *Kalmia latifolia*
Californian Bush-Poppy *Romneya coulteri*
Californian Lilac *Ceanothus thyrsiflorus*

Cane-Apple *Arbutus unedo*
Cantabrian Heath *Daboecia cantabrica*
Carolina Allspice *Calycanthus florida*
Carolina Buckeye *Aesculus pavia*
Cat Rush *Euonymus europaeus*
Cat's Claw *Bignonia unguis-cati*
Chaste Tree *Vitex agnus-castus*
Checkerberry *Gaultheria procumbens*
Cherry-Laurel *Prunus laurocerasus*
Cherry Woodbine *Lonicera alpigena*
Chickasaw Rose *Rosa bracteata*
Chilean Fire-bush *Embothrium coccineum*
China Rose *Rosa indica*
Chinese Box *Euonymus japonicus*
Chinese Gooseberry *Actinidia chinensis*
Chinese Guelder-Rose *Hydrangea maritima*
Chinese Pieris *Pieris forestii*
Chinese Plumbago *Ceratostigma wilmottianum*
Chinese Wisteria *Wisteria sinensis*
Chittam Wood *Rhus cotonoides*
Christ's Thorn *Paliurus spina-christi*
Cinnamon Bush *Ribes aureum*
Cinnamon Rose *Rosa cinnamomea*
Cinquefoil *Potentilla fruticosa*
Coral-Berry *Berberidopsis corallina*
Coral-Berry *Symphoricarpos orbicularis*
Coral Vine *Bignonia (Doxantha) capreolata*
Cornel *Cornus mas*
Cornelian Cherry *Cornus mas*
Cornish Heath *Erica vagans*
Corsican Heath *Erica terminalis*

Creeping Wintergreen *Gaultheria procumbens*
Cross-leaved Heath *Erica tetralix*
Cross Vine *Bignonia (Doxantha) capreolata*

Daisy-Bush *Olearia* spp.
Damask Rose *Rosa damascena*
Death-Alder *Euonymus europaeus*
Devil Weed *Osmanthus delavayi*
Devil's Twister *Clematis vitalba*
Dog-berrie Tree *Cornus sanguinea*
Dog-Rose *Rosa canina*
Dogwood *Cornus* spp.
Dogwood *Euonymus europaeus*
Dorset Heath *Erica ciliaris*
Duke of Argyll's Tea-Tree *Lycium chinene*
Dutchman's Pipe *Aristolochia macrophylla*
Dwarf Almond *Prunus tenella*
Dwarf Bay *Daphne collina*
Dwarf Bay *Daphne mezereum*
Dyer's Greenweed *Genista tinctoria*

Eglantine *Rosa rubiginosa*
Elder *Sambucus niger*
Elisha's Tears *Leycesteria formosa*
Epiphany Tree *Hamamelis mollis*
Evergreen Privet *Phillyrea* spp.

Fairy Rose *Rosa lauranceana*
Fat-Headed Lizzie *Fatshedera lizei*
Father Time *Clematis vitalba*
Fetterbush *Pieris floribunda*
Fiery Azalea *Rhododendron calendulaceum*
Fig-Leaf Palm *Fatsia japonica*
Fishbone Cotoneaster *Cotoneaster horizontalis*
Fish-Tree *Laurus nobilis*
Fire-Bush *Embothrium coccineum*
Firethorn *Pyracantha* spp.
Fleece Vine *Polygonum aubertii*
Florida Allspice *Calycanthus occidentalis*
Florida Dogwood *Cornus florida*
Flowering Currant *Ribes sanguineum*
Flowering Nutmeg *Leycesteria formosa*
Flowering Plum *Prunus triloba*
Flowering Quince *Chaenomeles japonica*
Flowering Raspberry *Rubus odoratus*
Fountain Buddleia *Buddleia alternifolia*
French Rose *Rosa gallica*
French Sage *Phlomis fruticosa*
French Tamarisk *Tamarix gallica*
Fuji Cherry *Prunus incisa*
Furze *Ulex europaeus*
Fuzzy Deutzia *Deutzia scabra*

Garland-Flower *Daphne cneorum*
Gatteridge Tree *Euonymus europaeus*
Giant Dogwood *Cornus coutroversa*
Gold Plant *Aucuba japonica*
Golden Bells *Forsythia viridissima*
Golden Currant *Ribes aureum*
Gorse *Ulex* spp.
Grape-Pear *Amelanchier canadensis*
Gravid Flower *Enkianthus quinqueflorus*
Griffith Rhododendron *Rhododendron griffithanum*
Grig *Calluna vulgaris*

Guelder-Rose *Viburnum opulus*
Guild-Tree *Berberis vulgaris*
Gum-Cistus *Cistus ladaniferus*
Gypsy's Bacca *Clematis vitalba*

Hag-rope *Clematis vitalba*
Hardy Orange *Poncirus trifoliata*
Hearts-a-bustin' *Euonymus americanus*
Heath *Erica* spp.
Heather *Calluna vulgaris*
Heavenly Bamboo *Nandina domestica*
Hedge Feathers *Clematis vitalba*
Holly Rose *Cistus* spp.
Honey-bind *Lonicera periclymenum*
Honeysuckle *Lonicera* spp.
Horse-chestnuts *Aesculus hippocastanum*
Horse-Tongue *Ruscus hypoglossum*
Hortensia *Hydrangea* vars.

Incense Bush *Ribes sanguineum*
Indian Currant *Symphoricarpos orbicularis*
Irish Heath *Daboecia contabrica*
Irish Ivy *Hedera hibernica*
Irish Whorts *Daboecia cantabrica*
Italian Ivy *Hedera chrysocarpa*
Italian Jasmine *Jasminum humile*
Ivy *Hedera helix*
Ivy *Kalmia latifolia*

Japan Rose *Camellia japonica*
Japanese Bitter Orange *Poncirus trifoliata*
Japanese Bush Cherry *Prunus japonica*
Japanese Rose *Kerria japnoica*
Japanese Skimmia *Skimmia japonica*
Japanese Wineberry *Rubus phoenicolasius*
Japonica *Chaenomeles* spp.
Jasmine *Jasminum officinale*
Jaundice-berry *Berberis vulgaris*
Jerusalem Sage *Phlomis fruticosa*
Jerusalem Willow *Elaeagnus angustifolia*
Jessamine *Jasminum officinale*
Jew's Mallow *Kerria japonica*
Joshua Tree *Yucca brevifolia*

Kiwi *Actinidia chinensis*
Knee-holme, -holly or -hulver *Ruscus aculeatus*

Lady Banks' Rose *Rosa banksiae*
Lady Larpent's Plumbago *Ceratostigma plumbaginoides*
Lady's Bower *Clematis viticella*
Lady's Eardrops *Fuchsia* spp.
Lambkill *Kalmia angustifolia*
Lantern Tree *Tricuspidaria lanceolata*
Laurel *Laurus nobilis*
Laurel *Prunus laurocerasus*
Laurustinus *Viburnum tinus*
Leather Flower *Clematis viorna*
Lemon Verbena *Aloysia triphylla*
Lilac *Syringa* spp.
Lily among Thorns *Lonicera periclymenum*
Lily-Tree *Magnolia conspicua*
Ling *Calluna vulgaris*
Louseberry *Euonymus europaeus*
Love-roses *Viburnum opulus*

Macartney's Rose *Rosa bracteata*
Madeira Broom *Genista virgata*
Makebate *Jasminum fruticans*
Maracoc *Passiflora* spp.
Matilija Poppy *Romneya coulteri*
Mediterranean Heath *Erica mediterranea*
Mexican Evergreen *Abelia floribunda*
Mexican Orange-Blossom *Choisya ternata*
Mezereon *Daphne mezereum*
Missouri Currant *Ribes aureum*
Mock Orange *Philadelphus* spp.
Mock Privet *Phillyrea angustifolia*
Monthly Rose *Rose damascena* var.
Moss Rose *Rosa muscosa*
Mound Lily *Yucca gloriosa*
Mount Etna Broom *Genista aethnensis*
Mountain Elder *Sambucus racemosa*
Mountain Laurel *Kalmia latifolia*
Mountain Sweet *Ceanothus* spp.
Mountain Tea *Gaultheria procumbens*
Moustache Plant *Caryopteris* spp.
Musk Rose *Rosa moschata*
Myrtle *Myrtus communis*

New Jersey Tea *Ceanothus americanus*

Oak-leaved Hydrangea *Hydrangea quercifolia*
Old-fashioned Wiegela *Wiegela florida*
Old Man's Beard *Clematis vitalba*
Old Man's Woozard *Clematis vitalba*
Oleander *Nerium oleander*
Oleaster *Elaeagnus angustifolia*
Orange Ball Tree *Buddleia globosa*
Oregon Grape *Mahonia aquifolia*

Palmetto Royal *Yucca gloriosa*
Paradise Plant *Daphne mezereum*
Park-Leaves *Hypericum androsaemum*
Passion-flower *Passiflora* spp.
Persian Lilac *Syringa persica*
Petty Whin *Genista anglica*
Pinchushion-Tree *Viburnum opulus*
Pinxterbloom *Rhododendron nudiflorum*
Pipe Vine *Aristolochia macrophylla*
Pipperidge-Bush *Berberis vulgaris*
Poet's Ivy *Hedera chrysocarpa*
Poison-Berry *Kalmia angustifolia*
Poison Ivy *Rhus radicans*
Poison Oak *Rhus toxicodendron*
Pomegranate *Punica granata*
Portugal Broom *Cytisus albus*
Portugal Heath *Erica lusitanica*
Portugal Laurel *Prunus lusitanicus*
Prickly Heath *Pernettya mucronata*
Prickly Pettigrue *Ruscus aculeatus*
Prickwood *Euonymus europaeus*
Prim-print *Ligustrum vulgare*
Privet *Ligustrum vulgare*
Provence Rose *Rosa centifolia*
Provins Rose *Rosa gallica* var. *officinalis*
Purging Periwinkle *Clematis flammula*
Purple-Leaved Vine *Vitis vinifera* var.

Queen's Needlework *Spiraea salicifolia*